Enemy Waters

Royal Navy, Royal Canadian Navy, Royal Norwegian Navy, U.S. Navy, and Other Allied Mine Forces Battling the Germans and Italians in World War II

When Britain declared war on Germany in September 1939, the Royal Navy was deficient in minelayers needed to try to hold enemy forces at bay and out of its home waters. Turning first to the Merchant Navy, it requisitioned a liner and two ferries for this use, and a dozen destroyers and submarines were also converted to carry mines. Later, six fast minelaying cruisers joined the force. When Italy entered the war on the Axis side in June 1940, the situation became dire. As U-boats continued to sink shipping in the North Sea and around the British Isles, the Italian Fleet and German and Italian Air Forces controlled the central Mediterranean. Royal Air Force Bomber and Coastal Command planes took up mining, as did old Swordfish biplanes of the Fleet Air Arm. Joining in the fight were units of exiled navies, including the Dutch minelayer *Willem van der Zaan*, Free French submarine *Rubis*, and the Norwegian 52nd Motor Launch Flotilla. U.S. Navy mine forces supported the invasion of French North Africa in late 1942, subsequent landings in Italy, and the invasions of Normandy and southern France. The Canadian 31st Minesweeping Flotilla was at Normandy, and joined in later operations. *Enemy Waters* puts readers in the heart of the action. One hundred and forty-five photographs, maps, and diagrams; appendices; and an index to full-names, places and subjects add value to this work.

Dedication

To the members of the Allied Mine Forces in World War II, and particularly those who fought in exile, their native countries being occupied by enemy forces. One of these individuals was Finn Christian Gysler. Escaping Norway on skis at night under the penalty of death, he embarked on a very long, convoluted, and surreptitious journey by plane, train and ship before finally arriving in London to join remnants of the Royal Norwegian Navy operating from there to fight the enemy.

Enemy Waters

Royal Navy, Royal Canadian Navy,
Royal Norwegian Navy, U.S. Navy,
and Other Allied Mine Forces
Battling the Germans and
Italians in World War II

Cdr. David D. Bruhn, USN (Retired)
and
Lt. Cdr. Rob Hoole, RN (Retired)

HERITAGE BOOKS
2019

HERITAGE BOOKS
AN IMPRINT OF HERITAGE BOOKS, INC.

Books, CDs, and more—Worldwide

For our listing of thousands of titles see our website
at
www.HeritageBooks.com

Published 2019 by
HERITAGE BOOKS, INC.
Publishing Division
5810 Ruatan Street
Berwyn Heights, Md. 20740

Copyright © 2019 Cdr. David D. Bruhn, USN (Retired)
and Lt. Cdr. Rob Hoole, RN (Retired)

All rights reserved. No part of this book may be reproduced or transmitted in any form or by any means, electronic or mechanical, including photocopying, recording or by any information storage and retrieval system without written permission from the author, except for the inclusion of brief quotations in a review.

International Standard Book Number
Paperbound: 978-0-7884-5872-9

Heritage Books by Cdr. David D. Bruhn, USN (Retired)

Battle Stars for the "Cactus Navy":
America's Fishing Vessels and Yachts in World War II

Enemy Waters:
Royal Navy, Royal Canadian Navy, Royal Norwegian Navy,
U.S. Navy, and Other Allied Mine Forces Battling the
Germans and Italians in World War II
Cdr. David D. Bruhn, USN (Retired) and Lt. Cdr. Rob Hoole, RN (Retired)

Eyes of the Fleet:
The U.S. Navy's Seaplane Tenders and Patrol Aircraft in World War II

Ingram's Fourth Fleet:
U.S. and Royal Navy Operations Against German Runners,
Raiders, and Submarines in the South Atlantic in World War II

MacArthur and Halsey's "Pacific Island Hoppers":
The Forgotten Fleet of World War II

Home Waters:
Royal Navy, Royal Canadian Navy, and U.S. Navy
Mine Forces Battling U-Boats in World War I
Cdr. David D. Bruhn, USN (Retired) and Lt. Cdr. Rob Hoole, RN (Retired)

Nightraiders:
U.S. Navy, Royal Navy, Royal Australian Navy, and
Royal Netherlands Navy Mine Forces Battling the
Japanese in the Pacific in World War II
Cdr. David D. Bruhn, USN (Retired) and Lt. Cdr. Rob Hoole, RN (Retired)

We Are Sinking, Send Help!:
The U.S. Navy's Tugs and Salvage Ships in the African,
European, and Mediterranean Theaters in World War II

Wooden Ships and Iron Men:
The U.S. Navy's Ocean Minesweepers, 1953–1994

Wooden Ships and Iron Men:
The U.S. Navy's Coastal and Motor Minesweepers, 1941–1953

Wooden Ships and Iron Men:
The U.S. Navy's Coastal and Inshore Minesweepers,
and the Minecraft that Served in Vietnam, 1953–1976

Contents

Foreword by Dr. Edward J. Marolda	xiii
Foreword by Rear Adm. Paul J. Ryan, USN (Retired)	xv
Foreword by Rear Adm. Paddy A. McAlpine, CBE, RN (Retired)	xvii
Foreword by Cdr. Fraser M. McKee, RCNR (Retired)	xix
Acknowledgements	xxi
Preface	xxiii
1. Operation MINCEMEAT	1
2. British Minelaying Forces	9
3. Initial British Minelaying Efforts	23
4. Mining and Mine Countermeasures	37
5. Mining of Norwegian Waters and Aerial Minelaying	49
6. Dutch Minelayers Escape the Netherlands	63
7. Minelaying Forces Suffer Loss of Seven Vessels in 1940	69
8. Mediterranean 1941	87
9. The "Rats of Tobruk"	105
10. RN and RNN Coastal Forces	121
11. British Offensive Mining in Northwest Europe in 1942	139
12. French Minelaying Submarine *Rubis*	147
13. U.S. Army and Navy Minelaying Forces	153
14. American Forces Land in French North Africa	163
15. Invasion of Sicily – Operation HUSKY	183
16. Coastal Forces Operations in 1943 and early-1944	191
17. British Fast Minelayer Force Decimated in 1942-1943	205
18. Operation MAPLE	209
19. U.S. Navy Readies Auxiliary Minelayers for Flagship Duties	217
20. Royal Canadian Navy Mine Force	221
21. D-Day at Normandy	237
22. Consolidation of the Beachhead	257
23. Opening the port of Cherbourg	265
24. Southern France – Operation DRAGOON	277
25. Victory in Europe	289
Postscript	293
Appendices	
A. World War II British and German Sea Mines	307
B. Mine Warfare Terms	313
C. HMS *Adventure* and *Blanche* Casualties	319
Bibliography	321
Notes	323

Index 353
About the Authors 383

Photos and Illustrations

Contents-1: Dan buoys being loaded aboard HMAS *Orara* — xii
Preface-1: HMS *Ark Royal* with Swordfish aircraft overhead — xxiv
Preface-2: Otto Burfeind, captain of the SS *Adolph Woermann* — xxxi
Preface-3: German M-boats under way — xxxiii
Preface-4: R-Boats off the French coast — xxxiii
Preface-5: RNPS member George Hunt — xxxvi
Preface-6: British Empire and Commonwealth poster — xxxix
Preface-7: Painting *Battle of Taranto* by Richard DeRosset — xlvii
1-1: Minelaying cruiser HMS *Ariadne* — 1
1-2: Leghorn (Livorno) on the northwest coast of Italy — 2
1-3: Free French cruiser *Léopard* — 4
1-4: Moored mines aboard HMS *Abdiel*, ready to be laid — 6
2-1: Minelayer HMS *Adventure* at sea — 11
2-2: Coastal minelayer HMS *Miner VI* — 17
2-3: HMS Vernon's mining tender *Skylark*, later *Vernon* — 18
2-4: HM Trawler *Vernon* (ex-*Strathcoe*) — 19
2-5: Controlled minefield cabling loaded in boats — 19
3-1: HMS *Express* configured as a fleet destroyer — 25
3-2: German coastal submarine *U-16* on the surface — 28
3-3: Cutting jaws of a paravane, used for minesweeping — 30
3-4: Paravane used for minesweeping — 30
3-5: *U-52*, a Type VIIB submarine — 33
3-6: German Heinkel He 111 aircraft — 35
4-1: HMS Vernon in July 1955 — 38
4-2: Lt. Comdr. John Ouvry with King George VI — 39
4-3: German GC magnetic mine on a trolley at HMS Vernon — 40
4-4: German destroyer *Wilhelm Heidkamp* — 43
4-5: British cruiser-minelayer HMS *Adventure* — 44
5-1: Painting by Richard DeRosset of HMS *Glowworm* — 56
5-2: Sunken ship in Rombak's Fjord, Norway — 58
5-3: A mine being loaded aboard a RAF Hampden bomber — 60
5-4: A RAF Bristol Beaufort preparing to make a landing — 60
5-5: A RN Fairey Swordfish aircraft dropping a torpedo — 62
6-1: Dutch minelayer HNLMS *Willem van der Zaan* — 63
7-1: Members of British Expeditionary Force at Dunkirk — 71
7-2: Conning tower of British submarine HMS *Porpoise* — 73
7-3: Arado 196 on catapult of German battleship *Gneisenau* — 76

7-4:	British destroyer HMS *Express* under way	83
8-1:	Submarine HMS *Porpoise* alongside HMAS *Australia*	88
8-2:	Fast minelayer HMS *Latona* at Alexandria, Egypt	93
8-3:	RN Fleet Club on Boxing Night Alexandria, Egypt	94
8-4:	Admirals rowing Lord Louis Mountbatten at Malta	98
8-5:	Italian torpedo boat *Generale Antonio Cantore*	102
9-1:	Members of the 2/10th Australian Infantry Battalion	108
9-2:	Painting by Edward Tufnell of HMS *Hydrangea*	111
9-3:	Ship with bomb damage in Tobruk Harbour	113
9-4:	German Air Force Stuka Ju87 aircraft in flight	116
9-5:	Sunken minesweeper alongside a wrecked ship	117
9-6:	Norwegian whaler *Hector VI* in 1929	120
10-1:	Trawler HMS *Vulcan* with 1st MTB Flotilla alongside	121
10-2:	British Motor launch HMS *ML 187*	124
10-3:	British Motor Torpedo Boat HMS *MTB-263*	126
10-4:	German Sperrbrecher, location and date unknown	130
10-5:	German R-Boats operating off Norway	131
10-6:	Norwegian sailors in Nova Scotia	132
10-7:	Officers and men of the Norwegian 52nd ML Flotilla	134
10-8:	HMS Vernon Women's Royal Naval Service members	135
10-9:	Norwegian motor launch *ML 125* after striking mine	137
10-10:	Norwegian sailor Olav Martin Stromsoy	138
11-1:	German battleships *Scharnhorst* and *Gneisenau*	139
11-2:	USS *DD-939*, the former German destroyer *Z-39*	142
11-3:	Swordfish being readied for a mining mission	145
12-1:	Free French submarine *Rubis* in difficulty off Norway	147
13-1:	Mines aboard Army mine-planter *Gen. E. O. C. Ord*	156
13-2:	USS *Chimo* passing movies to USS *Obstructor*	157
13-3:	USS *Terror* at sea	159
13-4:	USS *Salem* under way off Norfolk, Virginia	161
14-1:	Adm. Royal Eason Ingersoll, USN	166
14-2:	A U.S. destroyer passes astern of the carrier USS *Ranger*	168
14-3:	Destroyer USS *Hogan* before conversion to a minesweeper	172
14-4:	Minesweeper USS *Auk* off the Norfolk Navy Yard	174
14-5:	USS *Terror* loading Mk 6 mines at Yorktown, Virginia	179
14-6:	Casablanca Harbour on 16 November 1942	180
15-1:	Minelayer USS *Keokuk* near the Norfolk Navy Yard	187
15-2:	Generals Montgomery and Patton studying a map of Sicily	189
16-1:	Warden Hotel in Dover (HMS Wasp in WWII)	196
16-2:	German heavy naval artillery bombarding Dover	197
16-3:	Finn Christian Stumoen, his wife, and the mayor of Dover	204
17-1:	Troops of the 1st Airborne Division aboard USS *Boise*	208

18-1: Vosper *MTB 23* launching two torpedoes 211
19-1: Auxiliary minelayer USS *Barricade* at anchor 217
20-1: Canadian minesweeper HMCS *Fundy* under way 223
20-2: Canadian auxiliary minesweeper HMCS *Bras d'Or* at anchor 224
20-3: Canadian minesweeper HMCS *Kenora* 226
20-4: Minesweeping trawler HMCS *Anticosti* under way 228
20-5: Ferry *Sankaty* and as the minelayer HMCS *Sankaty* 230
20-6: Coastal minesweeper HMCS *Llewellyn* under way 230
21-1: Canadian Minesweeping Flotilla 31 237
21-2: Battleship USS *Arkansas* conducting shore bombardment 247
21-3: Minesweeper USS *Osprey* under way 250
21-4: Minesweeper USS *YMS-406* conducting builder's trials 253
21-5: German-built sea wall on Utah Beach at Normandy 254
21-6: Commanding Officers receiving the Bronze Star Medal 254
22-1: Painting *The Tough Beach* by Dwight C. Shepler 259
22-2: Minesweeper USS *Tide* sinking off Utah Beach 260
23-1: Aerial view of port of Cherbourg, France 266
23-2: YMS minesweepers exploding mines near Cherbourg 267
23-3: Minelayer USS *Miantonomah* near Norfolk, Virginia 274
24-1: *Algerine*-class minesweeper HMS *Persian* 282
24-2: Allied shipping in Cavalaire Bay, France 285
25-1: V-E Day celebrations on Bay Street in Toronto, Canada 289
Postscript-1: 'P' Party members performing their various duties 293
Postscript-2: Draper L. Kauffman as a Lt., RNVR 295
Postscript-3: Lt. Haderlie, USN, and Lt. Wilkinson, RNVR 296
Postscript-4: Gene Haderlie, USN, in Mine Recovery Suit 297
Postscript-5: Lt. Comdr. Gosse, RANVR, with 'P' Party 1571 299
Postscript-6: Australian Lt. Comdr. George Gosse, RANVR 299
Postscript-7: Soviet cruiser *Ordzhonikidze* in the Baltic Sea 302
Postscript-8: Commander Crabb's gravestone at Milton Cemetery 304
Appendix A-1 German ground mine, destined for HMS Vernon 308
Appendix B-1 German mine disposal continues in 21st Century 314
Appendix B-2 RNVR officer defuzing a parachute land mine 316

Maps and Diagrams
Preface-1: British and German Mine Areas, 1939-40 xxx
1-1: Mediterranean Basin 3
3-1: The Dover Strait area 24
3-2: Jade Bight and Heligoland Bight 26
3-3: England's east coast 31

3-4:	Scotland	32
3-5:	North Sea area	34
3-6:	German East Frisian Islands	36
4-1:	Royal Navy 'LL Sweep'	41
4-2:	Southern portion of the Thames Estuary	43
5-1:	Portion of Scandinavia	52
5-2:	Danish Straits	61
8-1:	Malta archipelago	89
8-2:	Eastern Mediterranean	95
9-1:	Eastern Mediterranean and surrounding areas	107
10-1:	England's south and southeast coasts	123
10-2:	The West Deep, Netherlands coastal area	127
10-3:	Northwest coast of France	128
10-4:	Nova Scotia	133
11-1:	Bay of Biscay	140
14-1:	French Morocco and French Algiers	164
14-2:	Navigation track of the Western Naval Task Force	167
14-3:	Fedala and Casablanca, French Morocco	171
14-4:	Minefield laid by *Terror*, *Miantonomah*, and *Monadnock*	181
15-1:	Allied invasion of Sicily	185
16-1:	Northwest coast of France	192
17-1:	Central Mediterranean	206
19-1:	Azores	219
19-2:	Ireland and other areas of the UK	220
20-1:	East coast of Canada	232
21-1:	Allied sea and air routes to the Normandy beaches	239
21-2:	Minesweepers operating in echelon formation	244
22-1:	German mines swept by Allied forces in Seine Bay	257
24-1:	French Riviera of southeast France	279

Photo Contents-1

Dan buoys being loaded aboard the Australian minesweeper HMAS *Orara*, circa November 1940. Lines of Dan buoys, affixed in place by mooring cables and anchors, identified naval and merchant shipping channels, anchorage areas, and assault lanes to hostile beaches swept clear of mines—the ultimate achievement of the mine forces. Australian War Memorial photograph 003950/02

Foreword

The heroic and routinely selfless wartime efforts of the U.S., British, Canadian, French, and other Allied minesweep and minelayer sailors and mine clearance divers rarely get the attention so often focused on the exploits of carrier pilots, submariners, and surface ship warriors. Through his comprehensive works, however, David Bruhn has brought to light the absolutely essential story of the mine warfare community and its sailors in the wars of the 20th century. His *Wooden Ships and Iron Men* trilogy, as well as his *Home Waters*, *Nightraiders*, and *Enemy Waters* books have brought to the notice of many readers critical information on this vital aspect of naval warfare. These books are essential for scholars, and for those who served in the mine warfare community and made it possible for the Allied navies to execute their offensive and defensive mine warfare missions. Bruhn's deep research and insightful analysis of the subject will benefit the study of this branch of naval history for many years to come.

Edward J. Marolda
Naval Historian

Foreword

Enemy Waters tells the fascinating story of the Allied aircraft, ships and submarines that participated in mining and mine clearing during World War II. Once again Commanders David Bruhn and Rob Hoole have provided exquisite detail about the tremendous efforts and sacrifices of our sailors and airmen, primarily in the European theater of operations.

Offensive mining is conducted at great risk to the minelaying platform when performed in enemy waters. Aircraft, surface ships and submarines have been lost while implanting minefields. Defensive mining is conducted in home waters, but not without other risks: grounding, pre-mature detonation as mines are being released over the side, and inadvertent collisions with "friendly" mines. This book provides historical examples of these risks.

Commanders Bruhn and Hoole put readers in the thick of mining and minesweeping operations in the North Atlantic, English Channel, North Sea, and Mediterranean. Many of the events and associated action are little known to the public, even to devotees of naval history. For example, HMS *Rorqual*, a British submarine minelayer, implanted over 450 mines in 10 sorties in the Mediterranean between June 1940 and May 1941, damaging or sinking approximately 15 ships.

While many World War II history books gloss over the contribution of Free French forces, *Enemy Waters* doesn't. The French submarine *Rubis* was in the UK when France surrendered to Germany in June 1940. The crew of *Rubis* joined the Free French forces and continued the fight, operating out of the UK. *Rubis* conducted minelaying operations off the coast of Norway and southwest France, implanting 684 mines in 23 sorties, responsible for the sinking of 23 ships. Besides conducting mining operations, *Rubis* also conducted torpedo attacks. As a submariner who appreciates the hazards of wartime submarine operations, I was particularly happy to see that *Rubis* survived the entire war and was decommissioned in 1949.

I personally appreciate the numerous charts and pictures that bring the book to life. Commanders Bruhn and Hoole have included over one hundred forty photographs and illustrations to help the reader appreciate the ships, aircraft, submarines, mines, minesweeping gear and geographic environments.

None of us has a crystal ball. Over 300,000 Allied mines were deployed in the Atlantic and Mediterranean campaigns. Approximately 250 minesweepers supported the Allied landings at Normandy on June 6, 1944 and additional minesweepers were in theater standing by to open other continental ports. At the same time, several hundred minesweepers were engaged in the Pacific theater. Today's allied mine inventory is probably in the single digit thousands. The number of allied minesweepers is about 150. The emphasis over the last 15-20 years has been on getting sailors out of the minefield, i.e., using remote controlled and unmanned systems to clear mines rather than historical minesweeping. While modern systems are improving mine detection and neutralization capability, the overall capacity of our current assets is significantly less than at almost any time since World War II. I hope we don't have to relearn the outstanding history documented by Commanders Bruhn and Hoole in their well-researched trilogy of books, *Home Waters*, *Nightraiders*, and *Enemy Waters*.

Paul J. Ryan
Rear Admiral, USN, retired

Foreword

Enemy Waters, the third book in the excellent Mine Warfare trilogy by Commanders Bruhn and Hoole will evoke many memories for those who have served at sea and experienced the challenges posed by sea mines, whether they are modern and newly laid or still highly dangerous WWII relics.

Having joined the Royal Navy to dive in an attempt to emulate my childhood hero Jacques Cousteau (that amazing military leader, scientist, film maker and inventor of the aqua lung), I eventually found myself as a newly qualified Mine Clearance Diving Officer (MCDO) and Operations Officer on board HMS *Chiddingfold*, a *Hunt*-class mine countermeasures vessel. We were preparing to deploy to the Persian Gulf in a small task group that included HMS *Berkeley*, commanded by co-author of this excellent book, Rob Hoole. A hugely experienced and capable MCDO during his active service, Rob has gone on to thrive as a successful author as well, a widely acknowledged and a highly respected naval historian, specialising in mine warfare and diving.

Working up in the middle of the English Channel before we deployed to relieve the other RN Minehunters that were keeping the Strait of Hormuz open at a time of increasing tension, we started the day off as we usually did with System Operator Checks or "SOCs." A chance to turn all the gear on and test it with the maintainers ready to finely tune it up was always welcome. We slowed down to mine hunting speed, turned on the 193M Sonar and, purely by chance, discovered a very solid sonar contact on which to run out the submersible. As soon as the Chief Petty Officer Mine Warfare saw the sonar contact, his mine hunting antenna was immediately raised. "That's a mine!" We launched the submersible, got it into position and with about ten metres to run, turned on the searchlight and video recorder. Into view came a fully intact, German WWII mine. After the usual reactions and preps, the divers went down to neutralise the threat.

As interesting a start to the morning as this was, it was not the first and certainly was not to be the last intact WWII sea mine found in the English Channel during that year and in every year since. They continue to turn up today during similar chance encounters, or in trawlers' nets, or are uncovered on the beaches at extremely low tides; and mine warfare and diving specialists from the navies on both sides of the Channel continue to regularly deal with old WWII ordnance.

Mines, old or new, remain a potent weapon in wartime and a threat to all shipping once the shooting ashore has ended. During WWII the Channel was strewn with them and for very good reason. This relatively cheap-to-manufacture weapon always achieves a high return on investment if offered a target. Countering the mine threat must be part of every amphibious plan to open the approaches to chosen landing areas for the amphibious forces and provide access to the ports and harbours for supporting shipping and follow-on forces once the battle has moved inland. This book not only reinforces this point but also reminds those designing the navies and amphibious forces of the future that ignoring the lessons of the past would be folly.

Paddy McAlpine CBE
Rear Admiral Royal Navy (Retired)

Foreword

Forewords rarely begin with a personal note, but my modest wartime experience is illustrative of Canada's ill-preparedness for either the First or Second World Wars. I joined its RCNVR as a seaman in March 1943 and after training served in an ex-American armed motor yacht. Our navy went to war with only six destroyers and four wooden hulled minesweepers. I was commissioned a year later, ended the war in an RCN *Algerine*-class ship, designed as an RN fleet minesweeper, although we used our twelve just as ocean escorts. I stayed in the Reserves until 1978, specializing in anti-submarine warfare. This included the basics of mining and minesweeping, although I was never involved directly, and ended my naval career in NATO control of merchant shipping (NCS) exercises. In the 1970s, from a personal concern about lack of any current mine countermeasures (MCM) capability, I researched and wrote a major paper for a naval association on the subject, urging a return to its attention—even in ships manned mostly by our willing Reserves. When the MCDV's (Maritime Coastal Defence Vessels, designed in part for the MCM job) were finally introduced in the mid-1990's, that is exactly who manned them to a large degree—and most successfully.

Canada, unlike most European countries like the United Kingdom, France, Belgium, and so forth, has never had to take mining very seriously. While anticipating enemy sea mining in two world wars, it never seriously occurred. On only two occasions, in 1943, were U-boat moored magnetic mines laid off Halifax and St. John's, Newfoundland, costing four ships sunk. Our own defensive fields were as much a hazard as any enemy's. The Japanese never even tried, as far as we knew. The 'Regular Force Navy,' despite becoming embroiled in both conflicts with a miniscule but quickly expanding surface navy, in both cases concentrated on enlarging that "fighting force" and coping with the Battle of the Atlantic, leaving mine warfare to the speedily recruited volunteers of the RCNR (ex-merchant seamen and fishermen) and the RCNVR, the enthusiastic, largely untrained or experienced amateurs. We were fortunate the enemy—Germany in the first war and the same plus Italy and Japan in the second—never took advantage of a pretty obvious opportunity to aggressively mine Canada's coastal ports and estuaries. Under current conditions of world unrest, that—it is claimed by some of us—will not be the case another time.

Canada's major sea ports all face the open ocean – Halifax, Sydney, St. John, and St. John's in Newfoundland on the Atlantic; Vancouver, Victoria, and now Prince Rupert on the Pacific. Access to the inland Great Lakes ports is via the narrow Cabot Strait leading to the St. Lawrence River highway—easily, and inexpensively, mineable. As the authors of this history point out, all our navies have expended much effort – during war years – countering potential and actual sea mining, from cobbled together make-do ships and crews, to specific shipbuilding, occupying major yard space and funding, to employing large portions of naval forces. Yet, as soon as the war ends and the dangerous job of sweeping up the enemy's and our own mine fields is dealt with, the naval staffs revert to concentrating on major ship warfare. It may change from battleships and cruisers to aircraft carriers and submarines, but mine warfare, the field during those wars for the Naval Reserves, is set aside once again. In Europe they are reminded acutely on occasion of the sea mine dangers even today when a fisherman dredges up a 75-year-old mine, some of which still cost those fishermen's lives.

The story in this book by Commanders Bruhn and Hoole serves not only to remind all of us what was valiantly faced by the Allies in two wars by opponent minelaying, but as a caution to maritime planners as to what is still largely ignored or at least assigned very minor attention now and in the future.

As one of the oldest naval weapons still very much in use – or at least potential use – the personal motto of Vice-Admiral Lord Nelson would seem still appropriate, and maybe serve as a caution:

Palmam qui meruit ferat – Give the prize to he who deserves it.

Fraser M. McKee, Cdr, RCN(R) (ret'd)
Toronto, Ontario, Canada

Acknowledgements

Brilliant and prolific maritime and aviation artist Richard DeRosset has created a stunning painting of the Battle of Taranto, which is used as the cover art. An explanation of the subject matter follows in the preface to the book. DeRosset, who has completed over 950 paintings in his illustrious career, has a great love of the sea borne of immense talent and his own experiences sailing the deep. After serving in the U.S. Navy, he worked aboard commercial fishing vessels, and later was master of the small cargo ship MV *Pacific Trojan*.

This book would not have been possible without the considerable involvement and expertise of co-author Rob Hoole and George Duddy. Canadian George Duddy is a retired Professional Engineer with a keen interest in maritime subjects, particularly those relating to the maritime history of western Canada and the Arctic. He has deep family roots in the British Isles. His father as a schoolboy took photographs near the end of World War I of surrendered German Battleships in the Firth of Forth; his great grandfather was a pioneering Leith steamship owner and his great great grandfather, as Master in both sail and steam, ended his career as marine superintendent for the Leith, Hull & Hamburg Steam Packet Co. Duddy offered many suggestions for improvement during his technical review and editing of the manuscript.

Rob Hoole is a former Royal Navy mine clearance diving officer and commanding officer of HMS *Berkeley* (M40)—a *Hunt*-class mine countermeasures vessel. An acknowledged expert on mine warfare, he is a long-standing member of the Ton Class Association and a regular contributor to its publications. Hoole is also founding Vice Chairman and Webmaster of the Royal Naval Minewarfare & Clearance Diving Officers' Association, and holds key positions in related organisations.

Dr. Edward J. Marolda is one of four esteemed individuals who were kind enough to pen forewords, reflecting their unique perspectives and adding richness to the book. Dr. Marolda served as the Acting Director of Naval History and Senior Historian of the Navy. In 2017 the U.S. Naval Historical Foundation honored him with its Commodore Dudley W. Knox Naval History Lifetime Achievement Award. Marolda has written scores of books and articles. An expert on Vietnam and Mine Warfare, he taught courses on the Cold War in the Far East and the Vietnam War at Georgetown University in Washington, D.C.

Rear Adm. Paddy A. McAlpine, CBE, RN (Retired), and Rear Adm. Paul J. Ryan, USN (Retired) provided in their respective forewords, unique perspective borne of senior command experience at sea and ashore. Rear Admiral McAlpine was previously the Royal Navy's senior member of the Mine Warfare and Clearance Diver Officer community. During his career, McAlpine commanded the frigate HMS *Somerset* and the destroyer HMS *Daring*. Following promotion to commodore, he flew his pennant from the Fleet Flagship, HMS *Bulwark*, as Commander United Kingdom Task Group during both the 2012 (Mediterranean) and 2013 (Gulf) COUGAR deployments and was then appointed in Command of the Portsmouth Flotilla in 2014. His final assignment was as Deputy Commander Naval Striking and Support Forces NATO in 2015. McAlpine retired in 2017.

Rear Admiral Ryan, a nuclear engineer and career submariner, commanded the attack submarine USS *Philadelphia* and the submarine tender USS *L. Y. Spear*. Ashore, he served in a variety of assignments while assigned to the Navy and Joint Staffs in the Pentagon, and on the staff of Commander-in-Chief, U.S. Atlantic Fleet. In February 2002 he assumed command of Mine Warfare Command in Corpus Christi, Texas, where he was responsible for the deployment of mine warfare forces in support of Operation Iraqi Freedom in the Persian Gulf. He retired in December 2003.

Fraser McKee's assessment of Canada's association with mine warfare is based on his unique perspective. McKee joined the RCNVR as a seaman in March 1943 and received an officer's commission a year later. He served aboard one of Canada's twelve *Algerine*-class ships at war's end which, although designed as Royal Navy fleet minesweepers, were used as ocean escorts in the U-boat-patrolled North Atlantic. These ships were not fitted for, nor equipped with, minesweeping gear. Remaining in the Reserves after the war, McKee specialised in anti-submarine warfare, which included basic mining detail. His final duty was in NATO operations for the control of merchant shipping (NCSO). McKee retired in the rank of commander in 1978, and has remained deeply interested in Canadian naval matters since, having written or co-written six books on Canadian naval history.

Finally, a tip of the hat to Lynn Marie Tosello, the final editor. In addition to the mastery of prose and diction one would expect, she also lent sophistication and eloquence to the project. Of particular importance, Tosello was quick to identify to the authors cases in which they had provided insufficient explanations of particular naval or mine warfare-related terms or subjects.

Preface

The nations not so blest as thee,
Shall in their turns to tyrants fall;
While thou shalt flourish great and free,
The dread and envy of them all.

Rule, Britannia! rule the waves:
"Britons never will be slaves."

—From the poem "Rule Britannia" by James Thomson (1700-48),
put to music by Thomas Augustine Arne (around 1740)
and sung as an unofficial British national anthem.

It would be idle to suggest that we made no mistakes, or to claim that the British mines were necessarily better than the German. But the cold fact remains that our mines defeated their minesweeping organization, while our minesweeping and degaussing organization defeated their mines. We were not merely 'swapping pawns', for the total effect expended by both sides, compared with the results, left a balance in our favour not far short of two to one.

—Capt. J. S. Cowie, RN, in his seminal book, *Mines, Minelayers and Minelaying*, published by Oxford University Press in 1949.[1]

This book takes readers vicariously to sea with mine forces of Allied nations—principally those of Britain, aided by the United States, Canada, Norway, South Africa, Australia, the Netherlands, and the Free French—as they combatted German and Italian forces in the European, African and Mediterranean Theatres in World War II.

As an example of WWII mine warfare activity in all theatres, over the course of the war, the United States and United Kingdom laid a combined 300,000 mines, 100,000 for offensive purposes. U.S. efforts were almost entirely against the Japanese in the Pacific, a theatre not covered in this book. British and Allied minelayers (including U.S.) sowed 263,850 mines in areas under British control. Of these, 77,312 mines were employed offensively—nearly 55,000 mines laid by aircraft, 11,000 by purpose-built cruiser-minelayers ("fast minelayers") and destroyers modified to lay mines; 6,500 by Coastal Forces; and 3,000 by submarines. German and Italian figures are not readily available, but the Axis powers laid 55,000 mines in the Mediterranean alone.²

EXPLANATION OF THE COVER ART

Those familiar with the fine art of Richard DeRosset which graces the covers of my books, might well expect to find for this one, a painting of a minelayer or minesweeper under fierce attack by enemy air or surface forces. In this case, the cover art is a superlative painting by Richard of the Battle of Taranto. During the night of 11 November 1940, twenty-one Fairey Swordfish torpedo-bombers launched from HMS *Illustrious* struck the battle fleet of the Italian Navy at anchor in the harbour of Taranto. The attack was carried out using aerial torpedoes, despite challenges to their employment. Torpedoes dropped by aircraft over shallow waters like those of the Italian harbour, tended to strike the bottom and fail to operate correctly. The British were able to overcome this problem, which led senior Japanese naval officers to study the attack closely—and use it as a model for the attack on Pearl Harbor.

Photo Preface-1

Aircraft carrier HMS *Ark Royal* with a flight of Fairey Swordfish aircraft overhead. Naval History and Heritage photograph #NH 85716

The attack on Taranto (in which three Italian battleships were hit by torpedoes, one was sunk, and two others seriously damaged) established beyond a doubt the potential of aerial-launched torpedoes, even in relatively shallow waters. Equally important, the raid by slow, virtually obsolete biplanes of the Fleet Air Arm shattered many greatly cherished beliefs about the dominance of battleships and naval gunnery. More immediately, the air raid benefited the British by knocking out of action a significant part of the Italian fleet. It also facilitated relocation of many of the remaining ships to ports farther north; safer from another attack, but more distant from their main area of operations.[3]

In contrast to World War I, where nearly all the mines had to be laid by ships or submarines, in WWII a majority were sown by aircraft, but ships, coastal forces, and submarines still had an important role. Royal Air Force twin-engine Hampden medium-bombers, and Coastal Command Beaufort torpedo bombers carried out most of the aerial-mining, aided by antiquated Swordfish torpedo-bombers of the Royal Navy's Fleet Air Arm, the type of aircraft depicted in the painting.

By 1940, the era of the biplane was long over, and the beloved Swordfish was a relic of the past with her agonisingly slow speed, open cockpit, fixed landing gear, and canvas skin. Yet, her importance during the war would rival that of any other combat aircraft owing to her great versatility. Because the lumbering biplane could carry a variety of ordnance, some wag likened it to a popular shopping bag used by British women known as a "string bag"—and the nickname stuck.[4]

The Swordfish could carry a 1,610-pound torpedo, or anti-ship mines, bombs, flares, or depth charges, and perform a variety of missions. These included reconnaissance over land or sea, aerial spotting of naval gunfire (to provide reports of the accuracy of falling rounds), convoy escort duty, attacks on U-boats, and dropping mines into enemy harbours. Mining of German ports was done at night when the slow biplanes could hide in the dark with relative impunity. Nearing the designated area, the pilot typically throttled back the engine and glided the encumbered aircraft silently down to the target.[5]

The biplanes could fly from land bases or be launched off a ship's catapult as floatplanes; but proved most valuable to the fleet operating from carriers such as HMS *Illustrious*. Despite many previous successes, the tragic loss on 12 February 1942 of all six Swordfish of No. 825 Squadron, while carrying out a torpedo attack on a group of enemy ships, exposed weaknesses of the aircraft. With a top speed of only 138 mph, an equally unimpressive rate-of-climb of 1,220 feet/minute, and for self-defence, a single forward firing Vickers machine gun and a single flexible Lewis machine gun for the rear gunner, the Swordfish had no chance.[6]

In a "suicide mission" on 12 February, the squadron was sent out to oppose a group of German warships—centred around the battleships *Scharnhorst* and *Gneisenau*, and heavy cruiser *Prinz Eugen*. In addition to the threat posed by the ships themselves, and E-boat escorts, two Staffels (squadrons) of Messerschmitt Bf 109 fighters were sent up to intercept the British planes. All six aircraft were shot down, with only five of the total eighteen aircrew members surviving. (A three-man aircrew consisted of a pilot, an observer, and a telegraphist air gunner.) Thereafter, Swordfish were relegated to mining operations only. Details about the breakout of these ships from Brest, France, and ensuing dash up the English Channel to Germany are provided later in the book.[7]

The lumbering aircraft's finest day was nine-and-a-half months earlier, when they were instrumental in the demise of Germany's greatest warship. Torpedo hits by Swordfish of No. 825 Squadron on the battleship *Bismarck* had jammed her rudder, leaving the formidable German battleship turning in circles, easy prey to the force of Royal Navy ships responsible for her ultimate destruction on 27 May 1941.[8]

BOOK PHILOSOPHY AND LIMITATIONS

Enemy Waters is the final book in a trilogy by the co-authors, begun with *Home Waters* and *Nightraiders*. *Home Waters* details the steadfast efforts in World War I of Royal Navy and U.S. Navy Mine Forces to battle German U-boats in British home waters. German ships and submarines embarked on an aggressive campaign to mine Britain home waters, with the goal of cutting off maritime commerce and forcing the island nation, dependent on the delivery of food, fuel, vital supplies and other materiel, to capitulate. In response, hundreds of fishing vessels from every port and harbour in Britain were pressed into minesweeping duties, and British minelayers in turn sowed fields to restrict the movement of and destroy German vessels.[9]

The efforts of British minesweeping flotillas enabled the powerful Royal Navy sufficient freedom of the sea to blockade the German Navy in port—except for occasional skirmishes, including the Battle of Jutland—and kept vital channels and harbours open. Following the war, Lord John Rushworth Jellicoe aptly described the critical importance of the fishing boats turned minesweeper, declaring that "the Royal Navy had saved the Empire, but it was fishermen in their boats who had saved the Royal Navy.[10]

Late in the war, U.S. Navy minelayers began work in conjunction with the Royal Navy to create a massive mine barrier stretching from the Orkney Islands to the Norwegian coast. Although not completed by war's end, it represented a partially successful effort to bottle up German U-boats in the North Sea and curtail continued attacks on shipping in the Atlantic. (A similar British-American effort had been made in 1918 to lay a massive mine barrage from the Orkneys to Norway at Bergen, the purpose being to close the North Sea exit to the U-boats.)

U-boat activity was not confined to Europe. In 1918, across the Atlantic, minesweepers of the U.S. and Royal Canadian navies cleared mines in waters off the eastern seaboard of North America laid by enemy submarines. Attacks by five U-boats against shipping had been undertaken in an unsuccessful effort to compel the American government to bring home Atlantic Fleet destroyers engaged overseas in hunting U-boats in British waters.

This book could be wholly devoted to operations by Royal Navy minesweeping flotillas in World War II, or to British offensive or defensive minelaying operations. There is more than sufficient material, and many such books exist, including for example: *Out Sweeps! Story of the Minesweepers in World War 2* by Paul Lund, and Capt. J. S. Cowie's *Mines, Minelayers and Minelaying*. *Enemy Waters* endeavours to present the intertwined relationship between offensive minelaying in enemy waters, defensive minelaying in home waters, and minesweeping in home and enemy waters to counter enemy mines. While the Royal Navy and Royal Air Force were responsible for the bulk of mining in African, Mediterranean, and European waters, and the Royal Navy for sweeping enemy mines, the authors have included discussion of the contributions made by other Allied Mine Forces. These include the provision of mine warfare ships by Allies and/or in the case of dominions, service of some personnel aboard these type and other Royal Navy vessels.

Absent from *Enemy Waters* is the level of detail about the mechanics of minelaying and minesweeping found in *Home Waters* and *Nightraiders*. The latter book is devoted to mine warfare against the Japanese in the Pacific in World War II. It opens with the British and Dutch laying minefields off Hong Kong and Singapore, and in the Netherlands East Indies, in a desperate attempt to prevent their capture by the Japanese. These efforts were unsuccessful. Enemy invasion forces advanced over land to take the British Colonies, while a two-pronged naval advance overwhelmed Dutch forces in the Netherlands East Indies.

Ships employed as minelayers were commonly referred to as "nightraiders" because carrying out missions under the cloak of darkness increased the odds of survival in enemy waters. As MacArthur, Halsey, and Spruance's forces advanced toward Japan, minesweepers worked with "nightraiders"—clearing waters off landing beaches, while minelayers worked to deny the enemy freedom of the sea. Australian seaplanes ("Black Cats") flew long, perilous night-missions to mine Japanese harbours, and British submarines and planes joined in attacking shipping. (There were also many U.S. Navy squadrons equipped with the same type planes—PBY Catalinas painted black for night operations—operating in the Pacific, and which had performed mining earlier in the war. But while the "Aussie Black Cats" were mining Japan's inner zone, they were used almost exclusively to carry out bombing attacks on enemy shipping and shore targets.) Late in the war, U.S. Army Air Force bombers ringed the Japanese home islands with thousands of mines.

USAGES OF MINES AND MINESWEEPERS

Bluntly put, the purpose of a sea mine is to destroy enemy warships and shipping—hopefully in a manner that does not accidentally backfire by destroying the proponent's own forces, commercial shipping and fishing fleets, and those of neutral nations. In a larger sense the deployment of mines, involves certain strategies both involving stealth actions and public notifications to protect non-combative forces including fishing fleets.

As an overview to this book, it is important to understand the offensive and defensive use of minefields, and the vital role of minesweepers in combating them. A prime use of a mine is as a sea-denial weapon; one employed to deny to an adversary the use of waters, ports or harbours—or to force enemy shipping (hugging coastlines under the protection of shore guns and shielded from attack by submarines owing to shallow waters) farther out to sea where it can be attacked. Mines are also used offensively to blockade enemy ships inside a harbour from gaining the open sea, and to prevent merchant vessels from entering a harbour to deliver vital food, fuel, supplies and other materiel to the enemy. This tactic is particularly effective if the enemy lacks sufficient resources to quickly clear the mines to open the harbour to shipping.

Current NATO definitions divide offensive mining as generally explained above, into Offensive and Tactical categories, and defensive mining into Defensive and Protective ones:

- Offensive: Mines laid in enemy territorial water or waters under enemy control to deny him free use of his sea lines of communication or access to his own ports, harbours and anchorages.

- Tactical: Mines laid for a specific purpose or operation or as part of a formation obstacle plan laid to delay, channel or break up an enemy advance.

- Defensive: Mines laid in international waters or international straits with the declared intention of controlling shipping in defence of sea communications.

- Protective: Mines laid in friendly territorial waters to protect ports, harbours, anchorages, coasts, and coastal routes.

Map Preface-1

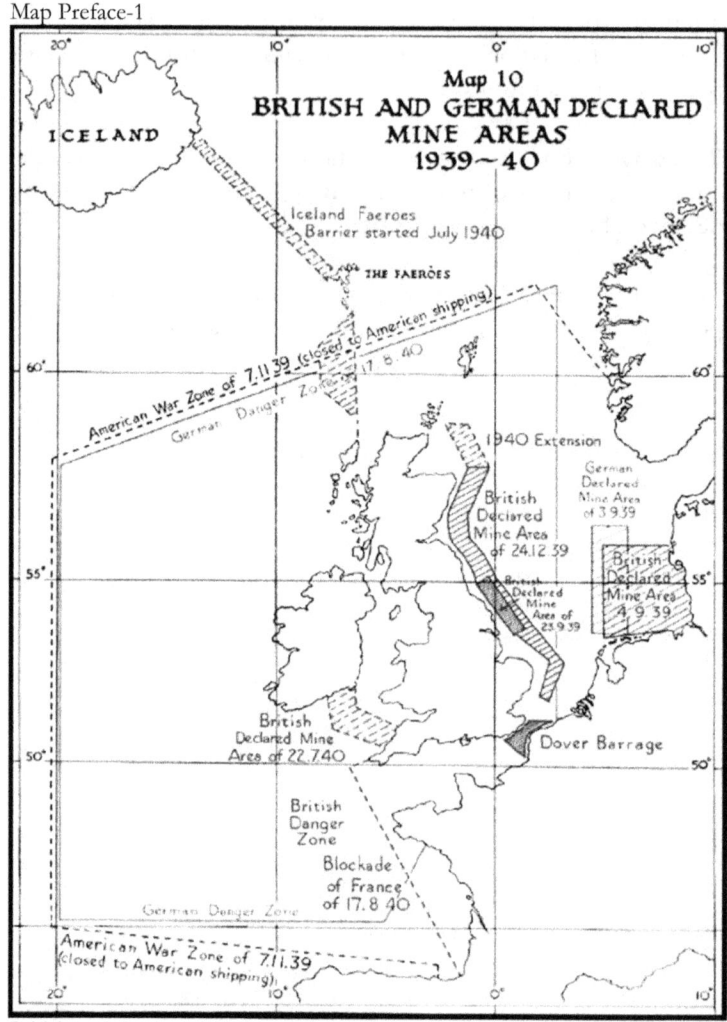

British and German Declared Mine Areas, 1939-40
S. W. Roskill, *The War at Sea 1939-1945* (London: HMSO, 1954)
http://www.ibiblio.org/hyperwar/UN/UK/UK-RN-I/UK-RN-I-6.html

Upon encountering offensive mines, the blockaded party typically despatches minesweepers, if available, to open the harbour or seaway. Mine clearance is painstakingly slow, as well as dangerous. As such, an urgent requirement to get to sea might force a ship's captain to try to run a suspected or known minefield. In recognition of the danger associated with trying to break out of a port blockaded by Royal Navy warships and/or mines, Germany authorised a special badge to be awarded to officers and men of the Kriegsmarine (German Navy) or Handlesmarine (Merchant Navy).[11]

Photo Preface-2

Otto Burfeind, captain of the blockade runner SS *Adolph Woermann*, who scuttled the merchant vessel (disguised as the Portuguese ship SS *Nyassa*) in the South Atlantic upon the approach of HMS *Neptune*. The British light cruiser *Neptune* picked up the crew and passengers and took them to England where they were interned at Seaton, Devon.
Courtesy of Dwight Messimer

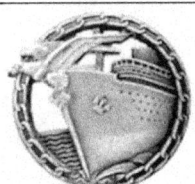

The "Badge for Blockade Runners" (Abzeichen für Blockadebrecher) was instituted on 1 April 1941 and first awarded on 1 July of the same year to the merchant vessel SS *Hugo Olendorff*

The criteria for awarding officers and men of German warships and merchant vessels attempting to run the blockade, and for some other notable actions, were:
- Distinction and good conduct, and
- Run a blockade and dock in a German port, or
- Service on a ship lost to enemy action, or
- To be wounded during an action at sea, or
- Scuttle a ship to avoid its capture, or
- For sinking an enemy ship in action, or
- Otherwise prevent the taking a German ship[12]

PROTECTIVE MINEFIELDS AND MINESWEEPERS

The purpose of protective minefields is to prevent enemy ships or submarines from slipping into harbour areas to attack friendly ships in supposed safe havens. If sufficient minesweepers are available to the enemy, his warships will most likely eventually leave harbour, perhaps to mine one's own shipping channels, ports, and harbours. In the case of enemy submarines, those configured to lay mines might do so, or torpedo or take ships under fire, or all three; being fitted with a deck gun and usually carrying some number of torpedoes. Defensive mining constituted the majority of mines laid by the Allies. As later enumerated, most were placed by surface ships.

IMPORTANCE OF MINESWEEPERS TO THE ENEMY

The scope of this book does not include German mine clearance efforts. It is important to note that Germany, like other maritime nations involved in the war, did not have sufficient numbers of minesweepers to combat her enemies. Comprising the German minesweeping forces were M-boats, R-boats, and a host of other types of ships that were pressed into service under the general designation Sperrbrecher. The larger M-boat and smaller R-boat classes were purpose-built minesweepers. The M-boats were called Minensuchboote or "Mine Search Boats," and the R-Boats Minenräumboote or "Mine Clearance Boats."[13]

Photo Preface-3

German M-boats under way
Courtesy of Dwight Messimer

Photo Preface-4

R-Boats off the French coast, probably 1941 or later
Courtesy of Dwight Messimer

The third type of minesweeper, Sperrbrecher, was not a class of ship, but instead vessels pressed into such duties, including merchant ships. The word Sperrbrecher translates as "barrier breaker" and "Sperrbrecher" was an assigned task rather than a specific type of ship. Sperrbrecher were employed as blockade runners and as vessels used to sweep mines—and many were lost during the war. F-boats, known as Geleitboote (escort boats), and their larger sisters, G-boats or Schnell Geleitboote (fast escorts), also fell into the Sperrbrecher category. The former vessels were heavily armed with up to ten anti-aircraft guns (20mm, 37mm, and 105mm).[14]

The G-boats carried up to sixteen anti-aircraft guns in the same 20 to 105mm bore sizes and were often called upon to lay mines as well as clear them. Additionally, four French boats captured at St. Nazaire while under construction, were completed as SG-boats. They were fitted with seventeen anti-aircraft mounts that carried guns from 20 to 105mm. At one time or another, all of these boats, F, G, and SG, acted as escorts for U-boats outbound from Heligoland and the German Bight (Bay) or returning, as well as sweeping or laying mines.[15]

As the war progressed, owing to extensive Allied mining (mostly British), the German Navy would be forced to devote more and more resources to mine clearance. In 1945 its minesweeping force involved 1,276 minesweepers and 46,000 personnel.[16]

The same held true for the Royal Navy, whose minesweeping force similarly grew to counter mines laid by the Axis. The number of British minesweeping vessels in commission on 3 September 1945 was 1,115—316 vessels having been lost. A comparison of minesweeper totals and cumulative losses in Home Waters and Abroad is provided in the table.

Numbers of British Minesweeping Vessels on 3 September 1945

Geographical Area	Vessels in commission	Vessel losses to date
Home Waters	654	185
Abroad	461	131
Totals	1,115	316[17]

The peak number of personnel directly engaged in British operated minesweeping was reached approximately at the time of the invasion of Normandy. The figure of 57,055 included the Royal Navy and its reserves, the Dominion navies and their reserves, and the officers and men manning ships of exiled allied navies.[18]

ROYAL NAVAL PATROL SERVICE (RNPS)

Unique silver badge awarded to those who served six months or more in the RNPS, worn on the sleeve of the recipient's uniform. Because a majority of RNPS members were Reservists, it became a "Navy within a Navy," one commonly referred to as "Harry Tate's Navy," "Churchill's pirates" and "Sparrows."

While Britain and other Allied forces were mining enemy waters, German aircraft, ships, and submarines were in kind mining British waters. Because of her expansive coastlines, Britain required, as she did in World War I, hundreds of minesweepers to allow the Royal Navy some freedom of movement and to keep her ports open to merchant shipping.

At the outbreak of war, the Royal Navy commandeered the Sparrow's Nest at Lowestoft, a private estate in Suffolk, to set up the headquarters for the Royal Naval Patrol Service. The Sparrow's Nest had been built as Cliff Cottage in the 19th century, a summer retreat for Robert Sparrow, the owner of a luxurious country house in nearby Worlingham. The estate became known as HMS Europa. (Unlike the names of vessels, the names of Royal Navy establishments preceded by HMS are not italicised because the 'ships' in question are 'stone frigates' to which shipboard terminology and customs still apply.) Located at the most easterly point of Great Britain on the North Sea, Lowestoft was then the closest military establishment to the enemy.[19]

Out of necessity, the Royal Naval Patrol Service utilised outdated and poorly armed vessels, mostly requisitioned trawlers crewed by ex-fishermen. The RNPS also helped man Royal Navy MMS (Motor Minesweepers otherwise known as "Micky Mouse") and BYMS (British Yard Minesweepers). The latter were the British-procured version of the famous U.S. Navy YMS minesweeper, to which the author's book *Wooden Ships and Iron Men Volume II: The U.S. Navy's Coastal and Motor Minesweepers, 1941-1953* is devoted.

The RNPS came to bear several unofficial titles that poked fun at it. One that gained prominence was "Harry Tate's Navy." This reference dated back to the First World War; it was jargon for anything clumsy and amateurish. It originated from an old music hall entertainer who portrayed a clumsy comic who couldn't come to grips with various contraptions, and whose act included a car that gradually fell apart.[20]

In WWII, regulars in the Royal Navy used this reference to poke fun at the trawlers and drifters of the Royal Naval Patrol Service. Nonplused, the members of the RNPS proudly adopted the title "Harry Tate's Navy" which, as the war went on, became a worthy synonym for courage. Because the peacetime crews of fishing vessels developed into Royal Naval Reserve seamen in the minesweeping fraternity, they quickly acquired a special camaraderie with one another. This amity continued throughout the war, though by the end most RNPS members were Royal Navy Volunteer Service "hostilities only" members with little previous connection to the sea. (An explanation about the differences between the RN, RNR and RNVR follows in a few pages).[21]

Photo Preface-5

George Hunt served aboard *MMS 1084* and HMS *Sobkra* in the Second World War. There is still a monthly gathering of RNPS veterans in Portsmouth, which he attends. Courtesy of Rob Hoole, photograph taken in June 2018

In keeping home waters swept, with some work abroad, the Royal Naval Patrol Service suffered the loss of over 250 vessels, more than any other branch of the Royal Navy. After the war's end, Winston Churchill sent the following message to the officers and men of the minesweeping flotillas in recognition of their sacrifices and vital contributions to the survival of Great Britain.

> Now that Nazi Germany has been defeated I wish to send you all on behalf of His Majesty's Government a message of thanks and gratitude.

The work you do is hard and dangerous. You rarely get and never seek publicity; your only concern is to do your job, and you have done it nobly. You have sailed in many seas and all weathers... This work could not be done without loss, and we mourn all who have died and over 250 ships lost on duty.

No work has been more vital than yours; no work has been better done. The Ports were kept open and Britain breathed. The Nation is once again proud of you.[22]

The officers and men of the RNPS fought in all theatres of the war, earning over 850 gallantry awards as well as over 200 Mention in Despatches. A Victoria Cross (VC) was won during the Norwegian Campaign by Lt. Richard Been Stannard RNR, the only one awarded to the unit. Stannard was commanding officer of HM trawler *Arab* of the 15th Anti-Submarine Striking Force. He received his VC for gallantry under air attack during operations off Namsos, Norway.[23]

Today, some of the few remaining original buildings of HMS Europa in Sparrow's Nest, Lowestoft, house the Royal Naval Patrol Service Museum and the War Memorial Museum. There is a memorial in Belle Vue Park, in the north end of Lowestoft, to commemorate members of the RNPS who died during 1939-1946 and who have no known grave other than the sea, as well as a few who died on shore and who also have no known grave.[24]

Importantly, some of the few remaining RNPS members continue to visit one another to share memories and remember their shipmates.

MINELAYING SERVICE RECEIVES FEW ACCOLADES

While the British public was very appreciative of the RNPS both during and after the war, the Minelaying Service and its members, to which the bulk of this book is devoted, were largely overlooked. This may have been due, at least in part, to a general belief that clearing the enemy's hated mines from home waters was good, while putting one's own into the sea (where innocent fishing vessels and merchant vessels could encounter them) was considered "odious." The British government promulgated notices identifying the boundaries of the waters it mined, but such postings were not always studied carefully by merchantmen and fishermen and losses occurred.

To illustrate this point, HMS *Latona*, one of six *Abdiel*-class "fast minelayers" (cruisers purpose-built to serve as minelayers) was not adopted by a community as part of a British Nation Savings WARSHIP WEEK campaign in 1941-1942. This may have been because of the nature of her primary mission. HMS *Abdiel* had been similarly slighted, as would be the other ships of the class. Ironically, these ships, apart from their minelaying functions, performed heroic, well-publicised deeds. Because the ships were very fast at 39 knots, some were routinely used to make overnight runs from Gibraltar to Malta, to supply fuel and food to the beleaguered British Colony. Others transported stores and personnel to Cyprus and Tobruk, also under siege by German and/or Italian forces. Apparently, even these accomplishments could not relieve their tainted status in the public eye.

BRITAIN NOT ALONE IN MINE WARFARE EFFORTS

At this point, some readers may be wondering about the involvement of the other countries whose Naval ensigns are displayed on the cover. The flags in the four corners of the group were those of the Royal Navy and dominions: The Royal Canadian Navy, Royal Australian Navy, and the South African Naval Force. South African minesweepers operated with the Royal Navy's Mediterranean Fleet. Officers and men from the RCN and RAN served aboard Royal Navy minesweepers, with some officers in command. Australia's minesweepers and minelayers were occupied with American ones in the vast Pacific. (Their activities are covered in great detail in Cdre. Hector Donohue, AM RAN (Rtd.) and Mike Turner's excellent book, *Australian Minesweepers at War*.)

The mine warfare ships of the Royal Canadian Navy were primarily engaged off Canada's east coast, combating German U-boats, as were ones of the U.S. Atlantic Fleet farther down the eastern seaboard of North America. U.S. Navy minesweepers and minelayers sailed for Europe in October 1942 to take part in Operation TORCH, the Allied invasion of French North Africa, and later participated in other operations, including assault landings in Italy and those at Normandy and in southern France. Canada's principal mine warfare involvement in Europe came in 1944, when it despatched sixteen *Bangor*-class fleet minesweepers to Britain to clear swept paths across the English Channel ahead of the Normandy assault forces.

Preface   xxxix

Following German occupation, in 1940, of France, the Netherlands and Norway, remnants of their navies fled to England and established headquarters in London. Discussion of actions by these exiled forces is largely limited herein to those of the Dutch RNLN minelayer *Willem Van der Zaan*, the FFNF (Free French) minelaying-submarine *Rubis*, and the Norwegian-manned 52nd Motor Launch Flotilla.

Photo Preface-6

Poster promoting the importance of the joint war effort of the British Empire and Commonwealth, 1939.
Lowe and Brydone Printers Ltd, London NW10 (printer)

BRITISH OFFENSIVE MINING IN WORLD WAR II

The British employed ships, submarines, planes, and coastal forces craft for offensive mining during the war. The largest effort by far, was that of aircraft which laid about 50,000 mines in northwest Europe, significantly impeding shipping traffic and naval movements. The purpose of an offensive minefield includes blockade, restricting the movement and attacks from enemy naval combatants, submarines, and merchant vessels; and endeavouring to destroy or damage such vessels in port, or at sea. No vessels would be sunk by an ideal field, because it would prevent the movement of shipping.[25]

As highlighted on the first page, in all theatres in World War II, including the Pacific (not covered in this book), over 76,000 British mines were laid in enemy waters. Of these, nearly 55,000 were laid by aircraft, 11,000 by fast minelayers and destroyers, 6,500 by Coastal Forces, and 3,000 by submarines. British losses of vessels and aircraft performing solely minelaying were one fast minelayer, two destroyers, four submarines, and four Coastal Force craft, and approximately 500 aircraft lost in 21,000 sorties. However, these numbers do not reflect true losses of resources. Not counted are minelayers sunk by the enemy while carrying out other tasking, with their killed and wounded personnel additional losses to the mine force and to the navy.[26]

The 11,000 mines laid by ships in enemy waters only accounted for a small portion (5½-percent) of the 199,002 mines sown. The remainder were put in defensive fields. Quantities by ship categories (with mines combined) are shown in the table.

Minelayer	Mines Laid	Remarks
Cruiser Minelayer (12,401 mines)		
HMS *Adventure*	12,401	Damaged by mines
Converted Merchant Vessels (133,334 mines)		
HMS *Agamemnon*	24,216	
HMAS *Bungaree*	6,800	
HMS *Menestheus*	22,866	Damaged by bombing
HMS *Port Napier*	6,331	Destroyed by fire
HMS *Port Quebec*	33,494	
HMS *Southern Prince*	23,762	Damaged by torpedo
HMS *Teviot Bank*	15,865	
Coastal Minelayers (17,525 mines)		
HMS *Plover*	15,327	
HNLMS *Van der Zaan*	2,198	
Converted Train-ferries (7,932 mines)		
HMS *Hampton*	5,190	
HMS *Shepperton*	2,742	

Converted Car-ferry (2,755 mines)		
HMS *Princess Victoria*	2,755	Sunk by mine
Fast (Cruiser) Minelayers (18,301 mines)		
HMS *Abdiel*	2,207	Sunk by mine
HMS *Apollo*	8,361	
HMS *Ariadne*	1,352	
HMS *Latona*	0	Sunk by bombing
HMS *Manxman*	3,111	Damaged by torpedo
HMS *Welshman*	3,275	Sunk by torpedo
Destroyer Minelayers (6,754 mines)		
HMS *Esk*	1,110	Sunk by mine
HMS *Express*	1,210	Damaged by mine
HMS *Icarus*	1,196	
HMS *Impulsive*	892	
HMS *Intrepid*	1,592	
HMS *Ivanhoe*	754	Sunk by mine
Total ship-laid mines[27]	199,002	

BOMB DISPOSAL AND RENDERING MINES SAFE

> *The RNVR (classic, wartime reservists known as 'Saturday night sailors') were gentlemen trying to be sailors.*
>
> *The RNR (professional seamen and part-time Navy officers) were sailors trying to be gentlemen.*
>
> *and The RN (regular Navy officers) were neither trying to be both.*
>
> —Old saying in the Royal Navy, courtesy of Rob Hoole

The emphasis of this book is surface ship, submarine, and coastal forces minelayers, consequently only a small portion is devoted to minesweepers and their stalwart crews, and practically none to "human minesweepers"—courageous divers who rendered mines safe. Co-author Rob Hoole, a retired RN Mine Clearance Diving Officer, offers in the postscript an overview of this area and associated bomb disposal in World War II.

The Naval Reserve Act of 1859 established the Royal Naval Reserve (RNR) as a reserve of professional seamen from the British Merchant Navy and fishing fleets, who could be called upon during times of war to serve in the regular Royal Navy. In 1862, the RNR was extended to include the recruitment and training of reserve officers, who wore on their uniforms a unique and distinctive lace consisting of stripes of interwoven chain. The Royal Naval Volunteer Reserve (RNVR)—the so-called "wavy Navy"—was created in 1903.

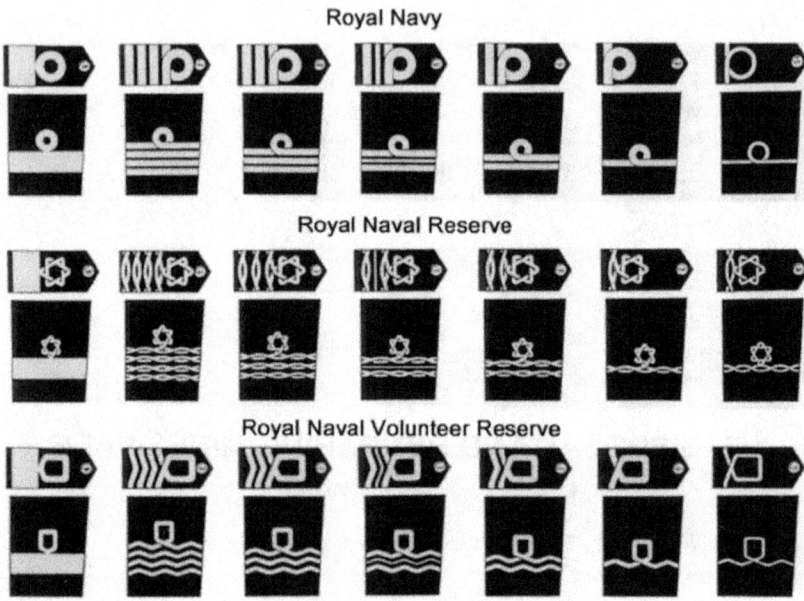

World War II British Naval Officer shoulder boards and sleeve insignia

While the Royal Naval Reserve was composed of personnel from the merchant marine and fishing communities, members of the Royal Naval Volunteer Reserve came from other civilian backgrounds. Another difference was the gold braid officers wore on their sleeves to denote rank. Those of the RNR were in intersecting waved pairs, while the pattern of RNVR braid was single waved lines—thus the RN moniker "wavy Navy" when referring to the latter officers.

The Royal Navy's Bomb Disposal/Render Mines Safe Community, like the Royal Naval Patrol Service, was mostly populated by RNR and RNVR members, and the same held true for the Dominion navies. At the onset of war, some regulars serving in the Royal Navy made fun of RNR and RNVR personnel, whom they termed amateurs. However, the amateurs soon became professionals in mine clearance, a warfare area generally underappreciated by "top brass" who'd gained promotion aboard battleships, cruisers and destroyers.

USEFUL INFORMATION

Before readers vicariously stand out to sea with units of the Mine Force, or enter murky depths with Clearance Divers, some explanation of language, officer ranks and awards for valour, nautical/naval terms, and the indexing of ship name entries might be in order; particularly to those not part of the Commonwealth. Those desiring more information about particular types of mines, or terminology associated with mines, minelaying and minesweeping may consult Appendices A and B.

The British spelling of particular words are used throughout as a nod to the Royal Navy and the British people to which a majority of the book is devoted. (They also apply to the Dominions generally.) The primary differences are the addition of the letter "u" in some words, and the use of "s" instead of "z" in others.

British	American	British	American
authorise	authorize	labour	labor
calibre	caliber	manoeuvre	maneuver
centre	center	metre	meter
colour	color	minimise	minimize
defence	defense	organisation	organization
despatch(es)	dispatch(es)	programme	program
draught	draft	publicise	publicize
endeavour	endeavor	recognise	recognize
favourable	favorable	utilise	utilize
harbour	harbor	valour	valor
honour(s)	honor(s)	vigourous	vigorous
italicise	Italicize		

COMPARABLE NAVAL OFFICER RANK STRUCTURE

The United States and Dominion navies were patterned after the Royal Navy and, sharing a common language, utilised a similar officer rank structure. (The Royal Canadian Navy represents other Dominions in the following table.) An RN sub lieutenant is the equivalent of a USN lieutenant (junior grade), and an RN midshipman the same as a USN ensign, because the Royal Navy does not use the latter rank. The rank of midshipman in the U.S. Navy and Imperial German Navy was below that of Ensign and Oberleutnant zur See, respectively.

Royal/Royal Canadian/U.S. Navy Rank		Kriegsmarine (German Navy) Rank	
Admiral	Adm.	Admiral	Adm.
Vice Admiral	Vice Adm.	Vizeadmiral	VAdm.
Rear Admiral	Rear Adm.	Kontreadmiral	Kadm.
Commodore	Cdre.	Kommodore	Kdre.
Captain	Capt.	Kapitän zur See	Kpt. z. S.
Commander	Comdr.	Fregattenkapitän	FKpt.
Lieutenant Commander	Lt. Comdr.	Korvettenkapitän	KKpt.
Lieutenant	Lt.	Kapitänleutnant	Kptlt.
Sub Lieutenant	Sub Lt. [RN]	Oberleutnant zur See	OLt. z. S.
Lieutenant, Junior Grade	Lt. (jg) [USN]	Oberleutnant zur See	OLt. z. S.
Ensign	Ens. [USN]	Leutnant zur See	Lt. z. S.
Midshipman	Mid. [RN]		
Midshipman	Mid. [USN]	Oberfähnrich zur See	Fähn. z. S.

The prefaces "A/" or "T/A" are associated with some officers' ranks in the book. In general, 'A' or 'Acting' higher rank was appropriate when an officer was assigned to perform the full duties of a post at a higher rank because no officer of the required substantive rank and branch was available to do so. Award of the 'Acting' higher rank attracted the pay of the higher rank but did not accrue seniority in the higher rank.

'T' or 'Temporary' commissions were granted by the Crown to many RNR and RNVR officers, in such rank as deemed appropriate, to suitable persons who volunteered their services in times of emergency. Temporary officers in the RNR and RNVR were entitled, while so employed, to the rank, pay and allowances of the corresponding permanent rank in the RNR and RNVR and were generally treated in all other respects, and were subject to the same regulations, as permanent RNR and RNVR officers. Such officers were liable to service either ashore or afloat, as directed, until their temporary commissions were terminated.

In general, regular RN officers were granted permanent commissions by the Crown. These lasted for the duration of their lives unless forfeited (e.g., owing to some serious misdemeanour) or resigned (e.g. to serve in another nation's forces).

A significant difference exists between references to officers in the Royal Navy and its dominions, and those of the United States Navy. Those of the former include "Sir," if knighted (Royal Navy), following an individual's military rank, and reference to military awards earned after surname.

Over the course of their careers, officers advance in rank and may receive additional awards. Since it is difficult to associate the latter with the former at any given point in time, the convention is to denote the final rank of an officer, and all awards they received in the first reference to that officer. So, the first reference to fictional Lt. John Smith, RN, would include in parenthesis after his surname (later Vice Adm. Sir John Smith, VC, DSO, DSC, CGM). In order to make the text easier to follow, particularly for those without naval backgrounds, this information is provided after the individuals' names in the index.

Photographs and descriptions of six of the awards most commonly referenced in *Enemy Waters* follows:

A Few British Military Awards

	Victoria Cross (VC): Highest award of the British honours system; for gallantry "in the presence of the enemy"		Distinguished Service Cross (DSC): For an act or acts of exemplary gallantry during active operations against the enemy at sea
	George Cross (GC): For acts of the greatest heroism or for most conspicuous courage in circumstance of extreme danger, not in the presence of the enemy		Conspicuous Gallantry Medal (CGM): Award for enlisted members for conspicuous gallantry in action against the enemy at sea or in the air
	Distinguished Service Order (DSO): for meritorious or distinguished service by officers during wartime, typically in actual combat		George Medal (GM): For gallantry "not in the face of the enemy" where the services were not so outstanding as to merit the George Cross

Of course, other awards existed; some with humorous acronyms describing how "other ranks" viewed those received by their seniors:

- MBE (Member) – My Bloody Efforts
- OBE (Officer) – Other Buggers' Efforts
- CBE (Commander) – Covers Bloody Everything
- Order of St Michael & St George which has three classes:
  CMG (Companion) – Call Me God
  KCMG (Knight Commander) or DCMG (Dame Commander) – Kindly Call Me God
  GCMG (Knight Grand Cross or Dame Grand Cross) – God Calls Me God

NAUTICAL/NAVAL TERMS

- Carley raft/float: An early life raft consisting of a large oval ring of copper tubing covered with kapok and waterproof canvas.
- Graving dock: A large basin with gates from which water can be pumped out; used for building ships or for repairing a ship below its waterline.
- Scuttle: To cause a vessel to sink by opening the seacocks or making holes in the bottom of its hull.
- Stoker: An engineering rating responsible for feeding coal into the firebox of a boiler providing steam to propulsion turbine. "Stoker" survives as an unofficial term for a marine engineering mechanic in the Royal Navy to this day.

FINDING INDEX REFERENCES TO A SHIP OR SHIPS

Former sailors, or relatives or friends of former sailors picking up a book such as this one often desire to ascertain whether or not it includes any references to a particular ship in which they or another individual served. In acknowledgement of this fact, an extensive index is included. To reduce its size, multiple ships listed on the same page or pages in the text are combined into a single entry. Entries for American ships are located under their associated ship type headings. For example, the battleship *Arkansas* can be found under Ships and Craft, and the sub-categories: United States, Navy, combatants, and battleships.

A reader searching for a particular ship of another country should review all entries under the heading for that country. For example, if searching for HMCS *Quatsino* under the sub-heading Canada, it would be found on the fourth line, as part of the entry for *Bellechasse, Chignecto, Courtenay, Kelowna, Miramichi, Outarde, Quatsino, Transcona,* 227, which are all listed on but a single page (227) of the text.

Photo Preface-7

Painting by Richard DeRosset of the Battle of Taranto in which, on the night of 11 November 1940, twenty-one Fairey Swordfish torpedo-bombers launched from HMS *Illustrious* struck the battle fleet of the Italian Navy at anchor in the harbour of Taranto.

1

Operation MINCEMEAT

However painful, the action we have already taken should be, in itself, sufficient to dispose once and for all of the lies and Fifth Column activities that we have the slightest intention of entering into negations. We shall prosecute the war with the utmost vigour by all the means that are open to us.

—From an address by Winston Churchill to the House of Commons on 18 June 1940 explaining his intention that the British Fleet sink the Free French Fleet at the port of Mers-el-Kebir, outside Oran, Algeria. This action was carried out on 3 July, following France's surrender, to prevent Germany's use of the ships to wage war.[1]

Photo 1-1

HMS *Ariadne* (M65), sister ship of HMS *Manxman* (M70); two of the six *Abdiel*-class minelaying light cruisers, then the fastest ships in the world.
Naval History and Heritage Command #NH 81892

In August 1941, British Vice Adm. Sir James Somerville, commanding Force H based at Gibraltar, conceived the idea that the fast minelayer HMS *Manxman* (M70) should disguise herself as a French cruiser, steam openly and alone along the Riviera coast in daylight, dash into the Gulf of Genoa by night, lay a minefield on the very doorstep of the great Italian port of Leghorn (Livorno), and then somehow find her way back to Gibraltar. *Manxman* was named for the people of the Isle of Man in the Irish Sea. She was one of the new *Abdiel*-class minelaying light cruisers, the fastest ships in the world and able to do the Gibraltar to Malta run overnight to supply fuel and food to the island. Powered by four Admiralty 3-drum boilers, coupled to geared turbines and two shafts (72,000 shaft horsepower), she had a top speed of 39 knots. *Manxman* was also heavily armed, boasting three twin mount 4-inch (100mm) guns, one quadruple mount 2-pounder (four 40mm guns), and two 0.5-inch (12.7 mm) quadruple mount Vickers machine guns (eight barrels). She could also carry up to 160 mines, transported along rails to a pair of doors in the quarters of the ship.[2]

Photo 1-2

Leghorn (Livorno) on the northwest coast of Italy, July 1944, showing damage inflicted by Allied bombers to docks and port installations.
Australian War Memorial photograph SUK12589

The code name for this daring plan was Operation MINCEMEAT. (This name was also used later in the war, for a famous deception that convinced the Germans that the Allies planned to attack Greece instead of Sicily.) During the long grim period between the fall of France on 25 June 1940, and the Battle of El Alamein, 23 October to 11 November 1942—which marked the culmination of the North African campaign between the British Empire and the German-Italian army—enemy forces reigned supreme in the central Mediterranean. At the eastern end of the Med, the British Fleet under Adm. Sir Andrew Cunningham guarded the Suez Canal. Force H, under Adm. Sir James Somerville, guarded the Straits of Gibraltar at the western end.[3]

All through the central Mediterranean, from Sardinia to Crete, the Italian Fleet and the German and Italian Air Forces had overwhelming superiority of numbers; resulting in enemy dominance of the waters of the Adriatic, the Riviera coast, and the Gulf of Genoa. No British ship, except a submarine, could enter those waters without its presence being sighted from the air and reported. Thus, there was no certainty of success. The operation might well end in utter failure with loss of life and perhaps the ship as well.[4]

Map 1-1

Mediterranean Basin
http://legacy.lib.utexas.edu/maps/europe/mediterranean_rel82.jpg

On 14 August 1941, *Manxman* was lying in the Kyle of Lochalsh—to the Navy simply "Port ZA"—on the rugged northwest coast of Scotland opposite Skye (an island of the Inner Hebrides). Rear Adm. Robert Lindsay Burnett, Flag Officer of Minelayers who commanded the squadron based there, was aboard the ship that morning meeting with her commanding officer, Capt. Robert Kirk Dickson, in his cabin. During this visit, Dickson's secretary came in and handed Burnett a secret telegram from the Admiralty. It conveyed that "For her next operation it is desired that *Manxman* should resemble the French cruiser *Léopard* as far as is reasonably practicable."⁵

Burnett had been well-known in his earlier years as a producer of naval theatricals, and this sort of thing tickled his fancy. After reading the missive, he told Dickson,

> Well, I don't know what their Lordships mean by reasonably practicable, but I'll give you just twenty-four hours to be the French cruiser *Léopard*. You can use the entire resources of this squadron, and of this base. Do what you like and make any signals you like in my name, but I don't want to see any of you again today. I'll come on board you at 11.30 tomorrow morning and I'll inspect you as a Frenchman.

With that he departed.⁶

Photo 1-3

Free French cruiser *Léopard* on 6 June 1942, following refit in the United Kingdom in which her forward stack and boilers were removed, fuel stowage increased and additional anti-aircraft guns fitted.
Naval History and Heritage Command photograph #NH 89001

With only the 1940 edition of *Jane's Fighting Ships* for guidance, *Manxman*'s crew set to work. From the photographs available in the book, there didn't seem to be any resemblance between the *Manxman* and the *Léopard* except that they both had three funnels (stacks). However, everyone had ideas and there was labour in plenty to put them into action. In addition to the ship's force, 120 personnel from the squadron had come aboard, each with a paint pot and brush and more ideas. The necessary changes began by raking the masts and funnels, not actually, of course, but giving them that appearance through the use of canvas, 40-foot spars, and black and white paint. *Manxman*'s flush deck was also given a false break, the upward trend of her forecastle corrected, and characteristic beveled French funnel tops fashioned out of sheet iron and installed.⁷

Work began at midnight to change the shape of the bow and stern. *Manxman*'s stem was straight, but base personnel had fabricated a false curved one of steel plates. It was a grand affair, but nothing would induce it to fit properly over the paravane gear. So, it was discarded; replaced by a substitute of shaped canvas, spread on wires between a bowsprit and heel fittings welded on each side of the stem at the waterline. Similar problems were experienced with a false stern, also of steel plates, which fell off the first night out from Gibraltar and had to be replaced by a structure of wood and canvas.[8]

With the ship's appearance altered as well as could be, attention was turned to other niceties. It would have been unfortunate if an aircraft inspecting the "French ship" found her upper deck thronged with British sailors. A hundred French uniforms were turned out, consisting simply of white singlets painted with blue horizontal stripes, and red bobbins to be pinned on top of blue caps.[9]

Rear Admiral Burnett arrived punctually at 1130 the following morning, inspected the French guard fallen in on the quarterdeck, and toured the ship. He was satisfied with her transformation and reported to the Admiralty that their instructions had been complied with. The false props were then unrigged and laid inboard. This action was taken to prevent anyone from seeing the disguise until it was necessary to use it, and partly because it would not have stood up to the Atlantic weather on the passage to Gibraltar.[10]

MEDITERRANEAN PASSAGE

HMS *Manxman* stood out of Kyle the next day, and after loading mines at Milford Haven on the southwest coast of Wales, sailed for Gibraltar on the evening of 17 August 1941. In an effort to avoid being identified by friend or foe, she made a big detour out into the Atlantic, and upon sighting other ships altered course to keep her hull down. Admiral Somerville's orders were to approach the Straits of Gibraltar after dark, and to make false identification signals upon arrival. Gibraltar was a nest of spies; thus, *Manxman* was not to enter harbour, or have any communication with the shore. She was instead to fuel from an oiler anchored out in the bay; and transfer to the same oiler all her secret books and cyphers to avoid the risk of carrying them into dangerous waters. Finally, *Manxman* was to be away into the Mediterranean and out of sight of land before daylight.[11]

While his ship was fueling in darkness in Gibraltar Bay, Dickson read the detailed operation orders which Somerville had aboard the oiler for him. It was then 0100 Friday morning, and *Manxman* was to lay the minefield at Leghorn in the early hours of Sunday morning. Somerville had just left Gibraltar with Force H—the battleship *Nelson*, aircraft carrier *Ark Royal*, the cruiser *Hermione*, and the destroyers *Nestor, Encounter, Foresight, Fury* and *Forester*. His intention was to operate southward of the minelayer and generally do all he could to distract the enemy's attention from her, while she slipped along the French coast and round the north end of Corsica.[12]

Photo 1-4

Moored mines aboard HMS *Abdiel* (M39), ready to be laid. The boxes upon which the mines rest are their anchors. Inside each anchor was a winding mechanism which allowed the cable to payout to preset lengths. Tarps were rigged to avoid disclosing the presence of mines aboard ship to enemy planes, passing vessels, or spies ashore.
Australian War Memorial photograph P03538.001

Manxman sailed from Gibraltar and Friday night, as she passed between the Balearic Islands and Spain, crewmembers rigged her French disguise and put on their French uniforms. At sunrise on Saturday, being then fifty miles northwest of Minorca—one of the Balearic Islands belonging to Spain—she hoisted French colours. Steaming north at moderate speed, as a French cruiser on the route between Oran and Toulon. Those aboard hoped that she appeared correct, in every detail, from the French pendant at the mainmast head to the washed clothes flapping on the quarterdeck.[13]

Then began the most intense period of the operation, in which captain and crew had to trust entirely to *Manxman*'s French disguise. They were under no illusions about what would happen if it failed while their ship was entirely alone in the northwestern Mediterranean. If any French, German, or Italian aircraft approached, the sailors on deck were to wave to them in a friendly manner. If a bomb was dropped, the plan was to signal "Attention! Vichy!" via searchlight to inform the attacker they were Vichy French. Should their ruse fail, Dickson intended to hoist the White Ensign, and engage the enemy in combat.[14]

Happily, this was not required. During the entire operation, only two aircraft were sighted. The first was a German Dornier and the second was probably French. By late afternoon on Saturday, *Manxman* was within forty miles of Toulon. With the Riviera coast well in sight to the northward, she turned east, and steered as if for the French port of Ajaccio in Corsica. When it was quite dark, Dickson ordered the false bow dismantled; to permit paravanes to be streamed to provide protection against any enemy mines that might be encountered, and the speed increased to 30 knots. Patches of mist made passage close round the north end of Corsica challenging. Once east of it the weather cleared, and there was enough diffused starlight to provide working light for men on deck to put mines into the water.[15]

MINING OPERATION

Manxman arrived in the Gulf of Genoa and the lay of 140 mines began at 0200. Conditions were ideal. The sky was thickly overcast, the sea glassy calm, and the mainland of Italy blacked out. The absence of lights ashore that mighty disclose the ship's presence was a saving grace. *Manxman* was near enough the shore that land was plainly visible, and the masked headlights of a motor car could be seen on the coast road just opposite her position. The black cloud shadows on the water were unsettling; but offered additional protection. Dickson later described the operation as having being uneventful, but not so the return passage:

> Once there was an explosion of some kind in the sky above Leghorn, twice flares were dropped over the land, and twice we thought we had been challenged by destroyers. But these alarms came to nothing and we completed the lay at 3-30 [0330].
>
> I remember the next three-quarters of an hour as being the most trying part of the whole performance, because we had to make our getaway past Gorgona Island, where we suspected the Italians of operating hydrophones. It was essential to reduce our propeller noise to the minimum when passing it, so the first part of the retreat had to be done at 10 knots, with daylight getting nearer and with all the horsepower of the fastest ship in the Navy waiting to be unleashed.[16]

By 0415, *Manxman* was clear of Gorgona and had increased engine revolutions to full speed. Soon after, a sighting was made of three ships in line ahead, but it was easy to avoid them and there was no indication that they had detected the minelayer. At daylight, *Manxman* reduced speed to 33 knots, so as to avoid making funnel smoke. An hour later, she encountered thick fog, but held that speed for six hours through mist which intermittently enveloped the ship, until west of Toulon. Dickson believed that the murk "really was the protecting hand of Providence, because fog in that part of the Mediterranean in August is very rare."[17]

Most of those aboard believed their adventures were behind them. But in the afternoon, while steaming down the Spanish coast, there appeared on the eastern horizon the masts and funnels of a ship which appeared to be a real French cruiser on passage from Oran to Toulon. There was nothing to do but continue onward. Fortunately, the other ship proceeded on her way, unalerted, eliminating the possibility of an engagement with the enemy. Much of the success of the mining operation may have been due to aircraft from the *Ark Royal* making a night attack on the cork woods at Tempio in Sardinia, with incendiaries, just as *Manxman* had been approaching Leghorn. Six hours after this diversion, an Italian fleet—three battleships, six cruisers and twenty-five destroyers—was reported near Sicily, steering west. Fortunately for Force H, the ships never came west of Sardinia.[18]

Manxman arrived back at Gibraltar after dark on Monday night. Wasting no time, after fueling in the bay as before, and picking up her secret books, she prepared for her homeward run. She was well clear and into the Atlantic before daylight, bound for Kyle of Lochalsh.[19]

AFTERMATH

Unfortunately, the heroics of *Manxman* came to naught. On 26 August, no less than 22 mines of the mines she had laid were sighted, and the entire field was eventually swept. This work was accomplished by the Italians with the assistance of three sets of German sweep gear capable of dealing with M Mk I mines; one of the latter was recovered intact. During clearance operations, the 109-ton Italian sweeper *Carmelo Noli* was sunk on 23 September 1941.[20]

Robert Kirk Dickson received the DSO (Distinguished Service Order) on 22 May 1942, and CB (Companions of the Order of the Bath) on 8 June 1950. He retired as a rear admiral on 22 April 1952.[21]

2

British Minelaying Forces

British minelaying operations were carried out from the day that hostilities commenced until 10 May 1945 when the submarine RORQUAL laid 44 mines off Thousand Islands in the Pacific. They were carried out in all types of weather and ships deployed in coastal waters had the additional hazard of enemy mines apart from encounters with hostile warships and aircraft.

—From a Royal Navy Minelaying Operations monograph prepared by Geoffrey B. Mason, Lieutenant Commander, RN (Rtd), for the Librarian of the Chatham Historic Dockyard Trust.[1]

When Britain declared war on Germany on 3 September 1939, the Royal Navy was deficient in minelayers. If three small, specialised indicator loop minelayers are set aside—*Linnet*, *Redstart*, and *Ringdove*, which could carry only ten mines—*Plover* was the single minelayer operational. In fact, that day she laid mines off Bass Rock near the Firth of Forth. The other minelayer, *Adventure* (the Royal Navy's first purpose-built minelayer), was in reserve, awaiting re-commissioning. A number of E- and I-class destroyers built in the 1930s with minelaying capability, were nearing conversion to this role at the time of the outbreak of hostilities. However, whereas *Adventure* could carry 280 mines and *Plover* 100, the converted destroyers were limited to sixty. There were also six S-class destroyers, commissioned 1918-1920, that could carry forty mines. This situation would later improve with the commissioning between April 1941 and February 1944, of six fast (39 knot) minelaying cruisers of the *Abdiel*-class, each capable of carrying up to 160 mines; and of four O-class destroyers in 1942, which could carry sixty apiece.[2]

To address the immediate shortfall, the Royal Navy turned to the Merchant navy, requisitioning, first, the liner SS *Teviotbank* (built 1938). Commissioned HMS *Teviot Bank*, after conversion to carry 280 mines, she could muster only 12 knots. Two subsequent acquisitions requiring minimal modification were the Southern Railway's "boat train ferries" SS *Hampton Ferry* and *Shepperton Ferry*, built in 1934. Their boilers exhausted through funnels athwartships, leaving completely open their main decks with four rail tracks and, aside from a low gate, sterns as well. With this deck arrangement, they were easily modified to carry 270 mines apiece. The former railway ferries quickly joined *Adventure* and *Plover* at Dover, in southeast England. The four ships began a massive six-day minelaying operation in the Dover Strait on 11 September 1939, effectively closing the English Channel to German U-boats heading for the Western Approaches.[3]

The factors that made the railway ferries attractive applied equally to the newly built Stranraer-Larne stern-loading car ferry *Princess Victoria*, another requisitioned ship. She was even more suitable being diesel-driven, as there was no requirement for frequent boiler cleaning; tragically, her service was short-lived. On 19 May 1940, she struck a mine off the entrance to the Humber River and sank quickly.[4]

The Royal Navy also acquired the Prince Line cargo liner MV *Southern Prince*, Blue Funnel Line sister cargo-passenger ships MV *Agamemnon* and MV *Menestheus*, and merchant vessels *Port Napier* and *Port Quebec* for conversion to auxiliary minelayers. The latter two ships were still under construction when requisitioned.[5]

The SS *Kung Wo* and SS *Mao Yeung* were acquired in the Far East to assist the Admiralty S-class destroyers *Scout*, *Stronghold*, *Tenedos*, *Thanet*, and *Thracian* with minelaying associated with the defence of Hong Kong and Singapore. Summary information about the Royal Navy's surface ship minelayers and submarine-minelayers follows. These totals do not include the considerable numbers of smaller craft of the Coastal Forces that engaged in laying mines in hostile waters.

RN MINELAYERS (50 SHIPS/22 SUBMARINES)

- Cruiser-minelayer *Adventure* (purpose-built)
- Eleven merchant ships converted to auxiliary minelayers
- Six *Abdiel*-class cruiser-minelayers
- Two E-class destroyer-minelayers
- Four I-class destroyer-minelayers
- Four O-class destroyer-minelayers
- Six Admiralty S-class destroyer-minelayers
- Coastal minelayer leader *Plover*
- Two *Corncrake*-class coastal minelayers
- Three *Linnet*-class indicator loop minelayers
- Eight *M 1*-class coastal minelayers
- Two mining tenders, *Vernon* (later-*Vesuvius*) and *Nightingale*
- Six *Porpoise*-class minelaying submarines

- Thirteen T-class minelaying submarines
- Three S-class minelaying submarines

CRUISER-MINELAYER HMS *ADVENTURE*

The abbreviation "Comm." in the ensuing tables refers to the date a ship was commissioned; and "No. Mines" to the number of mines it could carry, and the total number laid during World War II (if known). HMS is eliminated as part of ships' names for brevity.

Adventure-class Cruiser-Minelayer
(520 feet, 6740 tons, 28 knots, 395 ship's complement)

Ship	Comm.	No. Mines	Disposition
Adventure (M23)	2 Oct 26	280 mines/12,401	Damaged by mines, became a repair ship in 1944, sold 10 Jul 1947, broken up at Briton Ferry[6]

Photo 2-1

British Minelayer HMS *Adventure*, location and date unknown. National Archives photograph #80-G-165939

CONVERTED MERCHANT VESSELS

The information in parentheses in the "Former Ship Type" column refers to ship propulsion: MV (motor vessel) or SS (steamship), and date the ship was built, as opposed to when the RN commissioned it a minelayer.

Auxiliary Minelayers

Ship	Former Ship Type	Comm.	No. Mines	Disposition
Agamemnon (M10)	Blue Funnel Liner (MV/1929): 460 feet, 7593 tons, 16 kts	Oct 40	530/ 24,216	Became an amenities ship in 1944 for service with the Pacific Fleet, returned to owner on 26 Apr 1947
Hampton (M19)	Train-Ferry (SS/1934): 347 feet, 2839 tons, 16.5 kts	5 Sep 39	270/ 5,190	To the Ministry of War Transport (MoWT) in 1940
Kung Wo	River Passage Ship (SS/1921): 350 feet, 4636 tons, 10 kts	1941	248/ 224	Bombed by Japanese aircraft northwest of Pompong Island (near Lingga) and scuttled on 14 Feb 1942
Mao Yeung	Hong Kong River boat (SS): 371 tons	1939, 1941	100/ 476	Returned to owners in late Oct 1939, requisitioned again in Jan/Feb 1941
Menestheus (M93)	Blue Funnel Liner (MV/1929): 460 feet, 7494 tons, 16 kts	22 Jun 40	410/ 22,866	Damaged by bombing, amenities ship in 1944 for service with the Pacific Fleet
Port Napier (M32)	Mercantile (SS/1940): 503 feet, 9600 tons, 16 kts	12 Jun 40	550/ 6,331	Dragged anchor in gale, ran aground and set on fire at Kyle of Lochalsh on 27 Nov 1940
Port Quebec (M59)	Mercantile (SS/1940): 451 feet, 5936 tons, 14.5 kts	Jun 40	548/ 33,494	Repair ship *Deer Sound* (F99) in 1944, returned to owner on 20 Dec 1947
Princess Victoria (M03)	Car Ferry (SS/1939): 310 feet, 3167 tons, 16 kts	2 Nov 39	244/ 2,755	Sunk by mine off the mouth of the Humber River on 19 May 1940
Shepperton (M83)	Train-Ferry (SS/1935): 347 feet, 2839 tons, 16.5 kts	4 Sep 39	270/ 2,742	To the MoWT in 1940
Southern Prince (M47)	Cargo Liner (MV/1929): 496 feet, 10917 tons, 17 kts	15 Jun 40	560/ 23,762	Damaged on 26 Aug 1941 by torpedo from *U-652*, accommodation ship in 1945
Teviot Bank (M04)	Liner (SS/1938): 424 feet, 5087 tons, 12 kts	Dec 39	280/ 15,865	Returned to owners on 29 Mar 1946[7]

FAST MINELAYERS

The *Abdiel*-class cruiser-minelayers were very fast at 39 knots, could do the Gibraltar to Malta run overnight, and were used to supply fuel and food to the island. They were built in two groups with different armament. *Abdiel, Latona, Manxman,* and *Welshman* were fitted with six 4-inch (3x2) anti-aircraft guns, four 2pdr (1x4) AA and eight 0.5-inch (2x4) AA guns. *Apollo* and *Ariadne* had four 4-inch (2x2) AA guns, four 40mm (2x2) AA, twelve 20mm (6x2) AA guns, and 160 moored mines.[8]

Abdiel-class Cruiser minelayers
(418 feet, 2650 tons, 39 knots, 242 ship's complement)

Ship	Comm.	#Mines	Disposition
Abdiel (M39)	15 Apr 41	160/ 2,207	Sunk in Taranto Harbour, Italy, 10 Sep 1943, by mines laid by the German motor torpedo boats *S-54* and *S-61*
Latona (M76)	4 May 41	160/ 0	Sunk 25 Oct 1941 north of Ras Azzaz, Libya, by a German Stuka-dropped bomb
Manxman (M70)	20 Jun 41	160/ 3,111	Damaged by torpedo 1 Dec 1942, broken up at Newport in October 1972
Welshman (M84)	25 Aug 41	160/ 3,275	Sunk 1 Feb 1943 east-northeast of Tobruk, Libya, by two torpedoes from *U-617*
Apollo (M01)	12 Feb 44	160/ 8361	Arrived Blyth, England, Nov 1962 for breaking up
Ariadne (M65)	9 Oct 43	160/ 1,352	Broken up Jun 1965 at Dalmuir and Troon, Scotland[9]

DESTROYER MINELAYERS

The E-class destroyers *Esk* and *Express* were designed to be prepared in twenty-four hours for the minelaying role, by removing their 'A' and 'Y' 4.7-inch guns and all torpedo tubes and installing mine rails along their upper deck. Complete armament consisted of four 4.7-inch (4x1) gun mounts, eight 0.5-inch (2x4) machine guns, eight torpedo tubes (2x4), and 60 moored mines. The ships were capable of 36 knots, propelled by geared turbines producing 36,000 HP driving twin propellers.[10]

Impulsive, Ivanhoe, Intrepid and *Icarus*—I-class destroyers—were likewise fitted for rapid conversion. Two of their 4.7-inch guns had to be landed as well as both banks of torpedo tubes. The I-class ships were slightly smaller than those of the E-class. They could make 36 knots and had similar armament: four 4.7-inch (4x1) guns, two 0.5-inch (2x4) machine guns, ten 21-inch torpedo tubes (2x5), and 60 moored mines.[11]

E-class Destroyer-Minelayers
(329 feet, 1,370 tons, 36 knots, 145 ship's complement)

Ship	Comm.	#Mines	Disposition
Express (H61)	31 Oct 34	60/ 1,210	Damaged by mine, transferred to the Royal Canadian Navy as HMCS *Gatineau* on 3 Jun 1943, sold Vancouver 1956 and hulk sunk as breakwater
Esk (H15)	26 Sep 34	60/ 1,110	Sunk by a mine on 31 Aug 1940 northwest of Texel Island, off the Dutch coast[12]

I-class Destroyer-Minelayers
(323 feet, 1,370 tons, 36 knots, 138 ship's complement)

Ship	Comm.	#Mines	Disposition
Icarus (D03)	3 May 37	60/ 1,196	Sold for scrapping on 29 Oct 1946, broken up at Troon, Scotland.
Impulsive (D11)	29 Jan 38	60/ 892	Sold for scrapping on 22 Jan 1946, broken up in Sunderland.
Ivanhoe (D16)	24 Aug 37	60/ 754	Mined and damaged on 1 Sep 1940 northeast of Texel Island, off the Dutch coast, sunk later that day by the destroyer HMS *Kelvin*
Intrepid (D10)	29 Jul 37	60/ 1,592	Sunk on 26 Sep 1943 by German Ju 88 bombers in Leros Harbour, Dodecanese[13]

The O-class destroyers employed as minelayers—*Obdurate*, *Obedient*, *Opportune*, and *Orwell*—were armed with four 4-inch (4x1) guns, four 2pdr (1x4) AA, six 20mm (2x2, 2x1), eight 21-inch torpedo tubes (2x4), and 60 moored mines. Geared turbines producing 40,000 HP coupled to two shafts provided 37 knots top speed.

O-class Destroyer-Minelayers
(345 feet; 1,550 tons, 37 knots, 175 ship's complement)

Ship	Comm.	#Mines	Disposition
Obdurate (G39)	3 Sep 42	60/	Damaged 25 Jan 44 by a torpedo fired by *U-360*, arrived Inverkeithing 30 Nov 1964 for breaking up
Obedient (G48)	30 Oct 42	60/ 80	Arrived Blyth, England, 19 Oct 1962 for breaking up
Opportune (G80)	14 Aug 42	60/ 120	Arrived Milford Haven, Wales, 25 Nov 1955 for breaking up
Orwell (G98)	17 Oct 42	60/ 80	Arrived Newport 28 Jun 1965 for breaking up[14]

Thanet, *Tenedos*, *Scout*, *Stronghold*, *Thracian*, and *Sturdy* (remnants of a class of 67 destroyers ordered for the Royal Navy in 1917) were despatched to the Far East in 1939 to form local defence flotillas at Hong Kong and Singapore. When war broke out, *Sturdy* was on passage in the Mediterranean and was detained there. These Admiralty S-class destroyers were characterised by two funnels, tall bridge, long fo'c'sle, and heavily raked stem and sheer forward. Propelled by three Yarrow boilers and Brown-Curtis single-reduction geared turbines, they had a top speed of 31 knots. Armament consisted of three 4-inch (3x1) gun mounts, four 21-inch torpedo tubes (2x2), one 2 pdr AA gun, and 40 moored mines. Converting the destroyers to minelayers involved removing and landing ashore the after 4-inch gun, torpedo tubes, and depth charge racks and control units to enable mine rails to be fitted on either side. Twenty mines were carried in each set of rails.[15]

Admiralty S-class Destroyers
(265 feet, 905 tons, 31 knots, 90 ship's complement)

Scout (H51)	15 Jun 18	40/ 772	Arrived Briton Ferry 29 Mar 1946 for breaking up
Stronghold (H50)	2 Jul 19	40/ 1,174	Sunk 2 Mar 1942, while fleeing from Tjilatjap to Australia by a Japanese task group
Tenedos (H04)	11 Jun 19	40/ 266	Sunk 5 Apr 1942 by bombs from Japanese carrier aircraft while under repair at Colombo, Ceylon (now Sri Lanka)
Thanet (H29)	30 Aug 19	40/ 240	Sunk 27 Jan 1942 by Japanese destroyers *Fubuki, Asagiri, Yugiri, Shirayuki* off Endau (east coast of Malaya) while attacking Japanese transport ships in company with HMAS *Vampire*
Thracian (D86)	21 Apr 20	40/ 480	Scuttled by crew at Hong Kong on 16 Dec 1941, salvaged by the Japanese and repaired, and placed in service as IJN patrol vessel No. *101*[16]

COASTAL MINELAYERS

The steam-propelled *Plover* was designed to undertake mining trials, and to serve as the leader of other coastal minelayers. For anti-air defence, she had one 12 pdr 3-inch anti-aircraft gun, one 20mm Oerlikon, and two .303-inch machine guns; and for her primary mission, 100 moored mines.

ced *Plover*-class Coastal Minelayer Leader
(195 feet; 805 tons; 14.75 knots; 69 ship's complement)

Plover (M36)	27 Sep 37	100/ 15,327	Sold 26 Feb 1969 and broken up at Inverkeithing in April 1969[17]

The two *Corncrake*-class coastal minelayers were ex-*Fish*-class anti-submarine trawlers built by Cochrane & Sons Shipbuilders Ltd., Selby, United Kingdom. A single boiler sending steam to a single vertical triple-expansion reciprocating engine, propelled the ships at a pedestrian 11 knots. Their armament was equally modest: three 20mm Oerlikon gun mounts and two twin-barrel .303-inch machine guns.

Corncrake-class Coastal Minelayers
(162 feet, 830 tons, 11 knots, 35 ship's complement)

Corncrake (M82) ex-*Mackerel*	7 Dec 42	12 mines	Foundered in a storm in the North Atlantic on 25 Jan 1943
Redshank (M31) ex-*Turbot*	10 Jan 43	12 mines	Arrived at Sunderland for scrapping on 9 May 1957[18]

INDICATOR LOOP MINELAYERS

Unlike mines sown at sea (which could be either offensive or defensive in nature), the locations of controlled minefields, normally situated in the approaches to harbours, were chosen so that they could be under observation. One or more mines were manually detonated when a target vessel was observed to be within their effective range. For this reason, the mines were planted in predetermined locations with electrical cables connecting them to a centralised firing location. The mines, cables and junction boxes required maintenance. Specialised vessels undertook the hazards of planting and maintained the fields.

The three ships of the *Linnet*-class were the largest of a dozen indicator loop minelayers built for the Royal Navy immediately before and during World War II. The vessels were designed to lay controlled mines, used in coastal defences, as well as anti-submarine indicator loops to detect the presence of enemy submarines. They were very similar to the mine planters operated by the U.S. Army during the same era.[19]

Linnet-class Indicator Loop Minelayers
(164 feet, 498 tons, 11 knots, 24 ship's complement)

Linnet (M69)	18 Jun 38	12 mines	Arrived at Dunston to be scrapped on 11 May 1964
Redstart (M62)	1 Nov 38	12 mines	Scuttled at Hong Kong on 19 Dec 1941 to prevent capture by the Japanese
Ringdove (M77)	9 Dec 38	12 mines	Served as tender to HMS Vernon; sold to the Pakistani government in 1950 as a pilot vessel; sold in 1951[20]

The *M 1*-class coastal minelayers were ex-*Isles*-class trawlers converted for use in sowing controlled-minefields. Two Ruston & Hornsby 6-cylinder diesels, producing 60 bhp, propelled the diminutive craft at a modest top speed of 10 knots but, displacing only 8 feet, the minelayers could ply shallow waters. For anti-aircraft defence, the craft were fitted with one 20mm Oerlikon gun mount and two .303-inch machine guns.

M 1-class Coastal Minelayers
(122 ½ feet; 346 tons; 10 knots; 32 ship's complement)

M 1 (M19)	26 Oct 39	10 mines		Renamed *Miner I* in 1942, *Minstrel* 7 Sep 1962, sold in 1967
M 2 (M34)	19 Jan 40	10 mines		Renamed *Miner II* in 1942, expended as target on 18 March 1970
M 3 (M53)	16 Mar 40	10 mines		Renamed *Miner III* in 1942, sold Feb 1977, broken up at Sittingbourne
M 4 (M68)	12 Nov 40	10 mines		Renamed *Miner IV* in 1942, scrapped in May 1964
M 5 (M74)	26 Jun 41	10 mines		Renamed *Miner V* in 1942, cable layer *Britannic* 1960, expended as target 6 Jun 1970
M 6 (M94)	20 May 42	10 mines		Renamed *Miner VI* in 1942, sold mercantile 16 Aug 1988 Malta.
Miner VII (M88)	15 May 43	10 mines		Launched 29 Jan 1944 Dartmouth, *ETV VII* 1959, trials vessel *Steady*. Sold Mar 1980 for breaking up at Portsmouth, England
Miner VIII (M98)	31 Mar 44	10 mines		Launched 24 Mar 1943 Dartmouth, *Mindful* 7 Sep 1962, sold 22 Feb 1965 and renamed *Rawdhan*. Scrapped in 1981[21]

Photo 2-2

Coastal minelayer HMS *Miner VI* in April 1949.
Collection of Rob Hoole

Royal Navy Controlled Loop Minelayers also plied their trade outside the British Isles. HMS *Atreus* (base ship) and HMT *Alsey* (laying vessel) laid loop-controlled minefields at New Zealand in October 1942 and Brisbane, Australia in May/June 1942. The 443-foot *Atreus* was completed at the Scotts Greenock Yard No. 432 (Scotland), in March 1911 as a steam cargo vessel for the China Mutual Steam Navigation Company Ltd., Liverpool. The Royal Navy requisitioned her in November 1939 for use as a minelayer base ship but could also operate as a minelayer. She was returned to her owner in May 1946, and later delivered to Rosyth in 1949 for scrapping. *Alsey* was an Admiralty requisitioned motor launch built in 1940 and returned in 1945.[22]

MINING TENDERS

The Royal Navy's mining tenders (essentially Admiralty trawlers) included HMT *Nightingale* and HMS *Skylark*, built at the Portsmouth Royal Dockyard, in 1931 and 1932 respectively. Portsmouth is a port city on England's south coast, mostly spread across Portsea Island. *Skylark* was renamed *Vernon* in 1938—after the first *Vernon* was paid off—and later, *Vesuvius*. *Nightingale* spent her entire life in Portsmouth until sold out of service in 1958. These small vessels operated from HMS Vernon, a shore establishment in Portsmouth and home of the Royal Navy's Mining, Torpedo, and Electrical departments. They usually supported mining trials, but also performed some minelaying.[23]

Photo 2-3

HMS Vernon's mining tender *Skylark* being re-christened *Vernon* on 9 December 1938. Her name was changed to *Vesuvius* in April 1941 owing to difficulties with the postal arrangements.
Collection of Rob Hoole

Photo 2-4

HM Trawler *Vernon* (ex-*Strathcoe*)—tender to HMS Vernon for minelaying trials between 1924 and 1938—with controlled mines.
Collection of Rob Hoole

The diminutive 106-foot *Nightingale* displaced 255 tons, drew 6 ½ feet of water, and could make 10 knots. She was propelled by a single triple-expansion engine, producing 400IHP, coupled to a single shaft; and had no armament, other than the twenty mines she was capable of carrying. Her crew size was equally austere, only twelve men. *Vesuvius'* characteristics were very similar.

Photo 2-5

Controlled minefield cabling loaded in boats in Vernon Creek at HMS Vernon
Collection of Rob Hoole

COASTAL FORCES CRAFT

H.M. COASTAL FORCES

Coastal Force minelayers—comprised of a variety of craft—laid 6,900 mines in World War II. Coastal Forces was a division of the Royal Navy first established during World War I, and then again later in World War II under the command of Rear Admiral Coastal Forces. The last sailors to wear the 'HM Coastal Forces' cap tally were the ships' companies of HMS *Dittisham* (M2621) and HMS *Flintham* (M2628) on these inshore minesweepers being taken out of reserve in 1968.

Type Craft	Displ. (tons)	Speed (knots)	No. Mines
Fairmile B Motor Launches (ML)	75.5	17.7	9 moored, 8 ground
Motor Torpedo Boats (MTB) – various	37-40	31-38	4 ground
Fairmile D Motor Torpedo Boats (MTB)	102	29	10 moored
Motor gunboats (MGB)	28	39-42	4 ground[24]

MINELAYING SUBMARINES

As part of the following summary information, submarine displacement in tons and top speed in knots is given for both surfaced and submerged conditions.

Porpoise-class Minelaying Submarines
(289 feet, 1500/2157 tons, 15.75/8.75 knots, 6 O/53 E ship's complement)

Name	Comm.	No. Mines	Disposition
Porpoise (N14)	11 Mar 33	62/430 Mk 16, 71 Mk 2	Sunk by Japanese aircraft in the Malacca Strait on 19 Jan 1945
Narwhal (N45)	28 Feb 36	62/400 Mk 16	Believed sunk on 23 Jul 1940 off Aberdeen, Scotland by bombs from German aircraft
Cachalot (N83)	15 Aug 38	62/250 Mk 16	Rammed and sunk north of Benghazi, Libya, by the Italian torpedo boat *Generale Achille Papa* on 30 Jul 1941
Seal (N37)	24 May 39	62/50 Mk 16	Captured by the Germans 5 May 1940 after being damaged by a mine the day before
Grampus (N56)	10 Mar 37	62/50 Mk 16	Likely sunk 16 Jun 1940 off Syracuse by Italian torpedo boats *Circe* and *Cleo*
Rorqual (N74)	10 Feb 37	62/1292 Mk 16, 36 Mk 2	Went into reserve on 28 Jul 1945, sold 19 Dec 1945 and arrived Newport 17 Mar 1946 for breaking up[25]

T-class Submarine-Minelayers
(275 feet, 1090/1575 tons, 15.25/9 knots, 61 ship's complement)

Name	Comm.	No. Mines	Disposition
Tactician (P314)	29 Nov 42	12/12 Mk 2	Arrived Newport 6 Dec 1963 for breaking up
Tally-Ho (P317)	12 Apr 43	12/12 Mk 2	Arrived Briton Ferry 10 Feb 1967 for breaking up
Tantalus (P318)	2 Jun 43	12/12 Mk 2	Scrapped at Milford Haven in Nov 1950
Tantivy (P319)	25 Jul 43	12/12 Mk 2	Sunk as an anti-submarine target in the Cromarty Firth in 1951
Taurus (P339)	3 Nov 42	12/24 Mk 2	Loaned to the Royal Netherlands Navy on 4 June 1948 and commissioned into the Royal Netherlands Navy as *Dolfijn* the same day. Decommissioned and returned to the Royal Navy on 7 December 1953. Recommissioned into the Royal Navy as HMS *Taurus* on 8 December 1953. Scrapped at Dunston-on-Tyne in April 1960.
Templar (P316)	15 Feb 43	12/12 Mk 2	Sunk as a target in Loch Striven, Scotland in 1954. Salvaged on 4 December 1958. Arrived at Troon, Scotland on 17 Jul 1959 to be scrapped.
Thorough (P324)	1 Mar 44	12/24 Mk 12	Scrapped at Dunston on the River Tyne on 29 June 1962
Thule (P325)	13 May 44	12/12 Mk 12	Arrived Inverkeithing on 14 Sep 1962 for breaking up
Tradewind (P329)	16 Sep 43	12/12 Mk 12	Arrived Charlestown on 14 Dec 1955 for breaking up
Trenchant (P331)	31 Jan 44	12/12 Mk 12	Sold 1 July 1963 and arrived Faslane, Scotland, 23 Jul 1973 for breaking up
Trespasser (P312)	11 Sep 42	12/12 Mk 2	Arrived Gateshead on 26 Sep 1961 for breaking up
Truculent (P315)	31 Dec 42	12/12 Mk 2	Sunk in the Thames Estuary on 12 Jan 1950 after a collision with the Swedish merchant tanker *Divina*. Salvaged on 14 Mar 1950 and sold 8 May 1950 for breaking up at Grays, Essex.
Tudor (P326)	19 Dec 43	12/10 Mk 12	Sold 1 Jul 1963 and arrived Faslane 23 Jul 1963 for breaking up[26]

S-class Submarine-Minelayers
(217 feet, 715/990 tons, 14.75/9 knots, 6 O/42 E ship's complement)

Name	Comm.	No. Mines	Disposition
Sea Rover (P218)	7 Jul 43	8/8 Mk 2	Sold October 1949 and broken up at Faslane, Scotland from June 1950
Stoic (P231)	29 Jun 43	8/8 Mk 2	Sold July 1950 and broken up at Dalmuir, Scotland
Surf (P239)	18 Mar 43	8/16 Mk 2	Sold 28 October 1949 and arrived Faslane, Scotland July 1950 for breaking up[27]

MINELAYING AIRCRAFT

The scope of this book precludes much detail about the operations of minelaying aircraft. However, it's important to note that British and Dominion aircraft planted 54,194 mines, and that 500 planes went missing during these operations. Identified below are the types of aircraft employed, and their standard mine loads (number and types of mines they could carry).[28]

Royal Navy Fleet Air Arm

Aircraft	Mine Load, No. and Type
Swordfish, Albacore, Barracuda II, Avenger II	1-2,000lb (Mk 1-4 & Mk 6)

Royal and Dominion Air Forces

Aircraft	Mine Load	Aircraft	Mine Load
Beaufort I	1-2,000lb (Mk 1-4 & Mk 6)	Manchester I	4-2,000lb (Mk 1-4 & Mk 6)
Catalina PBY	4 mines/4,000 lbs maximum load	Marauder	2-1,000lb (Mk 5 & Mk 7)
Halifax III	4-1,000lb (Mk 5 & Mk 7)	Mosquito XIB	2-2,000lb (Mk 1-4 & Mk 6)
Hampden I	1-2,000lb (Mk 1-4 & Mk 6)	Stirling I	6-2,000lb (Mk 1-4 & Mk 6)
Lancaster III	6-2,000lb (Mk 1-4 & Mk 6)	Ventura (number and type of mines carried by RNZAF unknown)	3,000lbs bombs, six 325lb depth charges, 1 torpedo
Liberator I (VLR)	6-1,000lb (Mk 5 & Mk 7)	Wellington IC	2-2,000lb (Mk 1-4 & Mk 6)
Liberator II (LR)	6-1,000lb (Mk 5 & Mk 7)[29]	Wellington X	2-2,000lb (Mk 1-4 & Mk 6)

3

Initial British Minelaying Efforts

With Germany arming at breakneck speed, England lost in a pacifist dream, France corrupt and torn by dissension, America remote and indifferent... do you not tremble for your children?
—Winston Churchill, 1935

British ships, submarines, and planes engaged in three types of mining during the war. Defensive minefields were laid in the Dover Straits, along the East Coast of the UK, and in North Atlantic waters, including the Denmark Strait to protect vital ports and shipping routes from attack by enemy submarines and surface forces. Anti-submarine traps were similarly employed in the Northwest and Southwest approaches, the Irish Sea and the English Channel. The third type, offensive minelaying, particularly by aircraft, proved the more rewarding in terms of mines laid to damage inflicted. Fields laid in shallow coastal waters—especially in areas of busy maritime activity—disrupted shipping and made necessary employment of significant numbers of enemy mine countermeasures vessels.[1]

DEFENSIVE MINING IN HOME WATERS

It was very evident to the Admiralty that in the event of hostilities with Germany, the defence of the English Channel would be of vital importance, both for the safe passage and support of an expeditionary force on the continent and for the protection of convoys in the western approaches. Thus, a plan for the laying of deep and shallow mines across the Straits of Dover—Dover-Calais mine barrage—and deep anti-submarine fields in the Folkestone-Gris Nez area (between Folkestone, a port town in southeast England, and Cape Gris Nez in northern France) was worked out many months before the war.[2]

Map 3-1

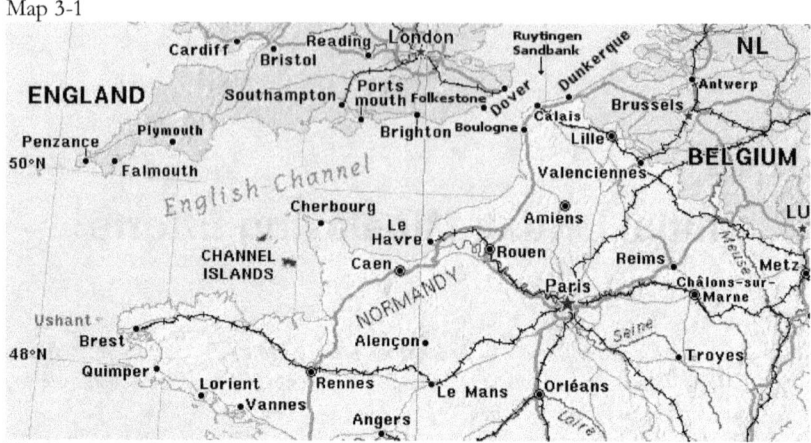

The Dover Strait and adjacent coasts of England and France

Speed of execution was of paramount importance, particularly in emplacing deep and shallow mine lines between the Goodwin Sands (a 10-mile long sandbank in the English Channel) and the Dyck Shoal near Boulogne, France. The minelayers available to carry out this work were the *Adventure*, the *Plover*, and the converted train-ferries *Hampton* and *Shepperton*. Before her Dover employment, *Plover* had the distinction of laying the first mine in the war by beginning the Bass Rock minefield, near the same-named island in the outer part of the Firth of Forth (an estuary on the North Sea near Edinburgh, Scotland) on 3 September. Upon completion of this effort the following day, she proceeded to Dover to take part in the laying of the cross-channel barrage.[3]

With the availability of the *Hampton* and *Shepperton*, each able to carry 270 mines, work began on the first stage, the Goodwins-Dyck Barrage (totaling 3,000 mines) on 11 September. The work was completed in five days. The type of mine used, as was the case for the Folkestone-Gris Nez fields (second stage, 3,636 mines), were moored mines of the H Mk II type. French authorities aided in the effort to deny U-boat passage through the Dover Strait into the English Channel by laying further minefields off Dunkirk. The final stage involved placing a double system of indicator loops on the seafloor between the two minefields to detect any U-boats which might attempt the passage of the Straits.[4]

OFFENSIVE MINING IN GERMAN WATERS

When Britain and France declared war on Germany on 3 September 1939, the Royal Navy's offensive mining capabilities were also very modest. However, efforts were under way to convert some E-class destroyers for minelaying, and HMS *Esk* and HMS *Express* rapidly took up these duties following these conversions. At Portsmouth (where her war complement of personnel reported aboard), *Express* "worked up" for operations as a minelayer on 7 September. She proceeded on the 9th to Immingham, on the North Sea, to take up her war station and embark mines. On 10 September, *Express* and *Esk* laid an initial minefield in Heligoland Bight (Operation AA), and two days later a second one in the same area (QQ).[5]

Photo 3-1

HMS *Express* (H61) configured as a fleet destroyer, circa 1942
Australian War Memorial photograph 302386

The mining was in conformance with plans made before the war to lay a series of small minefields along the probable enemy shipping routes in the Heligoland Bight, should Britain find herself at war with Germany. It was anticipated that German naval forces would likely use the same routes as in WWI—and this expectation proved true. The mining of the Bight began as soon as the state of the moon was favourable, the period of a new moon when the night was darkest. Such conditions improved the odds of a minelayer remaining undetected and of surviving a foray into enemy waters. Use of the night to cloak their activities gave rise to the moniker "nightraider" for these type vessels.[6]

The hazards of undertaking minelaying in enemy waters is amply illustrated in Chapter 1, which describes the exploits of HMS *Manxman* during one such clandestine operation. During Britain's 1939-1945 minelaying campaign, the first challenge minelayers and minelaying aircraft faced was getting to the appointed area and laying their mines with accuracy. The second, hopefully, assuming all went well and the mission was successful, was the return transit or flight out of enemy waters or air space. Stealth was critical to survival.[7]

Map 3-2

Jade Bight and Heligoland Bight, small bays off the German Bight (Bay).

FIRST U-BOAT CASUALTIES TO BRITISH MINES

In autumn 1939, the German U-boat arm possessed a relatively modest number of submarines, of the II, VII, VIIC, and IX types. The Type-II boats, by reason of their limited fuel capacity, were restricted to operational areas of the North Sea, English east coast, and Scotland's Orkney and Shetland Islands. The larger Type VII boats (a forerunner of the VIIC) could reach the west coast of England by transit around the Shetlands, and the Type VIIC operated off the north coast of Spain. Longer range Type IXC boats (a successor to the IX and improved IXB) could take the war to Gibraltar.[8]

The Type VIIC boat would be the German Navy's work horse (568 were commissioned), and the Type IXB its most successful boat (11 commissioned). The values for speed in knots and range in nautical miles listed in the table reflect submarine surface operations.

Representative Type U-boats

Type	Max Speed/Range	Torpedoes	Mines	Deck Gun
IIB	13 kts/3,100 nm at 8 kts	5	12 TMA	None
VIIC	17.7 kts/8,500 nm at 10 kts	14	26 TMA	88mm/45
IX	18.2 kts/10,500 nm at 10 kts	22	44 TMA	105mm/45
IXC	18.3 kts/13,450 nm at 10 kts	22	44 TMA	105mm/45[9]

The Type VIIC was a comparatively small and manoeuvreable boat of 517 tons with good range for its size and a comparatively high number of torpedoes (12-14). In the opinion of the U-boat command, it offered the ideal combination of tactical usefulness in attacks (light and easy to handle, difficult to see at night, small turning circle) and possessed the necessary fighting strength, expressed in range and armament. The Type IXC (about 740 tons), though more complicated to handle, had greater range and a higher number of torpedoes (22).[10]

Because outward and inward passage of the VII, VIIC, and IXC boats north of England consumed a considerable part of their action radius, the U-boat command attempted to send submarines through the English Channel on their way to the Atlantic. However, losses to mines in the narrow Dover-Calais Strait were too high, and this route was subsequently abandoned as too costly. Three submarines—two Type IIB 140-foot coastal boats, and one IX 251-foot ocean-going boat—fell victim to British mines in the second month of the war. Ironically, when sunk, they had either been seeding British waters with German mines; or were otherwise engaged, but capable of doing so.[11]

U-12, a 140-foot Type IIB coastal boat, was mined and sunk in the Dover-Calais mine barrage on 8 October, the first enemy submarine to fall victim to a defensive field. She had sailed from Wilhelmshaven on 23 September 1939 to operate in the English Channel. Nothing more was heard from her, and she was declared missing on 20 October after her failure to return. All hands were presumed lost when the body of Kptlt. Dietrich von der Ropp, her commanding officer, was found washed ashore on the French coast near Dunkirk.[12]

A much larger 251-foot, ocean boat, *U-40* (Kptlt. Wolfgang Barten) was sunk five days later, on 13 October in the English Channel east of Dover, by a mine in the British field C3. There were only three survivors. These type IX long-range submarines were fitted with six torpedo tubes (4 at the bow and 2 at the stern) and carried 22 torpedoes. They also had five external torpedo containers which stored ten additional torpedoes. As minelayers, the submarines could carry 44 TMA or 66 TMB mines. Secondary armament consisted of one large Utof 105mm/45 deck gun.[13]

A second coastal boat, *U-16*, was lost on 25 October; scuttled by her crew in the English Channel east of Dover, following heavy damage by a British mine. She had sailed from Kiel on the 18th to lay mines along the coast of Dover. Her commanding officer, Kptlt. Horst Wellner, had sent a radio message at 0415 that morning conveying his intention to scuttle the heavily damaged submarine. She was unable to return to German waters and would have had little chance on the surface. Like *U-12*, she had no deck gun; the space inside her small hull being used to carry five torpedoes, fired from three bow tubes, and twelve TMA moored-influence mines laid from these tubes.[14]

Photo 3-2

German coastal submarine *U-16* on the surface.
Naval History and Heritage Command photograph #NH 111239

British forces found the wreckage of the submarine aground on the Goodwin Sands the day of her scuttling, with her conning tower just awash and extensive damage forward. A salvage attempt failed due to poor weather and was not repeated because the wreck was full of silt. She later disappeared in shifting sands. No survivors were found. In the following weeks the bodies of nineteen crew members, all wearing lifesaving apparatus, were recovered or washed ashore, most between Dungeness and Hythe. The remains of the commanding officer and five men washed up on the French coast near Dunkirk and another on the Dutch island of Ameland.[15]

PRECEEDING HUNT FOR *U-16*

U-16 had been detected passing the St. Margaret's Bay indicator loop (located a short distance up the coast from Dover, on the seafloor of the bay) around noon on 24 October and as a consequence the 243-foot sloop HMS *Puffin* was sent to investigate. Concurrently, the small 373-ton ASW trawler HMS *Cayton Wyke* was patrolling the English Channel when Seaman John Cook, the sonar operator, detected the submarine. *Puffin* joined the *Cayton Wyke*, a former fishing trawler, upon receiving a contact report from the *Wyke*. (Many such trawlers were requisitioned by the Royal Navy during the war for conversion to warships and use principally as patrol vessels or minesweepers.)[16]

Together they had closed in for an attack on the submarine. *Puffin* dropped a pattern of three single depth charges on a contact she made in the vicinity of Goodwin Sands (sandbar) near Kent, followed up by a pattern dropped by *Cayton Wyke*. The two ships were initially credited with sinking *U-16* with these depth charge attacks, but extensive damage to the wreck found during investigation by Royal Navy divers, revealed that she had struck a mine in the Dover-Calais barrage, possibly after evading the aforementioned attacks.[17]

The Dover barrage apparently accomplished its purpose. Only one U-boat is known to have passed through the Straits successfully, and that was on the night of 11-12 September before the first stage of the barrage had been completed. Following the loss of the three submarines, it appeared that the Kriegsmarine (German Navy) did not make further attempts to send coastal submarines by the shortest route to the central and western Channel—waters through which all shipping approaching or leaving southern British ports had to pass.[18]

FITTING OF MINELAYERS WITH PARAVANES

In October, planned minelaying operations off the Dutch coast were suspended and *Express* and *Esk* were fitted with paravanes and taut-wire measuring gear (used to enable the positions, in which minefields were to be laid, to be located with greater accuracy). Because Royal Navy warships and merchant vessels had to venture outside coastal shipping channels for naval operations and trade, some form of self-protection was desirable. Additionally, while the channels themselves were swept daily to counter the presence of enemy mines, newly-laid "shipkillers" could appear at any time, particularly in a swept channel.[19]

This requirement in World War I had led to the development of the Burney paravane, so named after its inventor Lt. Charles Denniston Burney, Royal Navy. A pair of paravanes was towed in a v-shaped configuration from the ship's bow at the point where its stem met the keel. Each of the devices was streamed on its own wire at about a 50-degree angle off the respective ship's side. The ship's motion and the paravanes' shape caused them to stand out at a considerable lateral distance from the ship's sides, thus preventing a mine strike on its hull. When the mooring of a mine was encountered, it slid along the paravane tow wire, and entered a pair of jaws on the paravane, which immediately cut it.[20]

Photo 3-3

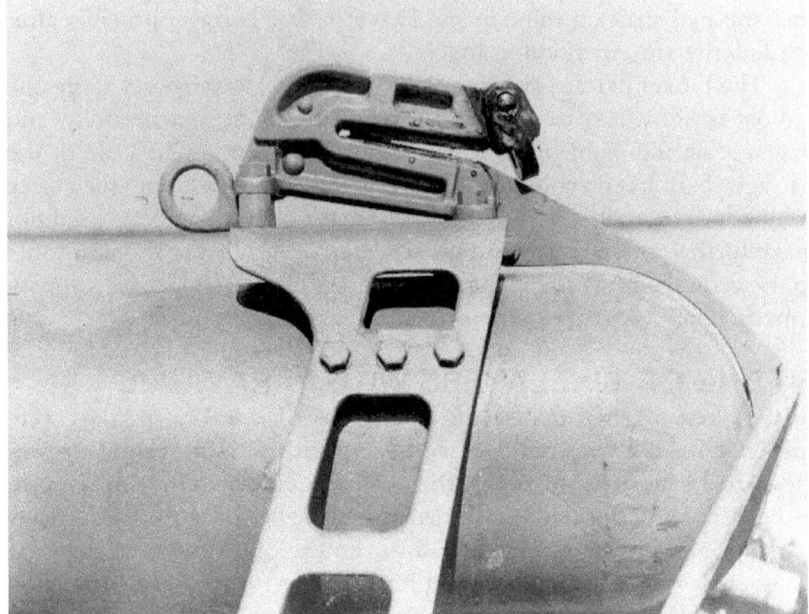

Cutting jaws of a paravane, used for minesweeping in World War I.
Naval History and Heritage Command photograph #NH 60758

Photo 3-4

A pair of paravanes, shaped somewhat like torpedoes, were dragged through the water by a ship. Mines cut loose from their moor by the sharp v-knife at the top floated to the surface and were exploded by rifle fire.
Naval History and Heritage Command photograph #NH 124026

FORMATION OF THE 20TH DESTROYER FLOTILLA

On 17 November, *Express* and *Esk* put to sea to lay a deep minefield in the approaches to the Thames Estuary, a part of the East Coast Barrier (Operation RG). This effort was completed on 29 November. Two weeks later, on 12 December, they joined the 20th (Minelaying) Destroyer Flotilla, on its formation at Harwich, on England's southeast coast. This designation was chosen out of sentiment for the destroyer minelayers of World War I, which were assigned to the 20th Destroyer Flotilla, subsequently decommissioned. Four I-class destroyers—*Ivanhoe*, *Icarus*, *Intrepid*, and *Impulsive*—also joined the flotilla. Based at Harwich, its units obtained mines at Immingham, farther up the coast, on the southwest bank of the Humber Estuary, northwest of Grimsby.[21]

Map 3-3

England's east coast

Later in December 1939, *Express* and *Esk* escorted the auxiliary minelayer *Princess Victoria* during a defensive minelay for the East Coast Barrier off Harwich (Operation OF). A converted car ferry, she could carry 244 mines. Destroyers *Brazen* and *Boreas* were at sea on 15 December to provide cover during the first day of the operation, and *Brazen* and *Codrington* on the 22nd, for the second portion. During this period, *Ivanhoe* and *Intrepid* of the new flotilla carried out their first minelaying operations. Offensive mining in Heligoland Bight on 18 December was followed on the 30th by a lay of defensive mines near the Farne Islands, off the coast of Northumberland in northeast England, as part of the East Coast Barrier.[22]

I-class destroyers *Icarus* and *Impulsive* completed their conversion to their minelaying role at Portsmouth on 24 January 1940. Following workups for her new duties, *Icarus* rejoined the flotilla on 26 February. *Impulsive*, which had been nominated for such duty in December but had remained assigned to the Home Fleet at Scapa Flow, in Scotland's Orkney Islands, until her refit, reported to the flotilla on 27 February.[23]

Map 3-4

Scotland

ANOTHER U-BOAT, AND TWO DESTROYERS SUNK

As a new year (1940) broke, operations by 20th Flotilla units continued. On 2 January, *Ivanhoe* and *Intrepid* once again ventured into enemy waters to lay mines in Heligoland Bight (Operation EW). Operations ID1 and ID2 followed on the 10th and 13th, as additional mines slide off the sterns of the two destroyers into the Bight. These fields were responsible for sinking the *U-54* and destroyers *Leberecht Maass* and *Max Schultz* the following month.²⁴

Photo 3-5

U-52, a Type VIIB submarine like the *U-54*.
http://www.iwmcollections.org.uk/

U-54, a 218-foot type VIIB submarine commanded by KrvKpt. Günter Kutschmann, was sunk north of Terschelling, one the Netherlands West Frisian Islands, with the loss of all hands. (VIIBs included many of the most famous U-boats of the war, including *U-48*, the most successful). No communications were made by her after leaving Heligoland. She had orders to operate off Cape Finisterre, a rock-bound peninsula on the west coast of Galicia, Spain. She was reported missing to authorities on 14 February 1940 when parts of two of her torpedoes were recovered by the German patrol boat *V-1101*.²⁵

KRIEGSMARINE OPERATION WIKINGER ENDS IN DISASTER AS A RESULT OF BRITISH MINES AND FRIENDLY FIRE

During the first months of the war, German naval forces were used mainly for conducting offensive mining close to the British coast and harbours and defensive mining to protect the German Bight against British raids. Small groups of two to four destroyers (each able to carry up to sixty mines) laid mines in Britain's main shipping lanes. Eleven such operations were carried out between October 1939 and February 1940. A variety of vessels—destroyers, torpedo boats, light cruisers, and converted ferries or transports—accomplished the defensive warfare. The results were several huge minefields in the North Sea with well-defined clear paths through which German ships could safely pass—at least that was the belief.²⁶

In February 1940, the Kriegsmarine requested several Luftwaffe air reconnaissance missions in the area of the Doggerbank, as several suspicious ships, fishing boats, trawlers and other small craft were operating west of the defence mine fields. It was not clear what activity they were engaged in, but to the Kriegsmarine they seemed to be involved in some military operations. On several occasions, reconnaissance aircraft reported that enemy submarines were meeting with those boats. A decision was made on 22 February to send the 1st Destroyer Flotilla—*Friedrich Eckoldt, Richard Beitzen, Erich Koellner, Theodor Riedel, Max Schulz* and *Leberecht Maass*—to intercept this group of vessels.²⁷

Map 3-5

North Sea area

Unlike typical destroyer operations, it was a moon-lit night, with an almost cloudless sky and a light wind from the southwest when the flotilla sailed on the 22nd to carry out Operation WIKINGER. The Luftwaffe had orders to provide fighter cover for the destroyers leaving their base and on their return one or two days later. But due to communication problems between the two branches of the German military (of which there was no love lost during the war), the aircraft requested to escort the six destroyers on their way out into the North Sea did not show up. It appears they had been assigned to a separate operation. The Luftwaffe X Fliegerkorps (10th Air Corps) planned to employ two squadrons of Heinkel He111 medium bombers to carry out attacks against merchant shipping off England's east coast.[28]

At 1900, the destroyers making over 25 knots entered the six-mile wide passage through the "Westwall" minefield protecting the German Bight. The ships proceeded in a line ahead formation on a west-northwest course so as to clear the minefield and arrive in the operation area as soon as possible. At 1945, a Heinkel He 111 aircraft attacked *Leberecht Maass*, believing her to be an enemy ship. Two bombs fell into the sea, just astern the destroyer. A third one made a direct hit between her forward superstructure and forward funnel and exploded.[29]

Leberecht Maass slowed and left the formation to starboard, while signaling "Habe Treffer, brauche Hilfe" (Being hit, need assistance). The bomber carried out another attack on her thirteen minutes later. Aircrewmen witnessed two of the four 50kg bombs dropped score direct hits, followed by a huge fireball. *Leberecht Maass* then broke into two parts and sank. They did not then know that their "friendly fire" (bombs) had sunk a German ship and not a British one.[30]

Photo 3-6

Propeller motors and bomb aimer's compartment of a German Heinkel He 111 aircraft. Australian War Memorial photograph 129675

At 2004, a second huge explosion erupted into the night sky. Following reports of a sighting and sonar contact on a submarine from two of his remaining destroyers, the flotilla commander ordered rescue operations then in progress for survivors discontinued, and a search begun for an enemy submarine.[31]

Max Schulz did not answer radio calls as her sister ships tried to locate a possible British submarine, but it was then unknown that she had also struck a mine, and sank, or was sinking. As a result, all 308 officers and men perished; killed by the explosion, trapped inside the wrecked destroyer as it plunged into the abyss, or succumbed to hypothermia or drowning in the cold sea. Following an unsuccessful hunt for a suspected submarine, the remaining four destroyers returned to the site of the *Leberecht Maass* sinking, retrieved their boats, and learned of the loss of *Max Schulz*. The enormity of the disaster then became apparent. Two destroyers lost and 578 personnel killed; only sixty crewmembers of *Leberecht Maass* rescued, and none of the *Max Schulz*.[32]

In the weeks following the disaster, the German 1st Minehunter Flotilla swept the area of the supposed mine-free passage and found several British mines close to the area in which the destroyers were lost. *Leberecht Maass* and *Max Schultz* were lost northwest of the German Frisian Islands, and *U-54* was presumed lost in the same field. After the war, the Royal Navy disclosed that mines had been laid in an area about five miles around the sinking locations. Following bomb hits to *Leberecht Maass*, evasive action by her and *Max Schulz* apparently resulted in the destroyers running though the freshly-laid British mines.[33]

Map 3-6

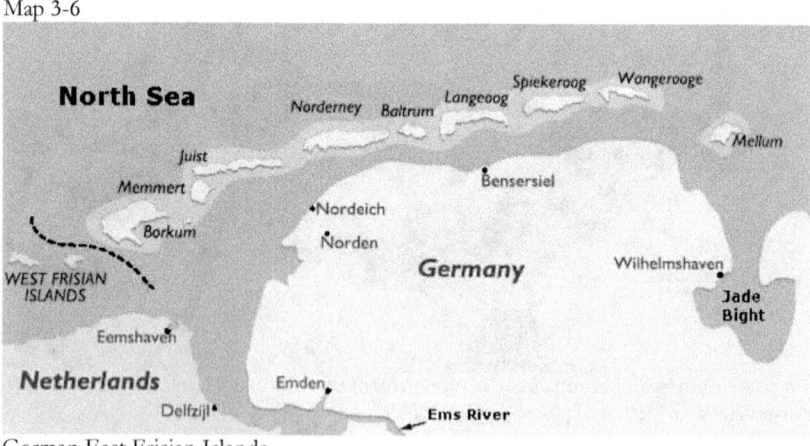

German East Frisian Islands

4

Mining and Mine Countermeasures

> *The suitability of the entire coast of the British Isles for minelaying...imposes a gigantic task upon the minesweeping organisation and...there are never enough minesweepers to meet the various commitments.*
>
> —The [British] Naval War Manual (1948)

While the Royal Navy was carrying out its initial minelaying efforts, Germany had not neglected either defensive or offensive minelaying. Like Britain, she mined and provided notification of a defensive area in the North Sea. Rectangle-shaped, it stretched north from Dutch waters for about 180 miles and was about 60 miles wide. Overlapping the British declared area, the purpose of the expansive mined waters was to deny the British an approach, from the west to German ports and bases on the North Sea coast.[1]

By the end of 1939, the British Admiralty believed that the areas mined within these waters by the enemy were accurately known. It accordingly decided to use the minelaying flotilla to place some small fields within unmined passages used by German naval forces to transit their declared area. To this end, the Royal Navy carried out a number of operations early in the new year to mine "swept channels." When placing their mines, the Germans had left a number of passages for their own use within the fields. They had hoped the British were unaware of them, but regularly swept them in case the British discovered them and were able to enter and mine them.[2]

GERMAN OFFENSIVE MINING

While all mines laid early in the war by the British were of the contact type, the Germans had developed and put into production the first of a series of magnetic mines. These type of "influence" mines are detonated not by actual contact with a passing ship, but rather by the disturbance to the earth's magnetic field that even small amounts of iron, steel, or other ferrous metal aboard a vessel creates. Detection of such a disturbance by a magnetic mine causes it to explode.³

Magnetic mines were not a new development. In fact, the British had first produced them in World War I and laid some in Belgian waters, in the mouth of the River Scheldt near the Port of Antwerp and off Zeebrugge, in 1918. This early type was not initially successful, but development continued intermittently until 1939, at which time the British standard magnetic mine was acceptable and ready for production. Concurrently countermeasures—including mine sweeps against magnetic mines that an enemy might employ—were being studied, but it was difficult, if not impossible, to design one in advance that would be capable of exploding mines fired by all the numerous variations in magnetic influence that an enemy might employ.⁴

Photo 4-1

HMS Vernon in July 1955

On the outbreak of war, the entire British minesweeping force then in service, and the large numbers of auxiliary minesweepers to be requisitioned and converted, were designed only to deal with moored-contact mines. As mentioned previously, research and development work had progressed to a point where production of a sweep for magnetic mines could be started, as soon as necessary intelligence could be obtained regarding the type of magnetic influence required to fire the enemy's mines. Ship losses off England's east coast in the first week of the war gave rise to suspicions that the Germans were using ground (resting on the seafloor) mines of the magnetic type, as well as contact mines. This was confirmed on 16 September, when the British passenger ship SS *City of Paris* was damaged by an underwater explosion (caused by a mine laid by *U-13* on 4 September) but her hull was not penetrated, as would be the case if she had made direct contact with a mine.[5]

Photo 4-2

19 Dec 1939 – Lt. Comdr. John Ouvry at HMS Vernon showing King George VI the German magnetic ground mine he rendered safe on the mudflats at Shoeburyness, at the mouth of the Thames Estuary, on 23 November 1939. The projections were intended to prevent the cylindrical mine rolling across the seabed.
Collection of Rob Hoole

About this time within the Admiralty, a special staff had been placed under Rear Adm. William Frederic Wake-Walker, RN, to expedite the production of countermeasures in collaboration with the mining department of HMS Vernon, and commercial firms with requisite expertise in this area. Located in Portsmouth on the site of the old Gunwharf, Vernon was a shore station ("stone frigate") responsible for mine disposal and mine countermeasures. The development work in progress was aided by the recovery on 23 November 1939 of a magnetic mine (Type GA) off the mudflats of Shoeburyness, where it had been dropped by an aircraft under fire by British shore batteries. It was rendered safe at great personal risk by Lt. Comdr. John G. D. Ouvry, RN.[6]

Ouvry, the officer with the most experience with non-contact mines, had gained experience with minelaying aboard the cruiser HMS *Inconstant* during the First World War. For this courageous deed, Ouvry was decorated with the DSO by King George VI at a ceremony on HMS Vernon's parade ground on 19 December 1939. Others decorated at the same time for this and other tasks where mines were rendered safe for recovery and examination, were Lt. Comdr. Roger C. Lewis (DSO), Lt. John E. M. Glenny (DSC), CPO Charles E. Baldwin (DSM) and AB Archibald L. Vearncombe (DSM). Of particular note, these were the first Royal Naval decorations of the Second World War.[7]

Photo 4-3

German GC magnetic mine on a trolley at HMS Vernon.
Collection of Rob Hoole

A magnetic mine resting on the seafloor in relatively shallow water could detect a ship passing above or near it. Since this type mine did not float, buoyancy was not an issue, and because it didn't require an anchor and mooring cable, it weighed less than a contact mine. These mines, carrying only an explosive and firing device, were smaller and lighter, perfect for laying from aircraft.[8]

Disassembly of the mine recovered from Shoeburyness revealed that it was triggered by changes in the earth's vertical magnetic field—caused by the magnetism of steel-hulled ships in the direction of the north-pole downward—a feature common to all ships in north latitudes. Once this was known, a relatively simple method of making ships more immune to these type mines presented itself—to magnetise them so that they had no downward-oriented north pole. This theory led to British vessels being demagnetised or degaussed as it was known.[9]

COUNTER TO MAGNETIC MINES FOUND, BUT SHIP LOSSES CONTINUE DURING THE DEVELOPMENT OF SWEEP GEAR

Following the early success of ground mines with simple magnetic fuses, Germany decided to significantly increase their production. Before this could occur, the Royal Navy was able to design an 'LL Sweep' based on newfound knowledge of the mine. When this means of countering magnetic mines was brought into service, a substitute mine was not available to the enemy. Further developments only gradually became operational, some of them for the first time in 1944, when the German "oyster" (pressure) mine and acoustic mine were fielded.[10]

The British 'LL' gear consisted of long leg and short leg buoyant cables that were towed together astern of a minesweeper. Its operation involved passing a powerful electric current through the cables at variable time intervals to produce a magnetic field close to a mine, of adequate strength to set off the firing mechanism and to detonate it at a safe distance from the minesweeper.

Diagram 4-1

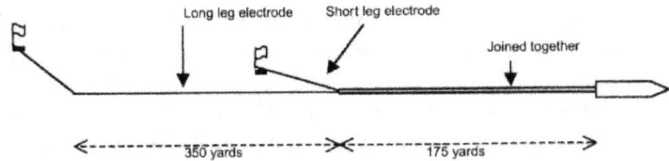

Royal Navy 'LL Sweep'
Collection of Rob Hoole

Implementation difficulties included: a great production effort to field significant quantities of 'LL' sweep gear, (2) obtainment of large numbers of vessels for use as minesweepers, and (3) provision of training for their crews before the minesweepers could put to sea in earnest. In the interim, mines sank or damaged scores of ships. By the end of October 1939, Britain had lost nineteen ships (59,027 tons) to mines, many of them sunk in five magnetic fields laid off England's east coast and in the Thames estuary. In the early months of the war, the Kriegsmarine despatched destroyers to carry out numerous bold and very successful minelaying operations in the mouth of the Thames and farther north, which included laying magnetic mines in the narrows of harbour channels.[11]

In November matters worsened; twenty-seven merchant ships of 120,958 tons and the destroyer HMS *Blanche* were sunk by mines and many more were damaged—including the cruiser HMS *Belfast* and the minelayer HMS *Adventure*. The quantities of German mines, gave the Nore Command great difficulty in identifying and marking safe channels in the Thames Estuary. In mid-month, only one of the three deep-water channels into the river was open, and it seemed that the enemy might succeed in completely halting the movement of shipping in and out of the Port of London. Fortunately, this did not occur, but many vessels were sunk.[12]

MINE STRIKES TO HMS *ADVENTURE* AND *BLANCHE* RESULT IN CASUALTIES, AND LOSS OF *BLANCHE*

> *ADVENTURE's War Organisation arranged for 50% of the Artisan ratings to sling at the after end of the ship. This proved to be a wise precaution because, not only was the number of casualties reduced, but Artisan ratings were available for Damage Control (There were 5 Shipwrights, 5 Electrical Artificers, and 10 Engine Room Artificers on the casualty list). Had this precaution not been taken all Key Ratings would have been sleeping in the part of the ship which was damaged and none would have been available afterwards, except those actually on watch at the time.*
>
> —Observation by Capt. Arthur Robert Halfhide, RN, *Adventure*'s commanding officer, regarding the good fortune that his ship's plumbers, painters, joiners, and coopers were sleeping aft, and therefore unhurt, when the minelayer suffered considerable damage, with many other ratings killed, from a mine.[13]

One of the German mine lays in British waters occurred the night of 12 November 1939, when the destroyers *Karl Galster*, *Wilhelm Heidkamp*, *Hermann Kunne* and *Hans Ludemann* laid a field of magnetic mines off Margate, in the South Edinburgh Channel of the Thames Estuary. This estuary, the largest of 170 such inlets on the coast of Great Britain and a major shipping route, extends from southwest London to where the Thames meets the North Sea.[14]

Map 4-2

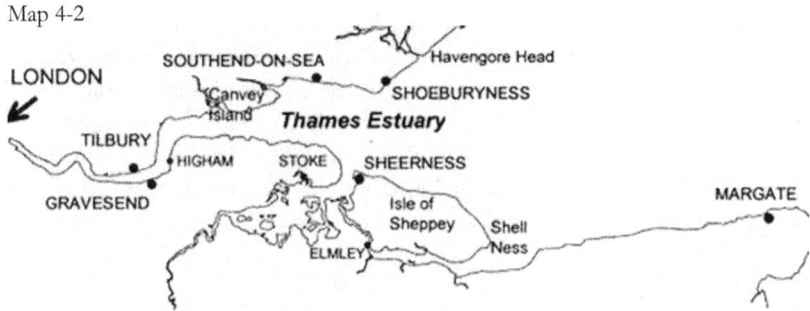

Southern portion of the Thames Estuary

Photo 4-4

German destroyer *Wilhelm Heidkamp*, circa 1939.
Naval History and Heritage Command photograph #NH 83980

The British minelayer HMS *Adventure* had sailed from Grimsby at 0932 earlier that day, in company with the destroyers *Basilisk* and *Blanche*, bound for Portsmouth, a port city on England's south coast. Visibility was poor proceeding down the east coast, between one and two miles. That night, unable to sight the Sunk Head and Northeast Gunfleet buoys, the ships anchored in Barrow Deep (a main shipping route past the shoals in the North Sea and outer Thames Estuary).[15]

The fog lifted in the early hours of 13 November and the minelayer and her escort weighed and proceeded at 0305. Capt. Arthur Halfhide, RN, *Adventure*'s commanding officer, decided not to stream paravanes while traversing the Thames Estuary. He considered that the danger of encountering moored mines in these restrictive waters was much less than the danger of losing the paravanes should it be necessary to alter course out of the narrow channel into shoal water, or stop, or go astern if meeting darkened ships either at anchor or under way.[16]

Photo 4-5

British cruiser-minelayer HMS *Adventure* in World War II, location and date unknown. National Archives photograph #80-G-165939

At 0506, speed was reduced from fifteen to nine knots, when close abreast the *Tongue Light Vessel*. Three minutes later the order was passed to "slip paravanes" (stream them off the port and starboard bows of the ship). The starboard one ran out correctly, but the towing wire of the port paravane fouled the heel fitting of an awning stanchion. It was necessary to recover the paravane and re-rig it. This was being done when at 0526 an explosion occurred just before the bridge on the port side. *Adventure* was then three miles from the light vessel with *Basilisk* and *Blanche* in column ahead in that order.[17]

Captain Halfhide believed (correctly) that explosion was due to a mine and not a torpedo, because he considered it too dark for a submarine to have attacked *Adventure*. His immediate concern was to stop the ship, which was turning slowly to starboard, so as to prevent her going full circle back into the minefield, and to avoid aggravating the damage she had sustained by bringing unnecessary pressure on weakened bulkheads. His next priority was to have a destroyer close by to render assistance in the event the damage to the ship was serious.[18]

These otherwise simple actions were hindered by degradation to ship control and communications equipment, identified below, and because all officers and men on the bridge were temporarily stunned by the force of the explosion or by blows received from falling instruments:

- The fore bridge was wrecked; the standard compass and Eversheds (target bearing indicators) were blown from their mounts and the bridge roof collapsed.
- All instrument lights on the bridge were out.
- No communication was available with the engine room, either by telephone or telegraphs through the lower steering position, as no answer could be obtained from the voice pipe to the lower steering position.
- The port siren was operating at full blast, rendering verbal communication on the upper deck almost impossible. (This was due to the rigging of the foretopmast having snapped off above the control top, fouling the siren lanyard).
- All bridge telephones were inoperative.[19]

Owing to the siren and failure of internal communications, it took seven minutes for Halfhide's order to stop engines to take effect. It also took considerable time to establish visual communication with the escort ships, owing to signaling gear on the bridge and flag deck having been strewn about and smashed by the force of the explosion. Meanwhile, boats and life floats were readied for abandoning ship should this be necessary.[20]

Halfhide eventually received reports of damage between Nos. 54 and 78 bulkheads, and that they were apparently sound. He also received word of a large number of wounded men and he ordered *Basilisk* alongside for the transfer of casualties to her. By 0715 all of the wounded able to be moved were aboard *Basilisk*, and *Adventure* was ready to proceed with rudder control at the after steering position. Halfhide ordered the destroyer to lead him into the Edinborough Channel, giving as wide a berth as possible to the location where she had been mined.[21]

Written reports by the commanding officer dated 16 and 21 November, concerning the considerable damage suffered by *Adventure*, described conditions in wrecked spaces, extending upward from the ship's hold to her superstructure:

> The space between Nos. 54 and 78 bulkheads...was opened up completely on the port side, and partially on the starboard side, from the keel to the upper deck, and contained a mass of twisted metal from frames, bulkheads, beams, pipes, lockers etc. Those portions of the decks which still remained were in some cases two feet deep in oil fuel. The whole of this space was filled with the gases of the explosion, fumes from the C.O.2 Room and oil fuel, and steam from heater pipes. These fumes were intensely irritating to the eyes and throat, and produced symptoms of lacrimation, coughing and nausea. About fifteen minutes after the explosion these fumes were less dense and irritating and the steam had been shut off. The only light available was supplied by hand torches which penetrated but a short distance into the fumes. Everything was smothered with oil fuel and many ladders were either shattered or unshipped. In this space were many wounded men, some trapped by fallen debris or twisted steel, and it was in these conditions that rescue work was carried out, with the constant danger to the rescuers of falling into the bottom of the ship.[22]

At 0725 on 13 November, *Adventure* went ahead at four knots, then increased speed to eight knots; accompanied by *Basilisk* and *Blanche*. An hour later, an explosion was observed at 0825 directly under the stern of the latter destroyer, and she was seen to lower a boat. At about the same time, a lifeboat approached from the direction of Margate. Halfhide was reluctant to leave *Blanche* unattended, but considered it unwise to send *Basilisk* to assist her lest she also be mined. He endeavoured to direct the lifeboat to *Blanche*; but was unable to establish communication with her either by semaphore, Morse, or International Code. The lifeboat finally turned away steering in the direction of Margate, upon which Halfhide ordered *Basilisk* to close the boat and send it to the *Blanche*, which had disappeared from view in the mist.[23]

At 0952, *Basilisk* overtook *Adventure* on her way to Sheerness on England's north coast to land casualties, and reported that *Blanche* was under tow by a tug. *Adventure* anchored at the Nore (a sandbank in the Thames Estuary, extending between Shoeburyness to the north and

Sheerness to the south) at 1128 on 13 November. She had proceeded there via use of the station-pointer (which plots position from horizontal sextant angles taken between two or more objects or geographical features) for navigation, the fore bridge for conning, and the after position for steering.[24]

AFTERMATH

HMS *Adventure*

Badge: On a black field, a silver anchor between two shields bearing the Cross of St George and the Irish Harp, respectively

Motto: 'Dare all'

Battle Honours: NORMANDY 1944

Twenty-five *Adventure* crewmen were immediately killed (or later perished), and two men aboard *Blanche* went missing and were presumed killed, as a result of the magnetic mine detonations. The identities of these casualties may be found in Appendix C. HMS *Blanche* sank one mile north of Spit Buoy while under tow by the tug *Fabia* en route to Sheerness; the first British destroyer to be lost to enemy action in the war. Later that day, the British cargo ships SS *Ponzano* and SS *Matra* were mined in the same area as *Adventure* and *Blanche* and were lost.[25]

Following temporary repairs at Chatham, in southeast England, *Adventure* moved to Plymouth on the south coast for refit. Arriving there on 20 December 1939, her crew was paid off and she was taken in hand for repair work which lasted through August 1940. With a new crew, she then carried out post refit trials in preparation for service and work-up at Scapa Flow. *Adventure* joined the 1st Minelaying Squadron at Kyle of Lochalsh on 17 October 1940.[26]

ENEMY AERIAL MINELAYING BEGUN IN LATE 1939; DIFFICULTIES ENCOUNTERED, AND NEW MEASURE BY BRITAIN HELP REDUCE HER VESSEL LOSSES

In mid-November 1939, German aircraft began mining British waters, but they lacked the means of fixing their positions accurately and were unable to drop mines where they would do the most harm. Accurate laying of mines by enemy submarines and ships through this point in the war had been greatly assisted by coastal lights kept illuminated for the benefit of British vessels. This practice changed on 21 November, when lights in the Thames Estuary were extinguished, and movement of all traffic west of a line between the Downs and Orfordness was stopped during the hours of darkness. This measure immediately helped to reduce vessel losses in the Thames approaches.[27]

In December, Germany shifted much of its minelaying northward from the Thames Estuary to the narrow channels off Norfolk County on England's east coast through which east coast convoys had to pass. The burden on the Nore Command and its minesweepers (responsible for defence of east coast convoys supplying the ports of northeastern England) continued to be heavy; but less so than in the preceding month. Thirty-three merchant ships were sunk in December by mines and eight others damaged, but there was a decline in the number of magnetic mines laid, due to enemy stocks having run low. Germany had only manufactured some 1,500 before the war; and was able to produce only a few more during the first months of the war. Stores of 20,000 contact mines had existed and it was, therefore, on the laying of that type that the enemy's campaign was initially chiefly dependent. However, the proportion of ship losses to magnetic mines far exceeded the numbers of those type laid thus far.[28]

SUMMARY OF GERMAN MINING CAMPAIGN IN 1939

During the first four months of the war, German mines caused the loss of seventy-nine merchant ships of 262,697 tons, significantly degrading British coastal shipping. One countermeasure denied the British military (on account of restrictive rules on air bombardment then in force) was bombing the seaplane bases from which German minelaying aircraft operated, or the naval bases used by submarine and surface minelayers. Additionally, despite the enemy mining campaign, the Cabinet had only just recently introduced control of German exports in retaliation for this method of waging war.[29]

5

Mining of Norwegian Waters, and British Aerial Minelaying

Offensive mining operations, although paltry in the first months of the war, built up gradually as new mines and vehicles became available and from mid-1940 onwards played an ever more important part in the general attack on enemy shipping. This form of attack, mounted in the main by aircraft but frequently supplemented by surface craft and submarine, not only caused grievous casualties amongst enemy controlled shipping and frequently affected movement patterns; just as significantly, it forced the enemy to devote a large proportion of his available manpower and material resources to a wide variety of mine countermeasures.

—British Mining Operations 1939-1945 Vol. 1.[1]

DEFENSIVE MINING IN BRITISH HOME WATERS

The Dover Barrage was largely completed in 1939 and, as previously discussed, was instantly successful in closing that route to German U-boats. Subsequent augmentation for anti-invasion reinforcement in 1940 brought its total contents to 9,897 mines. Elsewhere, a huge investment by the British in defensive mines did little to deter enemy operations. The East Coast Mine Barrier, begun in October 1939 and completed in November 1942, consisting of 38,045 mines, actually assisted the Germans in reaping a rich harvest of Allied and Neutral merchant shipping, especially during the first months of the war. In its original role, instead of providing intended protection to shipping, the barrier was responsible for creating a concentration of ships vulnerable to destruction by the enemy.[2]

A minefield in the St. George's Channel, a waterway connecting the Irish Sea to the north and the Celtic Sea to the southwest, was begun in July 1940 as an anti-invasion measure and reinforced thereafter with both shallow and deep fields up to June 1943. There is no evidence that this field, which absorbed 10,766 mines, claimed any victims. Had there been an invasion of the United Kingdom in that area, the mines could have imposed losses on the attacking sea-borne forces. As it was, while providing a measure of security, it restricted the movements of British forces and imposed a constant threat to friendly shipping from navigational error or drifting mines.[3]

Small defensive fields were laid in 1939 and 1940 in the Forth, the Clyde, and the Wash (a largely rectangular bay on England's east coast) totaling 907 mines. Finally, work began in July 1940 on the laying of the Northern Barrage. This vast effort consumed 80,352 mines by the end of 1942. It proved to be ineffective and was easily avoided by the Germans and did not result in much loss of commercial shipping, surface ships or submarines to them.[4]

During the autumn of 1940, another 9,195 mines were laid under the convoy routes in the Northwest approaches. Unfortunately, these came too late to offer any threat to prowling U-boats in search of merchant shipping, which by then had been driven farther seaward by British air patrols.[5]

AT LEAST ONE FIELD WAS MORE SUCCESSFUL THAN ORIGINALLY BELIEVED

In recent years, as divers continue to locate the wreckage of U-boats on the seafloor (often aided by precursor sonar searches), examination of hull damage has revealed that the demise of some submarines previously credited to aircraft or shipboard weapons, was actually due to mines. In 2006, divers uncovered the wrecks of *U-325*, *U-400*, and *U-1021* off the Cornish coast, lying in close proximity to each other seven miles off Newquay, where no U-boats were known to have been lost.[6]

Extensive research revealed they had been the victims of a secret "trap" minefield laid by the British to destroy such vessels. The Royal Navy had previously laid a defensive minefield between Cornwall, located in southwest England, and Ireland across the Celtic Sea. It contained a gap to allow supply ships to enter the harbours at Cardiff, Wales, and Bristol, England. The British intercepted a radio message from a U-boat commander to headquarters regarding his discovery of the gap, which was deciphered by codebreakers at Bletchley Park, in Milton Keynes, Buckinghamshire. The British then laid deep mines designed to allow surface ships to continue to use the gap, but destroy U-boats attempting to transit at depth.[7]

Following discovery of the mined submarines, which disappeared in late 1944 and early 1945, Eric Grove, a British naval historian and defence analyst, noted that the new information about the fate of *U-325*, *U-400*, and *U-1021* was revealing:

This shows the deep trap minefield was far more successful in killing U-boats than first thought. It was only recently the Government revealed the existence of this deep water minefield and the presence of these wrecked U-boats shows how effective it was. The U-boat crews were under orders to patrol coastal waters hunting Allied shipping at the end of the war because the Atlantic campaign was over. Admiralty records show two of the three U-boats in question were thought to have been sunk by depth charge in the Bristol Channel. This new research shows that was wrong and they actually struck mines off Cornwall.[8]

INITIAL OFFENSIVE MINING OF GERMAN WATERS

At the outbreak of war, the British reasoned that mining the Heligoland Bight (a bay which forms the southern part of the larger German Bight, itself a bay of the North Sea, located at the mouth of the Elbe River; see Map 3-2) would interrupt enemy movements on and under the water and cause casualties. The Royal Navy's initial efforts in September 1939 is covered in Chapter 3. However, subsequent mining in the area was delayed for several reasons. The German declaration of a mined area covering the approaches from the west, coupled with inadequate air reconnaissance to pinpoint the enemy's routes, postponed planned destroyer mining operations for a period of three months, after only two fields had been laid. One additional small field was laid by destroyers in December 1939; bringing the total number of mines laid offensively that year to 360, which caused neither casualty nor embarrassment to the enemy.[9]

As the new year, 1940, broke, the British Admiralty believed that it had sufficient intelligence of the enemy's routes for surface mining in the bight to resume, with the object of laying small fields of contact mines in the channels through the German mined area in the North Sea. However, by then, the 20th Destroyer Flotilla was largely employed for other purposes, including anti-submarine patrol and defensive mining. Two such minelays were carried out during the first half of January, but the destroyers did not return to minelaying until 3 March, when the *Express*, *Esk*, *Icarus* and *Impulsive* laid 240 mines in the enemy channel. Once available, emphasis on use of the flotilla shifted to the laying of mines in Norwegian territorial waters, planned for early April. This action was part of a long conceived and repeatedly postponed plan to disrupt the transport of Swedish iron ore through neutral Norwegian waters, necessary to sustain the German war effort.[10]

MINING OF NORWEGIAN WATERS

Britain decided early in 1940, to lay mines within Norwegian territorial waters, despite Norway being, like Sweden, a neutral country. The configuration of Norway's coastline was such that much of the 500-mile passage between Narvik and Stavanger could be made in the Inner Leads, a series of channels running inside an almost continuous chain of offshore islands. Narvik, shown on the following map, lies on the shore of Ofotfjord, an inlet of the Norwegian Sea located about 120 miles north of the Arctic Circle. During the war, iron ore was transported to this port by railway from Swedish mines in northern Sweden for export.[11]

Map 5-1

Portion of Scandinavia, including the west coast of Norway

By laying mines off Stadtlandet, Bud, and in the Vestfjord, ships carrying vital iron-ore to Germany would be forced farther seaward, where they would be subject to search and capture on the open seas outside Norway's territorial waters.[12]

OPERATION WILFRED/LOSS OF HMS *GLOWWORM*

There has been no greater impediment to the blockade of Germany than this Norwegian corridor. It was so in the last war, and it has been so in this war. The British Navy has been forced to watch an endless procession of German and neutral ships carrying contraband of all kinds to Germany, which at any moment they could have stopped, but which they were forbidden to touch.

It was therefore decided at last to interrupt this traffic and make it come out into the open seas. Every precaution was taken to avoid the slightest danger to neutral ships or any loss of life, even to enemy merchant ships, by the minefields which were laid and British patrolling craft were actually stationed around them in order to warn all ships off these dangerous areas.

The Nazi Government have sought to make out that their invasion of Norway and of Denmark was a consequence of our action in closing the Norwegian corridor. It can, however, undoubtedly be proved that not only had their preparations been made nearly a month before, but that their actual movements of troops and ships had begun before the British and French minefields were laid. No doubt they suspected they (the mines) were going to be laid. It must indeed have appeared incomprehensible to them that they had not been laid long before. They therefore decided in the last week of March to use the Norwegian corridor to send empty ore ships northward filled with military stores and soldiers concealed below decks, in order at the given moment to seize the various ports on the Norwegian seaboard which they considered to have military value.

—Speech made by Prime Minister Winston Churchill in the House of Commons on 11 April 1940, about the situation off the Norwegian coast, where naval battles were then taking place, and to justify Operation Wilfred.

On 5 April 1940, a large force of warships, which included elements of Operation WILFRED set out from Scapa Flow and proceeded toward the Norwegian coast. The plan was to lay two minefields, the first one at the mouth of the Vestfjorden—the channel leading to the port of Narvik where iron ore was shipped (Operation WV). The second was to be farther down the western Norwegian coast, adjacent to the

Stadtlandet peninsula (Operation WS). As a diversion, the simulated laying of a third field was to occur off the Bud headland south of Kristiansund (Operation WB).[13]

A portion of the force (Force WB and Force WS) detached on 7 April, to carry out operations to the south, as the main body continued en route to Narvik. Later that day, planned minelaying by the WS force off Stadtlandet was cancelled due to a report of German warships heading for the coast of Norway from Heligoland Bight. Early the next morning, 8 April, the British government informed the Norwegian authorities of its intention to lay mines inside their territorial waters. Soon after, Force WB as part of their diversionary procedure jettisoned empty oil drums into the waters off the Bud headland. They then patrolled the area to "warn" shipping of the mine danger.[14]

Operation WILFRED Naval Forces

Force WV (mouth of Vestfjord)
Battlecruiser HMS *Renown*
Destroyer HMS *Glowworm*
Destroyer HMS *Greyhound*
Escort destroyer HMS *Hardy*
Escort destroyer HMS *Havock*
Escort destroyer HMS *Hotspur*
Escort destroyer HMS *Hunter*
Minelaying destroyer HMS *Impulsive*
Minelaying destroyer HMS *Esk*
Minelaying destroyer HMS *Icarus*
Minelaying destroyer HMS *Ivanhoe*

Force WB (Bud headland)
Light cruiser HMS *Birmingham*
Minelaying destroyer HMS *Hyperion*
Minelaying destroyer HMS *Hero*

Force WS (off Stadtlandet)
Auxiliary minelayer HMS *Teviot Bank*
Minelaying destroyer leader HMS *Inglefield*
Minelaying destroyer HMS *Imogen*
Minelaying destroyer HMS *Ilex*
Minelaying destroyer HMS *Isis*[15]

Farther north, Force WV successfully laid its minefield in the mouth of the Vestfjord. The first mine went in the water at 0430; the Admiralty received word by 0445, and a warning statement was broadcast on Empire radio wavelengths at 0500 and again on the 0700 Home Service. This same warning was subsequently repeated in sixteen different languages on the foreign wavelengths.[16]

Britain had anticipated that Norway and Sweden would protest this violation of neutral rights, but the violence of the language in which their protests were made exceeded all expectations. However, fortunately, the laying of the mines coincided with the preliminary moves of the forthcoming German occupation of Norway. Its fleet was then already advancing up their coastline, and the day following the mining operation, Oslo was in enemy hands and Norwegian waters had ceased to be neutral.[17]

Meanwhile, as the southern Force (WS and WB ships) rejoined the Home Fleet, the northern WV force became embroiled in the early actions of the British attempt to thwart German landings in Norway. In an epic battle on 8 April 1940, the British destroyer HMS *Glowworm* was sunk by the German heavy cruiser *Admiral Hipper*.

HMS *GLOWWORM* SUNK BY *ADMIRAL HIPPER*

On 6 April, whilst screening the battle cruiser HMS *Renown* (part of the WV minelaying force) off the coast of Norway, *Glowworm* lost a man overboard in heavy weather conditions. After seeking permission from *Renown*, Lt. Comdr. Gerard Roope, *Glowworm*'s commanding officer, turned the destroyer around to look for him. During a fruitless search, she lost contact with the British force. As the weather worsened, she was forced to reduce speed to less than ten knots. At daybreak on 8 April, while *Glowworm* was searching to rejoin her force, she sighted an unidentified destroyer.[18]

When *Glowworm* challenged the new arrival to identify herself by flashing light, the ship replied that she was Swedish—then opened fire. She was in reality German, the destroyer *Bernd Von Arnim*, in company with a second destroyer, *Paul Jakobi*. Battle ensued, with the three destroyers manoeuvring at high speed and firing their guns.[19]

During violent manoeuvres, *Glowworm* lost two men overboard and several were injured by the roll of the ship, but her gunners scored a hit on the leading enemy destroyer. Shortly afterward, the German ships broke off the action; *Bernd Von Arnim* (packed with invasion troops) and *Paul Jakobi*, turned and fled into a rain squall. Roope suspected they might be trying to lure him into an ambush but, nevertheless, gave chase in an attempt to find the main German Invasion fleet, so that he could report its position to the Admiralty.[20]

After emerging on the other side of the squall, *Glowworm* came face to face with the 16,170-ton heavy cruiser *Admiral Hipper*—bristling with eight 8-inch, twelve 4.1-inch, and twelve 37mm guns. Recognizing that his ship had no chance of survival, nor of escape, Roope vowed to inflict as much damage as possible on the enemy before being sunk. The 1,345-ton G-class destroyer's modest armament consisted of four 4.7-inch guns, seven anti-aircraft machine guns, ten 21-inch torpedo tubes and depth charges. Roope reasoned if he could get close enough, his torpedoes might prove useful. However, before he could bring *Hipper* within gun range, she was pouring 8-inch rounds into *Glowworm*.[21]

As Roope continued to close the enemy, he ordered that smoke be made and an enemy sighting signal be sent. Using this smoke screen for cover, *Glowworm* carried out two torpedo attacks, firing ten torpedoes.

One missed the heavy cruiser by yards but none found their mark. All this time the destroyer was taking crippling hits—setting her ablaze. One of her guns was out of action, her range-finder had been hit, and the top of the mast collapsed across her siren pull cable, leaving the siren screeching amidst the blaze of battle and stench of cordite and black smoke. Out of options, Roope ordered a sharp turn to starboard and headed straight for the *Hipper*, giving the famous order "Stand by to ram." Going in under a barrage of 8-inch, 4-inch, and machine-gun fire, he steered for the enemy's starboard side.[22]

The *Hipper*, realising too late what was happening tried to turn and ram the destroyer but was much too slow. *Glowworm*, with guns firing and siren wailing, tore into the heavy cruiser's starboard side. Striking her amidships, with a grinding crunch as her bows crumpled against the *Hipper*, the destroyer tore away 100 feet of the heavy cruiser's armoured plating, damaging her starboard side torpedo tubes, killing one man at his gun and puncturing two fresh water tanks.[23]

Photo 5-1

Painting by Richard DeRosset of the British destroyer HMS *Glowworm* preparing to ram the German heavy cruiser RMS *Admiral Hipper* on 8 April 1940, during a losing sea battle in the Norwegian Sea, in which the much smaller surface combatant was sunk.

With her bows stove in, *Glowworm* drew away, and opened fire once more. Although her decks were swept by fire from *Hipper*'s 4.1-inch and close-range weapons, she still managed to get off another salvo hitting the heavy cruiser at a range of 400 yards. The salvo came from her only still firing single-mount, that of Petty Officer Walter Scott. Round after round from *Hipper*'s guns then found her. One passed through the wheelhouse; another burst in the Transmitting Station, killing most of the crew and all the staff of the Wireless Office. A third entered the ship's hull beneath the after-torpedo-tubes, passed through the ship and exploded against the forward bulkhead of the captain's cabin (which was being used as a first aid station), turning it into a shamble. Another round wrecked the after superstructure.[24]

The round that ruined the cabin also blew a huge hole in the destroyer's side and as water poured in, she heeled over to starboard. Her bows wrecked, a major fire raging amidships and all propulsion lost, Roope gave the order to abandon ship. There were many wounded and all that could be done was to put lifebelts on the injured men in the hope that they would float. Roope was one of the last to abandon, chatting with Petty Officer Townsend about the fact that they wouldn't play cricket for a while again, before he went to open the sea cocks to sink her. As *Glowworm* went down, men still aboard climbed onto her bow or dived into the stormy, freezing, oil covered water. She slipped under in late morning; her siren abruptly stopping caused a momentary eerie silence, until her depth charges blew up—killing men in the water.[25]

The *Hipper* stopped and her captain, Helmuth Heye, chivalrously remained for over an hour, picking up survivors. Personnel on deck, including embarked soldiers, helped to pull exhausted, oil-covered British sailors aboard. Ropes were thrown to others who, too exhausted to hold on to them, perished. Roope was in the water helping men to the ropes and to get life jackets on, before he finally took hold of a rope himself and was pulled some distance up the cruiser's steep, high side. Tragically, a combination of heavy seas and exhaustion loosened his grip and he slipped beneath the waves. Captain Heye told the survivors that their captain was a very brave man, and later sent a message through the International Red Cross, recommending him for the Victoria Cross; the only time in British History that the VC was advocated by the enemy.[26]

Of the ship's company of 149, only 31 survived; Lt. Robert Ramsey (the torpedo control officer) and thirty ratings. *Hipper* delivered these men to Trondheim, Norway, and later returned to Germany to enter drydock for the repair of battle damage. The survivors spent the rest of the war as POWs. It was only after the war, when Lt. Robert Ramsey told their story, that the events of that fateful day in April 1940 came to light. As a result of his gallant actions, Lt. Comdr. Gerard Broadmead Roope was awarded the Victoria Cross (posthumously). Lieutenant Ramsey received the Distinguished Service Order, and three ratings—Engine Room Artificer Henry Gregg, Petty Officer Walter Scott, and Able Seaman Reginald Merritt—the Conspicuous Gallantry Medal.[27]

AFTERMATH OF OPERATION WILFRED

While Operation WILFRED was in progress, a German invasion fleet was at sea. Force I (2,000 troops of the 3rd Mountain Division under

General Eduard Dietl aboard ten destroyers and supported by the battlecruisers *Scharnhorst* and *Gneisenau*) was heading toward Narvik. Before entering the approaches to Narvik, the battlecruisers were detached and sent off to the northwest. On the morning of 9 April, the destroyer force entered Ofotfjord, the western approach to Narvik, and sank the elderly Norwegian coastal defence vessel *Eidsvoll*. Three German destroyers then sailed into Narvik Harbour, and sank her sister ship *Norge*. Dietl landed, and was able to convince the Norwegian commander at Narvik to surrender without a fight.[28]

The following day a force of five British destroyers—flotilla leader *Hardy*, *Hotspur*, *Havock*, *Hunter*, and *Hostile*—under Capt. Warburton-Lee reached Narvik undetected. In the first battle of Narvik, they sank two enemy destroyers and damaged three more in the harbour. All but one of these ships, *Hostile*, had been a part of WILFRED.[29]

On their way out of the fjord, the British were attacked by the remaining five German destroyers, losing two ships themselves. *Hardy* was badly damaged, and had to be beached, while *Hunter* was sunk outright. Warburton-Lee was killed, and later awarded a posthumous Victoria Cross. *Hotspur* was also badly damaged; but rescued by the remaining two Royal Navy destroyers. Departing, the British were also able to sink the German ammunition ship *Rauenfels*.[30]

Photo 5-2

Sunken ship in Rombak's Fjord, Norway, photographed by a German cameraman after the naval battle near Narvik on 10 April 1940.
Naval History and Heritage Command photograph #NH 71407

Three days later the British returned to Narvik in much greater force with the battleship HMS *Warspite* and nine destroyers, and in the second battle of Narvik, 13 April, sank the surviving German ships. British and French troops landed at Narvik the following day to assist the Norwegians, pushing the Germans out of the town and almost forcing them to surrender. Further Allied landings took place between 18 and 23 April, but these efforts were all too little, too late, and the Norwegians surrendered on 9 June 1940.[31]

Operation WILFRED was essentially a failure as it did not prevent the Germans from gaining iron ore. But, ironically, once Norway's administration became pro-German she was no longer neutral, and British ships and aircraft were free to enter her territorial waters and air space to attack the enemy at will.

COASTAL COMMAND, BOMBER COMMAND, AND NAVY AIRCRAFT BEGIN MINELAYING OPERATIONS

Concurrent with the start of the Norwegian campaign, a new form of British offensive minelaying began, that of seeding the enemy's channels and estuaries from the air. Royal Air Force Coastal Command, Bomber Command, and naval aircraft all took part. The first lay was carried out by Hampdens of Bomber Command on the night of 13 April 1940. Two nights later, Coastal Command Beauforts followed suit; soon after Royal Navy No. 815 naval air squadron, which flew Swordfish aircraft and was then working under Coastal Command, joined the other minelayers. These operations developed rapidly into an important factor in disputing the control of the enemy's coastal routes, after magnetic mines and sufficient numbers of suitable aircraft became available to obtain good results. (The mines had been in development since before the war, but were not immediately available, delaying their deployment).[32]

The A Mk I ground magnetic mine had been designed for use by torpedo-carrying aircraft, but it soon became apparent that it would be impracticable to operate carriers in the southern part of the North Sea. The torpedo-carrying Beauforts of Coastal Command could reach the German North Sea ports and estuaries, but for operations farther afield, long-range bombers of Bomber Command were required. The use of RAF aircraft required adapting the mines for laying from greater heights and at greater speeds, which involved substituting a small parachute for the original ballistic tail. This change had the added benefit of reducing the overall length of the mine and thereby eased the problem of stowage in the planes' bomb racks.[33]

Photo 5-3

A magnetic mine being loaded aboard a British Royal Air Force Hampden bomber. Collection of Rob Hoole

Photo 5-4

Bristol Beaufort with wing flaps down, preparing to make a landing; 7 February 1941. Australian War Memorial photograph 005417

Map 5-2

Danish Straits. Three major passages traverse the Danish Straits: Little Belt to the west, Great Belt in the centre, and to the east "the Sound" separating Denmark from Sweden.

In their first mission on 13 April 1940, Hampdens of No. 5 Group laid mines in the Great Belt, the Little Belt, and the Sound, passages in the Danish straits, which connected the Baltic Sea to the North Sea through the Kattegat and Skagerrak. Two days later, Beauforts of Coastal Command laid mines off the Jade River, which flows into the Jade Bight, a bay of the North Sea, near Varel, Germany. Both types of aircraft carried only a single mine on these missions. At this time, Britain declared a large area of the Baltic dangerous. Soon after, mining operations were extended within this area, and also to Oslo, Norway, and to positions off the German North Sea estuaries, such as the Ems and the Weser. These latter locations were covered by the danger area originally declared in the Heligoland Bight, which then had been enlarged to cover the whole of the Danish coast, the Kattegat, the Skagerrak, and all Norwegian waters. A channel thirty-miles wide was left to allow access to the North Sea from Swedish waters.[34]

In April and May 1940, a total of 263 mines were laid in 385 aircraft sorties. Ten aircraft were lost, but twenty-four enemy ships were sunk and a further two ships damaged by the magnetic mines dropped. During the early days of these operations, a considerable number of the aircraft sent out returned without laying their mines. But as training improved and more planes became available, better results were achieved.[35]

Photo 5-5

Royal Navy Fairey Swordfish aircraft, dropping a practice torpedo in 1939.
Naval History and Heritage Command photograph #NH 94122

6

Dutch Minelayers Escape the Netherlands

Photo 6-1

HNLMS *Willem van der Zaan*, one of the units of the *Koninklijke Marine* (Dutch Navy). From a Dutch Navy poster

The Royal Netherlands Navy minelayer HNLMS *Willem van der Zaan* was laid down at Nederlandse Dok Mij, Amsterdam, on 18 January 1938. She was commissioned nineteen months later on 21 August 1939, with Lt. Comdr. Hendrik Dirk Lindner in command. The 246-foot ship was intended for double-purpose use as both a minelayer and training vessel, capable of accommodating forty cadets but also boasting robust armament. Comprising her main battery were two 120mm/50 Wilton-Fijenoord guns; and for anti-aircraft defence, two twin-40mm/56 Bofors machine guns with individual Hazemeyer directors, as well as two smaller twin-12.7mm/90 machine guns. When built, she could carry up to 120 mines on her covered mine deck (this number was reduced to 90 following modifications in the UK in 1940), and amidships there was an area for a seaplane, with an associated handling crane at the after part of her superstructure.[1]

Powered by two Yarrow 3-drum boilers and two Werkspoor triple-expansion-engines, producing 2,200hp, the Dutch ship could make 15.5 knots. Minelayers able to carry and sow large quantities of heavy mines (mostly converted merchant vessels) had very modest top speeds and were, accordingly, mostly employed placing defensive fields in home waters. "Nightraiders" capable of higher speeds (like the British *Abdiel*-class "fast minelayers) were typically employed to mine enemy waters. They were more able to transit to specified areas, carry out their missions, and be back in safer waters before sunrise, thus decreasing the likelihood of detection by enemy forces.²

Van der Zaan's first mining operation was defensive in nature, laying 98 mines near Den Helder on 3 September 1939. Later she added another 97 to the same field ("Schulpengat Buiten") on 22 September. Several months later on 7 May 1940, she took part in laying a minefield in the Boomkersdiep (27 mines) and "Zuider Stortemelk" (68 mines); and on that same day, laid the tactical barrages Northwest Vlieland (13 mines) and Southwest Vlieland (12 mines).³

GERMAN INVASION/EVACUATION TO BRITAIN

After Germany invaded Holland on 10 May 1940, the Dutch Navy was forced to flee when the bulk of the Dutch Army surrendered on 14 May. As a consequence, the government in exile established a new wartime headquarters in London. Because a majority of the fleet was in the Netherlands controlled East Indies, smaller offices were also set up in Ceylon (Sri Lanka), and Australia.⁴

The Dutch Navy suffered its heaviest war losses in an unsuccessful defence of the East Indies, which culminated in the Battle of the Java Sea in February 1942. Rear Adm. Karel Doorman went down on his flagship HNLMS *De Ruyter* along with most of his ships and 2,300 officers and men. As the ABDA (Australian-British-Dutch-American) fleet under his command no longer existed, Allied resistance in the East Indies ceased. (The heroic escape of some Dutch mine warfare ships to Australia, or Columbo, Ceylon, under Japanese naval and air forces attack on Java, is covered in *Nightraiders*, a companion book of this one.)⁵

Seven minelayers tried to leave Holland for Britain (of which six were successful). The identities of their commanding officers, and fate of the ships is provided in the following table. The rank of all of the commanding officers was luitenant ter zee 1e klasse (lieutenant commander), Royal Netherlands Navy.

Dutch Minelayers

Ship/ Date Com.	Length (ft.)/ Displ. (tons)	Commanding Officer	Ultimate Fate
Douwe Aukes 2 Nov 1922	180/687 (*Douwe Aukes*-class)	Joost Ruitenschild	Decom on 1 Feb 1962, sold for scrap 6 Jul 1962
Hydra 25 Jan 1912	192/995 (*Hydra*-class)	Johannes Adolph de Back	Scuttled on 15 May 1940 after being damaged by anti-tank gunfire
Jan van Brakel 25 Jun 1936	192/955	Jacobus Johannes Hogendoorn	Struck 1 Aug 1957 and expended as a target off Biak Island, Papua
Medusa 20 Dec 1911	163/657 (*Hydra*-class)	Johannes Marius Logger	Struck 4 Jun 1964, sold for scrap 25 Sep 1964
Nautilus 2 May 1930	180/951	Jan August Gauw	Sank 22 May 1941 due to a collision with the SS *Murrayfield* off Grimsby, England
Van Meerlant 22 Jul 1922	180/687 (*Douwe Aukes*-class)	Baron Thomas Karel van Asbeck	Mined in the Thames Estuary 4 Jun 1941
Willem van der Zaan 21 Aug 1939	246/1407	Hendrik Dirk Lindner	Struck 27 Feb 1970, sold for scrap 6 Oct 1970[6]

HNLMS *Willem van der Zaan*—the newest and most capable of the six Dutch minelayers that escaped Holland—sailed from Den Helder on 13 May 1940 bound for Britain. She arrived at Portsmouth the next morning with a cargo of ten torpedoes and other equipment for the cruiser *Sumatra*, which had accompanied her. Also aboard the minelayer were four downed German aviators plucked from the sea.[7]

On 14 May, the minelayers *Douwe Aukes*, *Jan van Brakel*, *Medusa*, and *Nautilus*; gunboat *Johan Maurits van Nassau*; and torpedo boats *G-13* and *G-15* evacuated Den Helder. In the afternoon, the group came under attack by German aircraft off the town of Callantsoog, eleven miles south of Den Helder. *Johan Maurits van Nassau*, the largest of the ships, was hit by bombs (with seventeen crewmen killed or who later died of their wounds) and was abandoned. The survivors were transferred to *G-13* and then put aboard the rescue ship *Dorus Rijkers* off Den Helder. The minelayers arrived at Portsmouth on 19 May.[8]

On 15 May, with the *van der Zaan* already safely at Portsmouth, and four other minelayers en route there, *Hydra* was lost to combat action off Holland. While bombarding enemy positions on Sint Philipsland, a village in the Dutch province of Zeeland, she was damaged by shore fire. Three crewmen were killed in the action, and *Hydra* was run aground to prevent sinking. The remaining minelayer, *Van Meerlant*, sailed from Vlissingen (Flushing), Netherlands, on 16 May, arriving at Portsmouth on the 19th via Ostend, Belgium, and Dunkirk, France.[9]

OLDEST MINELAYERS LAID UP FOR REPAIR WORK

Following arrival of the six Dutch minelayers at Portsmouth, the three longest in commission: *Medusa* (1911), *Douwe Aukes* (1922) and *Van Meerlant* (1922) were laid up at Falmouth in southwest England, crewed only with a few men to man the anti-aircraft guns. With little expected service life remaining, *Medusa* was not given a further active role—even at a time when anything that could float and fight was pressed into service. She was taken by tug *Empire Henchman* to Holyhead, Wales, where she served out the war as a training, depot, and accommodation ship from August 1940 to February 1944. Holyhead was the adopted homeport for exiled Dutch ships assigned to patrol the Irish Sea.[10]

Sister ships *Douwe Aukes* and *Van Meerlant* were returned to active service in July 1940 and proceeded for refit at Portsmouth. *Van Meerlant* engaged from 12 July to 3 August 1940 in laying a defensive minefield at Sheerness; and *Douwe Aukes* was similarly occupied from October 1940 to February 1941 at Southampton on England's south coast. Being coal-fired and with very short ranges, the elderly minelayers proved most suitable for local defence duties, for which they were used until transferred to the Royal Navy in 1941. These transfers freed their crews to be utilised aboard modern Dutch warships.[11]

NEWER SHIPS IMMEDIATELY JOIN WAR EFFORT

The three newer minelayers—*Nautilus* (1930), *Jan van Brakel* (1936), and *Willem van der Zaan* (1939)—were quickly pressed into service. *Nautilus'* initial duties included laying a defensive minefield at Hartlepool in northeast England. Subsequent assignments were to the Thames Local Defence Flotilla, and Boom defence at Sheerness. In February 1941, she was assigned convoy escort duties. Three months later on 22 May, *Nautilus* was sunk as a result of a collision with the British merchant ship SS *Murrayfield* on 22 May near Saltfleet (a coastal village in east central England) while escorting the commercial ships SS *Murrayfield* and Icelandic SS *Hekla*. Fortunately, her entire crew was rescued.[12]

Nautilus was primarily designed to carry out minelaying in coastal waters in conjunction with other minelayers, under the protection of coastal batteries and gunboats against enemy surface warships and submarines. For this role, deck space for a fixed number of mines and low speed sufficed, however her armament was insufficient to protect the ship against aircraft. Moreover, she did not have an enclosed mine deck below the main deck, necessitating that her mines be stored on the main deck, where they were exposed to the sea and weather.[13]

Jan van Brakel was very similar to the *Nautilus*, in that she too was a dual-purpose ship, with a pre-war secondary role of fishery inspection vessel. Following her arrival in the UK after having been strafed by German aircraft on 14 May (1 killed), *van Brakel* was assigned duties as a convoy escort on England's eastern coast. Sent to the Dutch West Indies in April 1942, she was similarly employed. On 9 June 1942, she rescued the 34 survivors of the American tanker SS *Franklin K. Lane*, which had been torpedoed and sunk north-northeast of La Guiara, Venezuela, by *U-502* while sailing in convoy. *Jan van Brakel* returned to the UK in November 1944; where she was converted to a mothership for the 126-feet class minesweepers of the Dutch 203rd Minesweeper Flotilla.[14]

WILLEM VAN DER ZAAN AND HM MINELAYERS

HNLMS *Willem van der Zaan* was the workhorse of the Dutch minelayers exiled in Britain. Although a purpose-built ship, she required alterations to enable her to lay British mines and, consequently, was towed to the Portsmouth Naval Yard. Her rebuild, which included being fitted with new paravanes and a smokemaker, was completed on 27 June 1940. To test her new minerails, she laid two dummy mines that day, before taking on fuel and ammunition for additional trials. After completion of the trials, on 1 July 1940 the *van der Zaan* came under the command of the Senior Officer, Minelayers, in Immingham.[15]

Willem van der Zaan and HMS *Plover* laid ten minefields in the East Coast Barrier in July and August, normally accompanied by minelaying destroyers of the 20th Flotilla, and on two occasions by the auxiliary minelayer HMS *Teviotbank*. These included a minefield south of the Doggerbank (a sandbar off the east coast of England), two fields near Dover, and the remainder in the East Coast Barrier (BS series).

Date	Mined Area	Minelayers Involved
5 Jul 40	Minefield AW south of Doggerbank	*Willem van der Zaan*, HMS *Plover*
6 Jul 40	Minefield increased to 250 mines	*Willem van der Zaan*, HMS *Plover*
12 Jul 40	Minefield near Dover	*Willem van der Zaan*, HMS *Plover*
13 Jul 40	Minefield MN in the Straits of Dover	*Willem van der Zaan* and HMS *Plover*, escorted by four MTBs
29 Jul 40	Minefield BS-29	*Willem van der Zaan*, HMS *Plover*, and destroyers *Esk*, *Intrepid* and *Impulsive* of the 20th Destroyer Flotilla

1 Aug 40	Minefield BS-32	*Willem van der Zaan* with *Plover* and *Esk, Intrepid, Impulsive, Express, Icarus* and *Ivanhoe*
2 Aug 40	Minefield BS-30	*Willem van der Zaan, Plover, Esk, Intrepid, Impulsive, Express* and *Icarus*
13 Aug 40	Minefield BS-31	*Willem van der Zaan, Plover, Intrepid* and *Impulsive*
20 Aug 40	Minefield BS-33	*Willem van der Zaan* with HMS *Teviotbank, Plover, Impulsive* and *Icarus*
25 Aug 40	Minefield BS-34	*Willem van der Zaan, Teviotbank, Plover, Impulsive, Icarus* and *Esk*
28 Aug 40	Minefield BS-36	*Willem van der Zaan, Plover, Intrepid* and *Icarus*[16]

At 2200 on 26 September 1940, *Willem van der Zaan* and HMS *Plover* stood out of Sheerness to lay Minefield MU. En route to the specified area, both minelayers ran aground. The tugboat *Lady Brassey* pulled them free, and they arrived back in Sheerness at daybreak without having laid any mines.[17]

A week later, the v*an der Zaan* entered the King George V graving dock at Southampton on 3 October for repairs. They were completed mid-month and, on 3 November, she joined a northbound convoy to arrive in Methil (Firth of Forth) on the 4th. The following day, she joined another northbound convoy, from which she was detached in the Pentland Forth, and arrived at Londonderry, Northern Ireland, on 7 November.[18]

On 13 November (after numerous missions and having laid 2,198 mines), she sailed from Britain for the Netherlands East Indies. Her passage took her to Freetown via Ponta Delgada, and from Freetown to Capetown via a stop at the South Atlantic island of St. Helena—a British protectorate, and one of the most inaccessible places on Earth, where Napoleon was exiled on 15 October 1815 (after ten weeks at sea on board HMS *Northumberland*; a 74-gun third rate ship of the line).[19]

Finally, on 14 January 1941, HNLMS *Willem van der Zaan* arrived at Tanjung Priok, Java, the main port of Batavia, Dutch East Indies, (today Jakarta, Indonesia). Just shy of four years later—spent operating in the Pacific and Indian Oceans—she entered the River Thames on 16 November 1944. (*Van der Zaan*, along with the submarine HNLMS *O-19*, had escaped the Netherlands East Indies on 1 March 1942, the only Dutch minelayers to do so.) The following day, *van der Zaan* went into dock (the Eastern Dock of the London Docks, Shadwell Basin) for repairs and maintenance.[20]

7

Minelaying Forces Suffer the Loss of Seven Vessels in 1940

The loss of the minelaying submarine SEAL *in May [1940] coincided with the general withdrawal of submarines from the Kattegat and Skagerrak, and her sister ship, the* NARWAHL, *was lost off the west coast of Norway in July, just as all submarines were being transferred southwards for anti-invasion patrol duties.*

On its next rare offensive task at the end of August – the mining of an enemy coastal route off Vlieland [island in the northern Netherlands] – the Flotilla encountered a newly laid enemy minefield and lost 50 percent of its strength at one blow, the ESK *and* IVANHOE *being sunk and the* EXPRESS *severely damaged. This debacle bought the [German Heligoland] Bight operations to an end, indeed, only one further surface offensive lay was carried out that year – in Seine Bay at the end of November.*

—From *British Mining Operations 1939-1945, Vol. 1*, a Naval Staff History published in London in 1973. In addition to the loss of the two submarine minelayers, two destroyer minelayers and damage to another, cited above, the British also lost in 1940 a third submarine and two auxiliary minelayers.[1]

The German invasion of Belgium, the Netherlands, Denmark and Norway in May 1940, and the fall of France toward the end of June, gave the enemy control over the entire west coast of Scandinavia and continental Europe. Seemingly, this expanse of German-controlled territory from North Cape to the Franco-Spanish border, was then wide-open to offensive mining. However, the close proximity of enemy forces to the British Isles and associated fear of an invasion added emphasis to the importance of defensive, versus offensive measures.[2]

Over the course of the year, British minelaying forces suffered the loss of seven vessels: the destroyers *Esk* and *Ivanhoe*; the submarines *Seal*, *Grampus* and *Narwahl*; and the auxiliary minelayers *Princess Victoria* (a former car ferry) and *Port Napier* (a cargo ship under construction when acquired). Before their loss, *Esk*, *Express* and *Ivanhoe* had comprised half of the 20th (Minelaying) Destroyer Flotilla.

After completing a minelay in the Heligoland Bight (a bay which forms the southern part of the German Bight, itself a bay of the North Sea) on 10 May, the destroyer flotilla was used almost exclusively until mid-August, laying defensive fields in the East Coast Mine Barrier. Diversions included some work on the Dover Barrage, and evacuation of Allied troops from Dunkirk, France. *Esk* was sunk and *Express* damaged on 31 August, and *Ivanhoe* sunk the following day all after hitting German mines off Texel Island in the Netherlands.

DUNKIRK

> ... *We were one of the first destroyers to take part, journey after journey from Dunkirk to Dover and back running almost without a stop for seven days and nights, being shelled and bombed incessantly but remaining to complete the final night of the evacuation, thousands of troops we must have carried, our own casualties being one killed and six wounded by shrapnel, [our] ship after running the gauntlet so much was badly damaged, but still seaworthy. The evacuation being completed we were ordered back to Portsmouth for repairs to our ship and 48 hours leave.*
>
> —Herbert Vaughan, a Communications Clerk Petty Officer aboard the destroyer minelayer HMS *Esk*, describing her participation in the evacuation of the British Expeditionary Force from Dunkirk[4]

Codenamed Operation DYNAMO, the evacuation of Allied soldiers from the beaches and harbour of Dunkirk, in the north of France, took place between 26 May and 4 June 1940. The well-known action of the successful effort by Royal Navy and Royal Canadian Navy destroyers, a variety of merchant ships, and a flotilla of hundreds of fishing vessels, yachts and other pleasure craft, and lifeboats called into service from Britain saved the bulk of the British army, and other Allied troops, surrounded by German troops at Dunkirk. In his "we shall fight on the beaches" speech on 4 June, Winston Churchill hailed their rescue as a "miracle of deliverance."[5]

Since so many articles, books, and films have been devoted to this event, coverage of it is not included herein. Amongst the many vessels awarded Battle Honour DUNKIRK for taking part in the evacuation were the six destroyer minelayers of the 20th Flotilla:

Battle Honour DUNKIRK (20th Destroyer Flotilla)		
HMS *Esk*	HMS *Icarus*	HMS *Intrepid*
HMS *Express*	HMS *Impulsive*	HMS *Ivanhoe*

Photo 7-1

Dunkirk, France. Members of the British Expeditionary Force firing at attacking low flying German aircraft during evacuation from the beach.
Australian War Memorial photograph 101171

SERIES OF TRAGEDIES MAR 1940

The loss of seven minelayers, and significant damage to an eighth one, HMS *Express*, began on 5 May 1940 with the capture of the submarine HMS *Seal* by German forces. A summary of the vessels listed in chronological order of loss is shown in the following table:

Date	Vessel	Commanding Officer	Fate
5 May 40	HMS *Seal*	Lt. Comdr. Rupert Philip Lonsdale, RN	Captured by German forces after being damaged by a mine the day before
18 May 40	HMS *Princess Victoria*	Capt. John Buller Edward Hall, RN	Mined and sunk in the North Sea, off the Humber River on the east coast of England
16 Jun 40	HMS *Grampus*	Lt. Comdr. Charles A. Rowe, RN	Most likely sunk off Syracuse in the Mediterranean, by the Italian torpedo boats *Circe* and *Cleo*
23 Jul 40	HMS *Narwhal*	Lt. Comdr. Ronald James Burch, RN	Believed sunk by a German aircraft (Do17) about 125 nautical miles east of Aberdeen, Scotland

31 Aug 40	HMS *Esk*	Lt. Richard John Hollis Couch, RN	Mined and sunk about 40 nautical miles northwest of Texel Island, the Netherlands
31 Aug 40	HMS *Express*	Comdr. Jack Grant Bickford, RN	Struck a mine about 40 nautical miles northwest of Texel Island, and badly damaged[6]
1 Sep 40	HMS *Ivanhoe*	Comdr. Philip Henry Hadow, RN	Mined and damaged about 40 nautical miles northeast of Texel Island; sank later that day
27 Nov 40	HMS *Port Napier*	Capt. (retired) John Norman Tait, RN	Destroyed at Kyle of Lochalsh, Scotland by an engine room fire on board[6]

HMS *SEAL* AND CREW CAPTURED BY THE ENEMY

We'll have to surrender, sir. Nothing for you to worry about, sir. God knows, you've done all you could.

—Voice of an anonymous crewmen shouting up from the control room of HMS *Seal* to the commanding officer on the bridge, whose submarine was taking on water and down by the stern; unable to dive and, with propulsion inoperative, drifting, helpless, and under attack by German aircraft with her two machine guns no longer functioning.[7]

HMS *Seal* was built at the Chatham Dockyard and commissioned on 24 May 1939. She was one of six *Porpoise*-class minelaying submarines, each designed to carry 50 mines in a long, high casing above the pressure hull. The mines were the standard British Mk XVI self-mooring type, carried aft by a chain mechanism and released from the boat through a hatch in the stern. Minelaying operations were normally safe and expeditious, but the high casing above the hull created a massive silhouette easy to detect when the submarine cruised on the surface.[8]

Seal was armed with one 4-inch deck gun forward, two Lewis light machine guns that could be mounted on the conning tower after surfacing, and six 21-inch torpedo tubes forward. The design allowed one reload for each tube. Two diesel engines driving one propeller could propel the submarine at a top speed of nearly 16 knots on the surface. Cruising on battery-power submerged was normally restricted to four knots. Ship's complement was fifty-nine officers and men.[9]

Seal had been engaged under Lt. Comdr. Rupert Philip Lonsdale, an experienced submariner, in continuous patrol and escort duties in the Atlantic and off Norway. On 29 April 1940, she departed port on her first minelaying operation, the purpose of which was to interdict the German shipping route to Norway by laying mines in the Kattegat.[10]

In addition to Lonsdale, there were five other officers aboard:

HMS *Seal*'s Officers

Commanding Officer	Lt. Comdr. Rupert Philip Lonsdale, RN
First Lieutenant	Lt. Terence B. J. D. Butler, RN
Navigator	Lt. Trevor Agar Beet, RN
Engineer	Lt. (E) Ronald H. S. Clark, RN
Sub. Lt.	Alexander R. L. Henderson, RN
Sub. Lt.	Philip William Hubert Boulnois, RN[11]

Photo 7-2

Conning tower of HMS *Porpoise*; the lead submarine in a class of six, which also included HMS *Cachalot*, *Grampus*, *Narwahl*, *Rorqual*, and *Seal*. Australian War Memorial photograph P00604.011

Lonsdale's mission, Operation FD7, was to enter the Strait of Skagerrak, round The Skaw (Skagen Denmark), and proceed into the much more navigationally difficult and better defended Strait of Kattegat—the stretch of shallow water that separates Denmark from Sweden—where she was to lay her minefield. (The Skagerrak is a rectangular arm of the North Sea separating the southeast coast of Norway, southwest coast of Sweden, and the Jutland peninsula of Denmark. It connects the North Sea and the Kattegat, which leads to the Baltic Sea.)[12]

PASSAGE TO ENEMY WATERS/MINELAYING

Seal entered the Kattegat in early morning darkness on 4 May, at 0130 on the surface. The strait was heavily patrolled by enemy aircraft and ships, and thus a dangerous place for a large minelaying submarine. In northern latitudes because daylight came early at this time of year, Lonsdale feared the silhouette of his submarine would not remain undetected for long. To minimise it, he ballasted down so that only the conning tower was awash. This effort proved fruitless. Lonsdale had to crash-dive at 0230, only minutes before full dawn broke, and just seconds before an approaching Heinkel He 115 reconnaissance seaplane detected his vessel. The plane dropped some bombs nearby, but caused only minor damage. (The aircraft was assigned to Küstenfliegergruppe 76 based at Aalborg, Denmark.)[13]

Later, Lonsdale, proceeding at periscope depth, sighted a group of anti-submarine trawlers at about 0800, searching the waters ahead of *Seal*, and blocking her path to the primary mining position. With only limited options available, he decided to drop the mines at an alternate position, off the island of Vinga, west of Gothenburg, just ahead and to one side of the area occupied by the trawlers. Minelaying which was begun at 0900 took only forty-five minutes to complete.[14]

PURSUIT BY GERMAN ANTI-SUBMARINE FORCES

Lonsdale then reversed course and undertook the hazardous task of extricating *Seal* from the Kattegat and bringing her safely home to Britain. Several hours of cat-and-mouse manoeuvres ensued, as trawlers searched for the submarine. Proceeding slowly on batteries at periscope depth, Lonsdale stopped whenever they did, reasoning that they were doing so to listen for underwater sounds on their hydrophones. Around 1500, Lonsdale sighted through the periscope to the northwest another group of hunters, German motor torpedo boats, each fitted with depth-charge throwers in addition to two torpedo tubes. His tactic quickly changed. The plan was now to evade both enemy forces until after dark, bring *Seal* to the surface, and run for safety out of the Kattegat using diesel power.[15]

However, with a weakening battery charge, Lonsdale was forced to adjust trim and settle *Seal* on a layer of dense salt water beneath the surface fresh water, at a depth of about 60 feet. As the submarine (rigged for maximum quiet) slowly drifted with the current, the terrifying sound of a wire scraping the hull was heard. At this point the submarine's hydroplanes became inoperable, but after their use was restored and nothing further was heard, *Seal* then resumed trim and it appeared that the wire had been shed.[16]

At 1855 after Lonsdale had ordered the evening meal be served, an explosion aft shattered their peaceful interlude. It was first believed that one of the vessels above had dropped a depth charge. However, when no more explosions followed, Lonsdale reasoned that the scrapping sound had been *Seal*'s entanglement in a mooring wire for a German mine, which with her movement had been pulled into the hull and had exploded.[17]

A quick survey revealed that the aftermost mining compartment was partially flooded, with perhaps 130 tons of sea water inside the hull. A first attempt to surface the submarine, then resting on the bottom, was made after dark at 2230. *Seal* rose forward, but her stern remained buried in the mud of the Kattegat, held there by suction and the weight of seawater inside her hull. Over the next few hours, other attempts were made unsuccessfully, as diminishing oxygen made crew members sluggish and weak.[18]

Amidst much desperation in one last attempt, *Seal* finally surfaced at 0130 on 5 May. Lonsdale opened the conning tower hatch and as fresh air rushed in, many crew members suffering from carbon dioxide poisoning were afflicted with severe headaches and began vomiting. Lonsdale sent an encrypted situation report to the Admiralty which was acknowledged. Fearing possible capture, he then ordered secret code books jettisoned over the side and directed that asdic (sonar) gear be smashed and charts with minefields plotted on them discarded in weighted bags. At about 0200, once these actions were completed, Lonsdale decided to run for the Swedish coast.[19]

However, only the starboard engine could be started and the rudder was found to be inoperative as a result of the mine explosion. The port engine was also finally started and Lonsdale tried to bring the bow of the submarine around to set a course for Sweden using the engines only. He ordered an ahead bell on one engine and a backing bell on the other, in an effort to twist the submarine, but was unable to turn her sufficiently by this method. Lonsdale then decided to proceed by backing astern. Hope of escaping vanished soon after when the starboard engine lost lubricating oil pressure and seized up completely.

Seal was then left to wallow on the surface, unable to dive and capable only of going in circles propelled by her port engine.[20]

ATTACK BY GERMAN AIRCRAFT

At 0230, the drone of an approaching German Arado Ar-196 from the 706 Coastal Defence Wing (*Küstenverteidigungsgeschwader 706*) shattered the quiet. Immediately after surfacing, Lonsdale had ordered the two Lewis guns brought up and mounted on the conning tower. He chose not to fire them as the Arado approached, hoping the German airmen would think that *Seal* was a disabled Swedish submarine returning to port. The Arado was piloted by Unteroffizier Heintz Bottcher with Leutnant Gunther Mehrens as observer in the back seat. Mehrens recognised British markings on the conning tower and ordered Bottcher to attack with machine guns, cannons and two bombs.[21]

Photo 7-3

Arado 196 seaplane on catapult of the German battleship *Gneisenau*.
Naval History and Heritage Command photograph #NH 83610

As a second Arado arrived (piloted by Unteroffizier Sackritz with Leutnant Karl Schmidt in the rear seat) joined the attack, *Seal* responded with Lonsdale himself manning one Lewis gun. The aircraft attacks took a toll among those on *Seal*'s bridge. The first lieutenant, Lt. Terence Butler, and a rating were wounded by shrapnel inside the conning tower and carried down into the control room. (In the Royal Navy, the First Lieutenant is the second in command of a smaller ship or submarine, known colloquially as 'No. 1', 'The Jimmy' or 'Jimmy the One').[22]

Lonsdale remained unscathed in the hail of enemy fire, but the other gun manned by Leading Seaman Mayes jammed, and cannon fire from the second Arado blasted holes in the sub's port side main ballast tank, allowing sea water to rush in, creating a pronounced list that could not be corrected. A larger Heinkel He-115 bomber (Leutnant Nikolaus Broili) arrived on the scene and began a new attack on the listing submarine, whose second Lewis gun now also jammed.[23]

SURRENDER OF BRITISH SUBMARINE

After considering the existing dire conditions, Lonsdale decided to surrender, believing that *Seal* would sink of her own volition before she could be salvaged by the enemy. Two depth charges fitted in the bilges were set to explode at a water depth of 50 feet, if the sub were scuttled. However, two considerations prevented him from intentionally taking this action and sinking the *Seal*. First, there were no vessels in view to rescue the crew, some of whom could not swim. Second, many crewmembers would be killed or injured if floating in the sea above the submarine when the depth charges exploded.[24]

Lonsdale asked that the white wardroom tablecloth be brought up to the bridge, and he waved it at the second Arado. The plane landed and its commander demanded that *Seal*'s skipper swim to the aircraft. Lonsdale turned over command to Lt. Trevor Agar Beet, the navigator, as Butler was incapacitated by his wounds, and swam to the Arado. The first Arado then landed alongside the surrendered submarine and took one petty officer aboard as an additional hostage.[25]

At 0630, with the He-115 bomber still circling ominously overhead, the German unterseebootsjäger *UJ-128* hove to near the foundering British submarine. The *UJ-128* was an anti-submarine vessel (ex-trawler *Franken*) commanded by Kptlt. Otto Lang. Her first lieutenant, Heinz Nolte, and three sailors boarded the submarine from a small boat. With *Seal* listing badly to port and settling by the stern, neither the Germans nor the British thought she would remain afloat. Her crew was transferred aboard the German ship as prisoners, and *UJ-128* began towing the sub toward Frederikshavn, a small port on Denmark's eastern coast at which the 12th UJ-Flotilla was based. (One crewman, Able Seaman Charles C. Smith, was missing and presumed killed.)[26]

Near the entrance to the harbour, the salvage tug *Seeteufel* (Sea Devil) took over the tow. In the early evening, she pushed *Seal* alongside the seawall in a secluded corner of the harbour and made up to the submarine to prevent her from sinking. Salvage operations began the following day. Workers patched *Seal*'s hull, repaired her high-pressure air system and ballast tanks, and pumped sea water inside her hull over the side, returning the submarine to near normal trim. The two scuttling depth charges were left in place as too dangerous to remove.[27]

With temporary repairs completed, *Seeteufel* took the sub under tow on the morning of 10 May, southbound through the Kattegat to German Naval headquarters at Kiel, heavily escorted by minesweepers, submarine chasers and aircraft. *Seal* arrived at Kiel, on the afternoon of 11 May, greeted by a host of dignitaries.[28]

She was to be restored and refitted as a U-boat, but efforts at Germaniawerft Shipyard, Kiel, to return *Seal* to service as an operational submarine ended in failure. Her diesel engines differed from those of German manufacture, and no spare parts were available. Additionally, German torpedoes and mines were also incompatible with their British counterparts. The former HMS *Seal* was commissioned as *U-B* on 30 November 1940, under the command of Fregkpt. Bruno Mahn. She would prove of limited value to the Kriegsmarine except for training and propaganda purposes.[29]

Lonsdale and his crew remained German prisoners until liberated by Allied forces in April 1945. A year later, the Royal Navy convened a court-martial in Portsmouth for Lonsdale and Beet, who had assumed command of the *Seal* upon Lonsdale's swim to the Arado seaplane. The five-officer court-martial board acquitted both Beet, who was tried first, and Lonsdale. The president of the court stated to Lonsdale, "I have much pleasure in handing you back your sword." In addition to the *Seal* crewmembers called as witnesses by the court, many others had traveled to Portsmouth at their own expense to attend the trial. These men then surged forward to congratulate their skipper and shake his hand.[30]

Shortly thereafter, Lonsdale resigned his commission and entered theological training for the Church of England, which he served for the remainder of his working life.[31]

LOSS OF MINELAYER HMS *PRINCESS VICTORIA*

> *MV Princess Victoria had an operational life of less than eleven months. She entered service as a cross-channel ferry in July 1939, was commissioned in November 1939 as HMS Princess Victoria, an auxiliary minelayer, and was sunk by a magnetic mine in May 1940.*
>
> —From the article "The Short Life and Sudden Death of HMS Princess Victoria" by Fraser G Machaffie[32]

HMS *Princess Victoria* (Capt. John B. E. Hall, RN) was despatched from Immingham on 10 May 1940 to lay a minefield off the Dutch coast (Ymuiden) to afford some protection behind ships escaping from Holland following a German invasion of the Netherlands. During Operation CBX1, she was under the escort of three units of the 20th Destroyer Flotilla—*Express*, *Esk*, and *Intrepid*—which had just completed a minelay in the Heligoland Bight.[33]

Princess Victoria's next excursion would be in home waters, placing mines in the East Coast Barrier on 18 May. Meanwhile, *Express*, *Esk*, and *Intrepid* sowed mines there on the 12th and 14th; and off the Dutch coast on 16 May—joined for the latter operation by flotilla mate *Ivanhoe*. On intervening days, the destroyers returned to Immingham to load mines for the next day's work. The code name designations of these operations are identified in the table, following their associated dates.

Minelaying Operations involving *Princess Victoria* and the 20th Flotilla

Date/Operation	Mined Area	Minelayers Participating
10 May 40 (ZMC)	Heligoland Bight	*Express*, *Esk*, *Intrepid*
10 May 40 (CBX1)	off Dutch coast	*Express*, *Esk*, *Intrepid*, *Princess Victoria*
12 May 40 (BS3)	East Coast Barrier	*Express*, *Esk*, *Intrepid*
14 May 40 (BS4)	East Coast Barrier	*Express*, *Esk*, *Intrepid*
16 May 40 (CBX3)	off Dutch coast	*Express*, *Esk*, *Intrepid*, *Ivanhoe*
18 May 40 (BS5)	East Coast Barrier	*Express*, *Esk*, *Intrepid*, *Ivanhoe*, *Impulsive*, *Princess Victoria*[34]

PEDIGREE OF THE AUXILIARY MINELAYER

The Admiralty had requisitioned the *Princess Victoria* (a Stranraer-Larne car ferry) for use as a minelayer on 13 September 1939. A product of William Denny & Bros Ltd, Dumbarton, Scotland, she was newly built, having been launched just a few months earlier on 21 April 1939. At a time when the steam turbine was the dominant power source for fast cross-channel ferries, *Princess Victoria* was unique in that she was propelled by two diesel engines. Her port and starboard engines (built by Sulzer and "Denny," respectively) produced a combined 6,160 shaft horsepower. During trials on 26 June, she achieved an official speed of 19.25 knots with the top run of 19.934 knots.[35]

Following modification for naval service, HMS *Princess Victoria* was commissioned on 2 November 1939. Although the 310-foot ex-ferry was but lightly armed with one 4-inch gun and two 12-pounder (2x1) anti-aircraft guns, she could carry an impressive 244 mines laid through two chutes at her stern. Ship's complement was 120 officers and men. *Princess Victoria* initially joined the First Minelaying Squadron and was later detached to operate from Immingham, for mining operations in the Heligoland Bight.[36]

HMS *PRINCESS VICTORIA*'S FINAL DAY

On the night of 18 May, *Princess Victoria* was returning to Immingham following a minelay in the East Coast barrier in company with flotilla destroyers, about forty miles east of Cromer, a town on England's north coast. HMS *Express* and *Ivanhoe* were serving as escorts after *Esk*, *Intrepid*, and *Impulsive* had been "dropped off" at Harwich. At 2315, off the entrance to the River Humber, a mine exploded beneath the *Princess Victoria* opening four compartments. The terrific blast beneath the starboard side of the ship, forward of the bridge, was believed to be caused by a magnetic mine probably dropped by a German aircraft.[37]

Crewmembers asleep or occupied by watch duties below deck had little opportunity to escape. Within 20 to 30 seconds, *Princess Victoria* listed 45 degrees. Inside two minutes she was nearly on her side, with list increased to 70 degrees. Touching bottom in the estuary, she straightened up and sank on a nearly even keel, leaving her mast, funnel and part of the bridge structure exposed above the water. As the destroyers HMS *Grafton*, *Gallant*, and *Foresight* rescued survivors (eighty-five), *Ivanhoe* and *Express* patrolled the estuary to warn shipping of the mine danger.[38]

Thirty-seven officers and men perished, including the commanding officer, Capt. John Buller Edward Hall. At the time of the explosion there were seventeen officers and ratings on duty on the bridge, in the chart house or in the wireless/telegraph office—of which only five survived. The force of the explosion likely flung many of the men into the sea where they drowned if not already dead from the blast. The bodies of some were swept south down the North Sea by the tides and currents, and later washed up on the beaches of the county of Norfolk. All these individuals were RN, RNVR, or RNR.[39]

Twenty of the casualties were Naval Auxiliary personnel (Merchant Navy) who worked below deck and were probably entombed when the ship sank. These men were part of a group of engineering officers and engine room personnel, cooks, and stewards who had been transferred to naval command and placed under King's Regulations when the Royal Navy requisitioned the MV *Princess Victoria* in 1939. Eight of the engine room crew were lost (the 5th Engineer Officer, six greasers and an engine room storekeeper) as well as six stewards and three cooks who were off duty and in their bunks on the lower deck. The remaining three were a boatswain, a carpenter, and a donkeyman (an engineer whose job was to operate steam engines, winches etc., not in the engine room).[40]

An initial casualty report designated the casualties listed in the table, as either Killed, or Missing and Presumed Killed.

HMS *Princess Victoria* Casualties (as reported on 19 May 1940)

Killed (4)			
James W. Burrows	Signalman	Thomas Rickett	Ordinary Seaman
Thomas Davidson	Telegraphist	Denis A. Self	Able Seaman
Missing, Presumed Killed (33)			
John F. Cockshutt	Ordinary Signalman	James McCalmont	Boatswain
Ernest G. W. Cookson	Ordinary Seaman	Alexander McIsaac	Greaser
James Coupethwaite	Carpenter	Ian G. McLean	2nd Cook
Sydney N. Darby	Assistant Steward	Alexander McNab	Assistant Steward
Sydney W. Dudley	Assistant Cook	Jack Mitchell	Able Seaman
Henry F. Evans	Leading Signalman	Felix T. Murray	Greaser
Patrick Flaherty	Greaser	Maurice H. O'Kelly	Leading Supply Assistant
William Graham	5th Engineer	Thomas R. Parker	Assistant Steward
John B. E. Hall	Captain	Harry Proudfoot	T/Sub. Lt. (E)
Henry A. Humphreys	Able Seaman	William J. Robinson	Engine Room Stores
Louis A. Lambert	Lt. Comdr.	Walter C. Scutt	Ordinary Seaman
Alexander Little	Donkeyman	Frederick H. Snowling	Chief Petty Officer
James Little	Greaser	Frederick A. Theobald	Leading Telegraphist
Alfred C. Logan	Greaser	William Underwood	Chief Steward
Louis C. Lowther	Greaser	Charles W. Wear	Ordinary Seaman
Albert H. MacDonald	Assistant Steward	Philip T. Willats	Chief Cook[41]
John Mackenzie	Assistant Steward		

HMS *GRAMPUS* LOST IN THE MEDITERRANEAN

Details about the loss of the submarine HMS *Grampus* on 16 June 1940, may be found in the following chapter.

HMS *NARWAHL* BELIEVED SUNK BY A BOMBER

HMS *Narwhal* departed the submarine base at Blyth on 22 July 1940 with orders to lay a minefield northwest of Kristiansund, off German-occupied Norway. (Blyth, a coastal port town in northeast England, is situated on the River Blyth some thirteen miles northeast of Newcastle.)

Narwhal failed to return to base following the patrol and was reported overdue on 1 August 1940. At the time it was assumed that she had been sunk in the vicinity of the Norwegian coast during her mission, taking her commanding officer Lt. Comdr. Ronald Burch, RN, and the other fifty-seven men aboard to a watery grave.[42]

After April 1940 the German *B-Dienst* ("Observation Service," a part of the Naval Intelligence involved with cryptanalysis and deciphering of enemy and neutral states' message traffic) had deciphered several signals addressed to HMS *Narwhal*. As a result, her route was known to the enemy prior to her departure, and the Luftwaffe was sent to intercept her. Lt.z.S. Bernhard Müller piloting a Dornier Do17 bomber, reported attacking a submarine on the afternoon of 23 July, about 125 nautical miles east of Aberdeen, Scotland. This attack was successful and her loss was later confirmed. Her loss was apparently the only instance of a British submarine being sunk as a direct result of enemy signal intelligence during the war.[43]

In 2017, Polish deep-sea explorers found the wreckage of what is believed to be the *Narwahl* sitting upright on the seabed about 300 feet down around 150 miles east of Edinburgh. The team was searching for the Polish submarine ORP *Orzel* ("Eagle"), which had been missing since escaping the German invasion of Poland in 1939. Bomb damage to the ruined hull—between the conning tower and stern—matched the description provided in the German bomber crew's combat report.[44]

TEXEL DISASTER (OPERATION CBX5)

> *The dead and wounded were being hoisted up from the foremost boiler room where the mine had most certainly struck us, also there were a few men lying on the deck, wounded and covered in oil fuel who had been rescued from the* EXPRESS. *Boats and carley rafts were being lowered, all around were cries of help from men in the sea, and from the other ships where they were imprisoned from the explosions, three ships mined almost on top of one another seemed incredible.*
>
> —Herbert Vaughan, a Communications Clerk Petty Officer aboard HMS *Esk*, describing first HMS *Express*, then HMS *Ivanhoe* and then his ship detonating mines, and the horror that followed.[45]

In mid-afternoon on 31 August 1940 a group of destroyers sailed from Immingham on a minelaying mission to the Heligoland Bight, off the Dutch coast north of Texel, a Dutch island. Five minelaying destroyers of the 20th Flotilla—*Express, Esk, Icarus, Intrepid,* and *Ivanhoe*—each loaded with twenty-six mines left the Humber at 1500. Capt. Jack Grant Bickford, RN, the flotilla's senior officer, commanded *Express*. It was a beautiful summer's day and being Queen Wilhelmina of Holland's birthday, each ship flew the Dutch ensign at its masthead. The sky was clear and after proceeding down river there was hardly a ripple on the sea when the ships passed the Humber Light vessel. They then steamed in a line ahead almost due east for the Dutch coast, escorted by members of the 5th Destroyer Flotilla. The destroyers HMS *Kelvin*, *Jupiter*, and *Vortigern* were along to provide protection in the event of an enemy surface attack.[46]

Photo 7-4

British destroyer HMS *Express* (H61) under way, location and date unknown. Australian War Memorial photograph 302386

20th Destroyer (Minelayer) Flotilla		5th Destroyer Flotilla	
HMS *Esk*	Lt. Comdr. Richard John Hollis Couch, RN	HMS *Kelvin*	Comdr. John Hamilton Allison, RN
HMS *Express*	Comdr. Jack Grant Bickford, RN	HMS *Jupiter*	Comdr. Derek Bathurst Wyburd, RN
HMS *Icarus*	Comdr. Colin Douglas Maud, RN	HMS *Vortigern*	Lt. Comdr. Ronald Stanley Howlett, RN
HMS *Intrepid*	Comdr. Roderick Cosmo Gordon, RN		
HMS *Ivanhoe*	Comdr. Philip Henry Hadow, RN		

During passage, aerial reconnaissance detected a German Naval force and as a consequence the task force was directed to intercept. This order, from Commander-in-Chief Nore, Adm. Sir Reginald Plunkett, RN, was based on an erroneous belief that the German ships were part of an invasion force. Herbert Vaughan, a Telman Petty Officer (communications clerk) aboard the *Esk* described receipt of this warning:

> At 8 p.m. action stations were sounded as we were approaching the danger area, myself going on watch in the W/T [Wireless Telegraph] office where everything was rather quiet. It is now beginning to get near dusk, and it was one of those moonless nights. At about 10.30 p.m. we received messages from the Admiralty saying enemy light forces had been sighted to the N.E. of us and detailing us to endeavour to intercept them on completion of our minelaying, other Naval units of cruisers and destroyers were also detailed to locate and engage the enemy.[47]

A short time after this order was received *Express* hit a mine at position 53°25'N, 03°48'E, which destroyed most of the ship forward. *Ivanhoe* approached *Empress*, lowered a boat to rescue survivors in the water, and was manoeuvring to go alongside her when she detonated a mine and propulsion was lost. *Esk* also closed to assist and almost immediately struck yet another mine. Approximately fifteen minutes later, a second explosion amidships caused her to break in two parts, both of which sank quickly at 53°23'N, 03°48'E.[48]

The mine strikes to the three destroyers, which it was later learned had stumbled into an unknown minefield, occurred in quick succession as Vaughan explained:

> At 11.7 p.m. I heard a violent explosion followed by continuous blasts of a syren, a messenger informed me it was the *EXPRESS* that had struck a mine, all remaining ships stopped engines, I waited patiently in the office for details or a signal giving information of the explosion, however nothing came. *IVANHOE* then proceeded slowly to pick up survivors from *EXPRESS*, I now learnt from outside that the whole fo'cle of *EXPRESS* from her bridge had been blown completely away, another terrific explosion shook the ship, this time it was the *IVANHOE* who had struck a mine whilst survivors were jumping aboard from the *EXPRESS*, whilst all these happenings were going on outside I remained on watch awaiting orders, the minutes seemed like hours, who would be the next one, not long were we kept in suspense for a mighty and terrific explosion shook the ship from head to stern, the deck underneath me seemed to rise up, [and] all lights went out so total darkness...[49]

On *Ivanhoe*, steam was restored two hours later, allowing an astern speed of seven knots. However, steam pressure could not be maintained and inspection revealed her keel was broken. A decision to abandon was made after it was learned that motor torpedo boat *MTB 15*, awaiting to embark survivors, was severely short on fuel and had to leave the area.[50]

Seacocks were opened before abandoning, but *Ivanhoe* did not sink. As a result of the mine explosion, one rating was killed, and seven others missing and presumed killed. Her drifting hulk was later found by the destroyer escort HMS *Garth*. *Ivanhoe* was sunk by a torpedo fired by *Kelvin* at 53°26'N, 03°45'E after an attempt by *Jupiter* to tow her had failed.[51]

Express, with bow blown off, was able to make sternway under her own power until taken in tow by *Kelvin*. After the tow fouled, it was reestablished by *Jupiter* until tugs (one the *St. Cyrus*, T/Lt. Peter Allan RNR) arrived to take *Express* under tow to Kingston-upon-Hull, a port city in Yorkshire on the northern bank of the Humber Estuary. Captain Bickford died later in hospital and was buried at sea. A summary of total casualties among the three destroyers follows.[52]

Minelayer	Killed	Missing Presumed Killed	Died of Wounds
HMS *Esk*	33	102	
HMS *Express*	10	46	2
HMS *Ivanhoe*	1	7	
Total[53]	44	155	2

Express was later towed to the Thames, where she underwent repair at HM Dockyard, Chatham from October 1940 to August 1941. She was recommissioned and carried out post-refit trials in September. After completing her refit on 3 October, she proceeded to Scapa Flow to rejoin the Home Fleet and was subsequently allocated for service with Force Z in Singapore.[54]

LOSS OF HMS *PORT NAPIER* TO A FIRE ON BOARD

In June 1940, five merchant ships which had been requisitioned by the Admiralty and under conversion for use as auxiliary minelayers, were formed into the 1st Minelaying Squadron at Kyle of Lochalsh, Scotland. They were not available for minelaying until October. Upon assuming command of the squadron in August, Rear Adm. Frederick Wake-Walker hoisted his flag in the *Southern Prince*. Rear Adm. Robert Burnett succeeded him in November 1940.[55]

1st Minelaying Squadron

Ship	Length (ft.)/ Displ. (tons)	Commissioning Commanding Officer	#Mines
Agamemnon	460/7,593	Capt. (retired) Neville Brevoort Carey Brock, RN	530
Menestheus	460/7,494	Capt. William Howard Dennis Friedberger, RN	410
Port Napier	503/9,847	Capt. (retired) John Norman Tait, RN	550
Port Quebec	451/5,936	Capt. (retired) Edward Clifford Watson, RN	550
Southern Prince	496/11,447	A/Capt. Edward Murray Conrad Barraclough, RN[57]	560

The squadron was later joined by the cruiser-minelayer *Adventure*, having anti-aircraft armament which eliminated the requirement for an escorting cruiser on minelaying missions.[56]

On 27 November 1940, the strength of the squadron (which could collectively embark 2,600 mines) was reduced by the loss of the *Port Napier*, destroyed by an engine room fire at Kyle of Lochalsh. Fortunately, there were no casualties. Although she had a full load of mines on board at the time, they had not yet been fitted with primers and detonators.[58]

20TH FLOTILLA MINELAYING OPERATIONS END

In April 1941, the remaining destroyers of the 20th Flotilla were withdrawn from minelaying service and returned to fleet duties. Surface minelaying was then carried out by fast minelayers and Coastal Forces craft with escorts from local destroyer flotillas.[59]

8

Mediterranean 1941

> *German intervention in this theatre [Mediterranean] produced a year of crisis, with Malta frequently out of action, the through route virtually closed to convoys and the fleet later reduced to negligible proportions by air and underwater attack. On land, Rommel's Afrika Korps advanced to the Egyptian border at the time of the ill-fated Greek campaign, only to be driven back into Tripolitania by a British offensive in November [1941].*
>
> —*British Mining Operations 1939-1945 Vol 1.*[1]

In the Mediterranean in 1941, the only British minelayers available at the start of the year were HMS *Rorqual* and the Swordfish of No. 830 Squadron, based at Malta. *Rorqual* (a *Porpoise*-class submarine built by Vickers Armstrong, Barrow-in-Furness) had been commissioned on 10 February 1937. She would become the most successful British minelaying submarine of World War II, sinking 57,704 tons of enemy shipping (35,951 with mines) during service in the Far East and Med— and would also be the only one of the six submarines of her class to survive the war. Before progressing to 1941, an overview of her preceding service is in order. *Rorqual* laid her first mines in the Mediterranean (fifty mines off Brindisi, Italy) on 14 June 1940, four days after Italy declared war on France and Great Britain. Two days later, HMS *Grampus*, a sister submarine, was lost off Sicily, contributing to the dearth of minelayers in the Mediterranean in early 1941.[2]

Spanning 289 feet, *Rorqual*'s armament included a 4-inch deck gun, two machine guns, six 21-inch bow torpedo tubes (with twelve reloads), and fifty mines. The mines were stored on a rail running nearly the entire length of the top of her pressure hull, enclosed by a non-watertight casing. When laying a minefield, large stern doors at the rear were opened and an endless chain (similar to a conveyor belt) inside the extra deep casing discharged mines into the sea. To accommodate the casing along the centreline of the upper hull, her conning tower and periscope were slightly offset to the starboard side.³

Rorqual's ship's complement was six officers and fifty-three ratings. Following construction, she was commissioned for service in the 4th Submarine Flotilla on the China Station—under the command of Lt. Comdr. Ronald Hugh Dewhurst, RN—where she was deployed at the outbreak of war in September 1939. The other flotilla members were HMS *Parthian*, *Phoenix*, *Proteus*, *Pandora*, and *Grampus*, supported by the submarine depot ship HMS *Medway*.⁴

Photo 8-1

British Submarine HMS *Porpoise* berthed alongside the heavy cruiser HMAS *Australia*, with HMS *Salmon* (an S-class submarine) outboard of her.
Australian War Memorial photograph P00604.012

Rorqual was nominated in February 1940 for transfer to the Mediterranean with the flotilla. The submarines took passage in March and arrived at Alexandria, Egypt, in April. *Rorqual*, along with *Parthian*, *Phoenix*, *Proteus*, and *Pandora* comprised the 1st Submarine Flotilla at Alexandria. *Grampus* was based at Malta—an archipelago in the southern Mediterranean, 50 miles south of Italy, 176 miles east of Tunisia, and 207 miles north of Libya. A British colony since 1815, Malta had long served as a critical way station for ships and the headquarters for the British Mediterranean Fleet. Malta would play an important role in World War II, receiving the George Cross for the bravery of its people in the face of an Axis siege. (Britain gave Malta its independence in 1964. It is today a republic, with the George Cross proudly displayed in the upper left corner of its red and white coloured national flag.)[5]

Map 8-1

Malta archipelago in the southern Mediterranean

ITALY ENTERS THE WAR IN JUNE 1940

First they were too cowardly to take part. Now they are in a hurry so that they can share in the spoils.

—Adolf Hitler remarking on Italy declaring war on France and Great Britain, on 10 June 1940. Franklin D. Roosevelt was equally critical of the action by Benito Mussolini, but for different reasons, saying during a speech in Virginia, "The hand that held the dagger has struck it into the back of its neighbor."[6]

In May, *Rorqual* was ordered to Malta to carry out exercises with Fleet units in anticipation of the outbreak of war with Italy. This possibility occurred on 10 June 1940 when, after withholding formal allegiance to either side in the battle between Germany and the Allies, Benito Mussolini, dictator of Italy, declared war on France and Great Britain. Although Hitler thought that perhaps the German occupation of Paris had spurred Italy to enter the war, Mussolini claimed that he wanted in before complete French capitulation only because fascism "did not believe in hitting a man when he is down."[7]

LOSS OF HMS *GRAMPUS* OFF SYRACUSE, SICILY

HMS *Grampus* badge
Motto: Grandis inter alios
'Great amongst others'

Emblem of the Regia Marina, the navy of the Kingdom of Italy from 1861 to 1946

Shortly before midnight on the same day that Italy declared war, 10 June, HMS *Grampus* left Malta at 2340 on a war patrol. She was, like *Rorqual*, a *Porpoise*-class submarine. Her orders were to patrol off the east coast of Sicily, and to lay a minefield off Augusta, which faced the Ionian Sea. After having done so, she was to depart her patrol area the evening of the 16th and proceed to Alexandria.[8]

Grampus (Lt. Comdr. Charles Alexander Rowe, RN) sent a signal at 1930 on 13 June, confirming that she had completed her minelaying mission: fifty mines in a shipping channel northeast of Augusta near Castello Maniace Light. Nothing more was heard from her. She was believed sunk, with the loss of all hands, by Italian torpedo boats off Syracuse, about ten nautical miles south of Augusta, three days later.[9]

In early evening on 16 June, the torpedo boat *Circe* sighted what appeared to be the conning tower of a submarine, 3000-4000 metres distant, and opened fire with her guns. The leader of the 13th Torpedo Boat Squadron, she had earlier sailed with sister ships *Clio*, *Calliope*, and *Polluce* from Syracuse to conduct an anti-submarine sweep. (The 273-foot *Spica*-class torpedo boats were of modest displacement, 1,040 tons, but possessed armament similar to that of destroyers. The 34-knot boats were fitted with three 100mm/47 calibre dual-purpose guns in single mountings, nine to eleven Breda 20mm cannons, and four 450mm torpedo tubes.) After *Clio* joined *Circe* in the attack, torpedo tracks were observed by sailors aboard the two vessels, and *Polluce* reported being missed by two torpedoes ahead.[10]

Following this desperate act by the submarine, the torpedo boats carried out depth-charge attacks; dropping munitions (*Circe*: 19, *Clio*: 13, *Calliope*: 10, *Polluce*: 19) in the area where she had disappeared. With high explosives detonating all around her, she would have had little chance. *Polluce* reported sighting a large oil slick and the wreckage of the submarine. Capitano di Fregata Aldo Rossi—the commanding officer of *Circe*, squadron leader, and commanding officer of the 1st Torpedo Boat Flotilla—was doubtful of the torpedo-track sightings, owing to the sea conditions, and was not convinced the submarine had been sunk. Accordingly, he ordered the sweep continued. Nothing was seen, and Italian authorities later assessed the sinking as "probable."[11]

As discussed in the previous chapter, the first *Porpoise*-class submarine casualty, HMS *Seal*, was lost on 5 May 1940 in Norwegian waters; captured by German forces after being damaged by a mine the day before. *Grampus* was the second *Porpoise* lost in 1940, and another soon followed. HMS *Narwhal* (Lt. Comdr. Ronald James Burch, RN) sailed from Blyth on England's northeast coast on 2 July to lay mines off Kristiansund, Norway. A German Dornier Do17 bomber attacked a submarine the following afternoon about 125 nautical miles east of Aberdeen, Scotland. *Narwhal* failed to return to base and was reported overdue on 1 August 1940.[12]

RORQUAL LAYS HER FIRST MINEFIELD IN THE MED

Rorqual sailed from Malta at 2307 on 10 June 1940, with orders to take up a defensive position off Malta. Less than three hours later, she received a signal ordering her to patrol the Straits of Otranto and to lay a minefield off Brindisi. Lt. Comdr. Ronald Hugh Dewhurst carried out these orders in late morning on the 14th: laying fifty mines at 40°39'N, 18°10'E.[13]

As indicated in the below table, this field may have accounted for the sinking of the Italian merchant vessel *Rina Croce* on 25 September 1940 and that of the sailing vessel *Peppino C.* on 17 August 1941. One or both may also have been victims of an Italian defensive minefield located in the same area. Between 14 June 1940 and 11 May 1941, the *Rorqual* laid a total of 450 mines in fields off Italy, Libya, and Greece, which sank or damaged several enemy vessels.

HMS *Rorqual* (Lt. Comdr. Ronald Hugh Dewhurst, RN)

Date	Field	Damage to Shipping
14 Jun 1940	50 mines off Brindisi, Italy 40°39'N, 18°10'E	May have been responsible for the sinking of the Italian merchant ship *Rina Croce* on 25 Sep 1940 and that of the sailing vessel *Peppino C.* on 17 Aug 1941. There was also an Italian defensive minefield in the same area.
21 Jul 1940	50 mines off Tulmaythah, Libya 32°45'N, 20°57'E.	On 24 Jul 1940 the Italian merchant ship *Celio* was sunk, and on 14 Aug 1940, the merchant ship *Leopardi* after hitting mines. This field was also likely responsible for the loss of the corvette HMS *Erica* on 9 February 1943.
17 Aug 1940	50 mines off Tulmaythah, Libya 32°40'N, 20°32'E	
4 Oct 1940	50 mines northwest of Benghazi, Libya 32°45'N, 20°00'E	The Italian schooner *Intrepido* (V71) likely sank in this minefield on 26 Oct 1940, her crew was rescued by another sailing vessel
5 Nov 1940	50 mines east of Misrata, Libya 32°25'N, 15°21'E	On 5 Dec 1940 the Italian torpedo boat *Calipso* sank about 6 miles northwest of Cape Misurata, after hitting two of these mines. On 23 Dec 1940 the Italian torpedo boat *Fratelli Cairoli* sank off Misrata, after hitting one of these mines. The same field may have been responsible for the losses of the Italian *Cadamosto* and the German *Spezia* on 22 Dec 1941.
9 Nov 1940	50 mines east of Misrata, Libya 32°40'N, 15°00'E	
28 Jan 1941	29 mines off Sansego Island (now Susak Island, Croatia) and later that day 21 miles near Ancona, Italy	On 27 Feb 1941, the Italian merchant ship *Ischia* was damaged when she hit a mine off Ancona. This field may have caused the loss of the Italian merchant ship *Pascoli* on 7 May 1941
25 Mar 1941	10 mines five miles northeast of Capo Gallo, Sicily, Italy	28 Mar 1941, the Italian torpedo boat *Generale Antonio Chinotto* sank in this minefield
26 Mar 1941	40 mines in two lines near Asinelli Rock Light	That same day, an Italian convoy ran into this field losing the Italian water tankers *Ticino* and *Verde*
11 May 1941	50 mines in the Gulf of Saloniki off Panomi Point, Greece	Likely responsible for the loss of the Romanian *Carmen Sylva* on 26 May and, the following day, that of the German *Helene* (ex-Greek *Eleni Kanavarioti*)[14]

ADDITIONAL RN MINELAYERS ARRIVE IN THE MED

The station mining capacity at Malta increased in April 1941 with the arrival of HMS *Abdiel*, followed by the submarine HMS *Cachalot* in May and HMS *Latona* in June. *Abdiel* and *Latona* were both bright and shiny, newly-built fast minelayers (commissioned on 15 April and 4 May 1941, respectively), as the first two of six cruiser minelayers that would comprise the *Abdiel*-class. HMS *Manxman*, whose exploits as a "French cruiser" are detailed in Chapter 1, was one of these type ships.[15]

Photo 8-2

Fast minelayer HMS *Latona* at Alexandria, Egypt, 27 September 1941.
Australian War Memorial photograph 020452

Abdiel, commanded by Captain the Hon Edward Pleydell-Bouverie, RN (second son of the Earl of Radnor) arrived at Alexandria on 30 April, joining the Mediterranean Fleet. On 21 May, she laid a field of 150 mines off Cape Dukato, the southern tip of Levkas Island in the Ionian Sea. The battleships HMS *Warspite* and *Valiant* with cruiser *Ajax* provided protection during the operation—code named MAT ONE.[16]

Intended to target shipping using the Corinth Canal—a four-mile long waterway separating mainland Greece from the Peloponnese—the deadliness of the field was validated later that day, by the loss of escorts and transports in convoy. Falling victim to its mines were the Italian destroyer *Carlo Mirabello*, the Italian gunboat *Pellegrino Matteucci*, and the German transports *Kybfels* and *Marburg*.[17]

Photo 8-3

Alexandria, Egypt, December 1941. The restaurant of the Fleet Club on Boxing Night. Australian War Memorial photograph 022877

THE BATTLE FOR CRETE

Two days after the minelay, *Abdiel* left Alexandria on 23 May with the sloops HMS *Auckland* and *Flamingo* in support of the military defence of Crete. Following German occupation of the Greek mainland in April 1941, the German High Command had decided to seize the island of Crete, to provide itself a good base in the eastern Mediterranean and to prevent British use for mounting operations in the Balkans. On Crete was a Commonwealth force comprised of New Zealanders, British, and Australians, hastily deployed to help the Greeks defend their homeland. Outgunned and outnumbered, they had been forced to undertake a fighting withdrawal south through Greece.[18]

During the last week of April more than 50,000 Allied troops were evacuated from the mainland, most sent to Crete. By month's end, there were more than 42,000 British, Commonwealth, and Greek soldiers on the island. Over 12 days in May, these troops desperately tried to fight off a massive airborne assault. Despite suffering appalling casualties, the parachutists and glider-borne troops who led German Operation MERKUR (Mercury) eventually gained the upper hand; forcing the evacuation to Egypt of the bulk of the Allied force.[19]

Map 8-2

Eastern Mediterranean

On 24 May, *Abdiel* delivered troops and stores to Suda Bay on the northwest coast of Crete, and the following day returned to Alexandria with wounded and prisoners. The 26th brought more of the same; passage back to Suda Bay with destroyers HMAS *Nizam* and HNS *Hero* carrying troops and stores. The fast minelayer disembarked 800 Commandos and stores and embarked 930 troops on 27 May, then made the return trek to Alexandria. *Abdiel* sailed once again on the 31st, this time with the cruiser HMS *Phoebe*, and destroyers *Hotspur*, *Jackal*, and *Kimberley*, carrying medical supplies and for the final evacuation of Allied troops. She took passage with 3,900 troops the following day for return to Alexandria, arriving the same day.[20]

OPERATION GUILLOTINE

HMS *Latona* (Capt. Stuart Latham Bateson, RN) joined her sister ship *Abdiel* at Alexandria on 21 June, following passage via the Cape of Good Hope and Red Sea. Like *Abdiel* (and later, the other four ships in the class), *Latona* was not adopted by a community as part of a British Nation Savings WARSHIP WEEK campaign in 1941-1942, possibly because of the nature of her primary mission. Ironically, because the spacious, fast minelayers proved ideally suited for the transport of stores and personnel, *Latona* was not used at all for mining after her arrival in the Mediterranean. Priority was instead given to deliveries to Cyprus and Tobruk.[21]

One day after her arrival, *Latona* departed Alexandria on 22 June, in support of military operations in the eastern Mediterranean. The following month, she and *Abdiel* embarked RAF personnel for passage to Cyprus to reinforce the garrison. The purpose of Operation GUILLOTINE (18 July to 29 August) was to transport the British 50th (Northumbrian) Infantry Division and units of the Royal Air Force to Cyprus from Egypt. Cruisers, the fast minelayers, and destroyers carried personnel, while four transports took the equipment and supplies. The defence of Cyprus was thus greatly strengthened against the possibility of another Crete. Cyprus became important as a supply and training base and as a naval station, but its use as an air base proved to be the greatest contribution to the Allied cause.[22]

SUPPORT OF TOBRUK GARRISON

Operating from Alexandria, the *Abdiel* and *Latona* (and other ships) provided support in the summer and autumn of 1941 of the Allied military garrison at Tobruk, Libya. (The following chapter, titled The Rats of Tobruk, describes the importance of this garrison to the defence of Egypt, and efforts of the Australian 9th Infantry Division and later, British troops, to hold it.) The contributions of the fast minelayers involved troop transport and ferrying of supplies.[23]

LOSS OF HMS *LATONA*

Motto: Vestigia nostra cavate:
'Beware our tracks'

Battle Honour LIBYA 1941

In early evening on 25 October 1941, *Latona* sustained a bomb hit in her after engine room—dropped by a plane unseen in the darkness—and went dead in the water. She was able to remain afloat for over an hour and appeared salvageable, until a further explosion, caused by the detonation of ammunition on deck, ignited by a fire started by the hit. Abandoned by crew and passengers, *Latona* was scuttled at 2230 by a torpedo from the destroyer HMS *Encounter*, at 32°15'N, 24°14'E.[24]

She and her escorts, HM destroyers *Hero*, *Hotspur*, and *Encounter*, had been twenty miles north of Ras Azzaz, Libya—and already under attack by a group of ten German Stuka and two Italian S.79 3-engined bombers—when she was bombed. *Latona* had earlier delivered 150 tons of cargo and 25 men to Tobruk as part of Operation CULTIVATE (replacing the 9th Australian Division garrisoning Tobruk with British soldiers); then embarked 450 troops and set out for Alexandria.[25]

Encounter and *Hero* rescued *Latona* survivors. Whilst alongside the minelayer to evacuate her crew, *Hero* was subject to further air attack; resulting in structural damage from three near misses. Forty members of *Latona*'s crew were lost. Reported "Missing, Presumed Killed," most were engineers who perished in the space hit by the bomb.

Name	Rank/Rating	Name	Rank/Rating
Colin R. Beale	Engine Room Artificer 4c	Albert E. J. Lipscombe	Stoker 1c
John Bretherton	Electrical Artificer 4c	Harold L. Mantle	Stoker 2c
Henry J. Brown	Chief Engine Room Artificer	Wilfred H. Murphy	Able Seaman
George F. W. Bruce	Commissioned Gunner (T)	Charles A. Peedle	Stoker 1c
Arthur F. L. Crisp	Petty Officer	Eric W. Pillinger	Ty/Lieutenant (E)
Joseph G. W. Hopgood	Act/Stoker Petty Officer	John Rutherford	Engine Room Artificer 4c
Sidney Hughes	Stoker 1c	Christopher A. Silvester	Chief Petty Officer Stoker
Ronald Jones	Stoker 1c	Percy Ward	Engine Room Artificer 4c
Ronald Kennedy	Midshipman	John H. Whincup	Stoker 2c
Frederick T. Kirkwood	Stoker 1c	Theodore G. B. Winch	Commander (E)[26]

FURTHER REINFORCEMENT OF CYPRUS

On 28 October, *Abdiel* left Alexandria for passage with ten destroyers to Beirut, to transport the 5th Indian Infantry Division to Cyprus to further strengthen it. Raised in 1939 in Secunderabad, British India, the division had, during 1940 and 1941, fought in the East African Campaign in Eritrea and Ethiopia. After moving to Egypt, Cyprus, and Iraq, it would be heavily engaged in the Western Desert Campaign and the First Battle of El Alamein in 1942. From late 1943 to the Japanese surrender in August 1945, the division fought continuously from India through the length of Burma.[27]

Over 14,000 Indian troops were embarked at Beirut. *Abdiel* sailed from there on 1 November and at completion of troop transport duties, returned to Alexandria on the 8th.[28]

ABDIEL TRANSFERRED TO THE EASTERN FLEET

Abdiel was ordered to the British Eastern Fleet in December 1941, and departed Alexandria bound for Trincomalee, Ceylon (Sri Lanka); arriving there on 10 January 1942. The fleet was newly constituted, having been formed on 8 December 1941 from the existing East Indies Squadron and China Squadron.[29]

SUPPLY OF MALTA BY SUBMARINES

Relatively little was accomplished in the Mediterranean in 1941 in the way of mining, owing to successive crises that demanded the use of fast minelayers and submarines for extraneous tasks. For their part, *Abdiel* and *Latona* performed numerous relief and trooping duties, including protracted operations in support of Tobruk—during which *Latona* was lost before laying a single mine operationally.[30]

Photo 8-4

Admirals rowing Lord Louis Mountbatten across Valetta Grand Harbour, Malta, to HMS *Surprise* on 15 December 1954, after he relinquished command of Allied Forces, Mediterranean, to become First Sea Lord of Britain. Mountbatten is seated in the stern of the boat; the admirals are, L to R: Louis Mornu of France, Marco Calamei of Italy, George Zepos of Greece, Sheres Karapiner of Turkey, Peter Cazalet of England, and James Fife of the United States.
Naval History and Heritage Command photograph #NH 62445

Following the commencement of hostilities with Italy, it became evident that the supply of Malta by surface ships would be a hazardous undertaking—one involving both Force H, based at Gibraltar, and the Mediterranean Fleet at Alexandria. From either base, a lengthy ship transit of some 1,000 miles was involved, requiring four days steaming, of which a considerable proportion would be in daylight, outside British air cover and within range of Italian air bases.[31]

HMS *Cachalot* (Lt. Hugo Rowland Barnwell Newton, RN) arrived at Alexandria from Holy Loch on 22 May 1941, joining sister submarine *Rorqual* for Mediterranean service. Aboard *Rorqual*, Lt. Comdr. Ronald Hugh Dewhurst turned over command of the submarine on 1 June to his successor, Lt. Lennox William Napier, RN. Napier, like Dewhurst, would carry out minelaying missions in the Med that year, but many less, with fewer resultant enemy casualties, as shown in the table:

Date	HMS *Rorqual* (Lt. Lennox William Napier, RN) Field	Damage to Shipping
25 Aug 1941	50 mines off Cape Skinari, Greece	
8 Oct 1941	50 mines in the Gulf of Athens 37°29'N, 23°53'E	19 Oct 1941, the Italian torpedo boat *Altair* ran into the minefield laid by *Rorqual*, was damaged, and sank the next morning while in tow. 20 Oct 1941, the torpedo boat *Aldebaran* sailed to the assistance of *Altair* but also ran into the field and sank in the Gulf of Athens.
21 Oct 1941	10 mines off Cavoli Island, off southeast Sardinia	
22 Oct 1941	40 mines off Cape Ferrato, Sardinia	Likely responsible for the loss of the Italian merchant ship *Salpi* on 9 Feb 1942
18 Nov 1941	29 mines west of La Rochelle, France 46°05'N, 01°57'W	
19 Nov 1941	45°56'N, 01°47'W[32]	

Supply of Malta occupied *Rorqual* and *Cachalot* in late spring and early summer. The two submarines made their first petrol delivery to Malta in May. In a larger operation, on 3 June, *Rorqual* and *Cachalot* departed Alexandria with the 1st Flotilla—HM submarines *Truant*, *Triumph*, *Taku*, *Tetrarch*, *Torbay*, *Regent*, *Rover*, *Osiris*, *Otus*, *Parthian*, *Perseus*, and *Pandora*—and five Greek submarines, for transport of petrol and stores to Malta. Aboard *Rorqual* were:
- 2 officers, 21 other ranks, 1 Maltese civilian
- 147 bags of mail
- 2 tons of naval stores including medical stores
- 1,478 cases of aviation fuel
- 15 tons of bulk aviation fuel, and 45 tons of bulk kerosene
- 2 coils of wire[33]

Chapter 8

Rorqual arrived at Malta on 12 June, disembarked her passengers and cargo, and embarked personnel, ammunition, stores and materiel for return passage to Alexandria:
- 2 officers, 3 dockyard officials, and 12 other ranks
- 46 cases of QF 4-inch HE (high explosive) shells
- 100 cases of 4-inch shrapnel shells
- 130 bags of mail, and 30 seaman's bags
- 2.5 tons of naval stores
- Several parts of machinery[34]

LOSS OF HMS *CACHALOT*

The fate of the crew of HMS Cachalot *was confirmed today in a report to the War Cabinet in London. The submarine was rammed on 1 July by an Italian torpedo boat as she was returning to Alexandria from [a] 'club run' to deliver urgent supplies to Malta. According to unconfirmed Italian reports, HMS* Cachalot *was sunk and her crew taken prisoner.*

The War Cabinet also heard how attacks from Malta continue to have an impact on Axis shipping and military installations on both sides of the Mediterranean.[35]

After sailing on 3 June, *Cachalot* had first called at Port Said, and then returned to Alexandria, before departing again on the 12th for Malta. Lieutenant Newton brought the submarine back to Alexandria on 28 June, ending her first storage trip. Several days later, she sailed on her second storage trip to Malta. After discharging her cargo, she left on 26 July bound for Alexandria. En route, she received orders to patrol to the northwest of Benghazi, Libya, to intercept an important Italian transport ship.[36]

Cachalot was in position on 30 July, lying in wait for the enemy. Newton wanted to avoid a night action on the surface due to the large silhouette of the submarine. He expected to intercept the transport at 0230 and had given orders that he be called thirty minutes prior. Upon sighting the ship, Newton planned to turn away from it and proceed along its projected track, while remaining on the surface to charge batteries. He further intended to be eight miles ahead of the enemy at dawn, and to then dive and attack submerged.[37]

Newton was asleep on the bridge, in anticipation of a "busy day," when at 0155, the officer of the watch, T/Lt. R. D. C. Hart, RNVR, sounded the night alarm and called him. Hart reported to Newton that there was a destroyer nearby, and *Cachalot* dived. Proper trim was obtained with great difficulty; there were also problems with the hydrophones and hydroplanes; and due to an error, the batteries were almost depleted. (Trimming a submarine to balance the forces acting on it when it reaches a desired depth, is accomplished by adjusting the amount of water in variable ballast tanks.)[38]

When *Cachalot* surfaced at around 0250, Newton thought the enemy must have passed and ordered a course steered along its expected track at full speed to try to catch up. However, because of the low charge on the batteries, he intended to give up the chase if the transport was not sighted within an hour.[39]

Around 0335 the starboard lookout reported an enemy tanker bearing green 120 degrees (abaft the starboard beam). It was also seen by the others on the bridge, and *Cachalot* turned to follow, with all lookouts searching visually for escort vessels of which none had been seen. A twenty-minute chase through patches of mist followed and Newton considered that if there were one or more escort ships, they might have lost touch with the tanker.[40]

He decided that in order to keep the enemy in sight, in the reduced visibility, she must be slowed down as soon as possible. He ordered the deck gun manned, and four rounds were fired on bearing green 30 (30 degrees on the starboard bow) for a range of 1,500 yards. After the fourth round the gun crew sighted the ship and fire was continued 'independently.' The eleventh round appeared to be a hit and dense clouds of smoke appeared amidships the target.[41]

The enemy then appeared to alter course toward *Cachalot* as if to ram. The submarine turned to counter this threat, but almost at once lost sight of the tanker in the smoke. One minute later, a torpedo boat was sighted only 800 yards distant, closing at high speed. By the time preparations were made to dive, the enemy was only 300 yards away. Newton, recognizing that he could not escape, and wishing to avoid being rammed, gave the order to abandon ship.[42]

The torpedo boat, realising that *Cachalot* was not diving, and not wishing to collide with her at high speed, backed down full, but still rammed the submarine from astern at a speed of about 4 knots, puncturing 'Z' ballast tank. She then remained stopped twenty feet astern with all her armament trained on *Cachalot*, whose single deck gun would not bear. Submarine crewmembers, meanwhile continued to abandon ship.[43]

Photo 8-5

Italian torpedo boat *Generale Antonio Cantore* (formerly classified a destroyer), a sister ship of the *Generale Achille Papa*, which was responsible for the loss of HMS *Cachalot*. German Federal Archives Bild 101I-185-0116-02A

Newton went below to inspect for damage and found that while 'Z' tank had been holed, the pressure hull remained intact. He decided that he would attempt a static dive, together with the First Lieutenant and with key ratings only on board—giving the Italians the appearance of having scuttled the ship—and hope to slip away on a main ballast trim. He returned to the bridge to find all the hatches open to facilitate abandonment. While this complication was being rectified, the Italians opened fire with anti-aircraft armament. Fortunately, all the shots were high, causing no casualties.[44]

Realising that a diving attempt was futile, Newton ordered main vents opened and *Cachalot* sank bow first in 200 fathoms of water. No debris escaped to the surface. All seventy of her passengers and crew, with the exception of a Maltese steward, Giuseppe Muscat, were saved. They were well treated by the Italians and, upon learning that one of *Cachalot*'s passengers was missing the Italian commanding officer, Tenente di Vascello Gino Rosica, ordered a search of the area.[45]

Newton learned from him that the ship sighted in the murk and believed to be a tanker, had in fact been the *Generale Achille Papa*, which he and his crew were now aboard. The torpedo boat had been heading northward to rendezvous with the *Capo Orso* arriving from Brindisi. The smoke believed due to a hit by *Cachalot*'s gunfire had been the Italian warship laying a smoke screen from which she emerged to counter-attack and ram her attacker.[46]

AFTERMATH

Generale Achille Papa was one of six 240-foot *Generale*-class destroyers, constructed in 1921-1922. They were the last ships of the Regia Marina fitted with three stacks and, in 1929, being obsolete, were reclassified as torpedo boats. Three of the torpedo boats would be lost to British mines in the war. The remaining ones were scuttled by their crews following the surrender of Italy to the Allies on 8 September 1943.

Fate of the Italian *Generale*-class Torpedo Boats (ex-Destroyers)

Torpedo Boat	Disposition
Generale Antonio Chinotto	Sunk on 28 March 1941 by a mine laid by the British submarine HMS *Rorqual* off Palermo
Generale Antonio Cantore	Lost on 22 August 1942 to a mine laid by the British submarine HMS *Porpoise* off the coast of Cyrenaica.
Generale Marcello Prestinari	Lost on 31 January 1943 to a mine laid by the British minelayer HMS *Welshman* in the Sicilian channel
Generale Antonio Cascino	Scuttled by crew on 9 Sep 1943 at La Spezia
Generale Carlo Montanari	Scuttled by crew on 9 Sep 1943 at La Spezia; salvaged by Germans but never commissioned and sunk at La Spezia on 4 Oct 1944 by Allied aircraft
Generale Achille Papa	Scuttled by crew on 9 Sep 1943 at Genoa; salvaged by Germans, originally designated TA7 (later SG20), but never commissioned[47]

Lt. Hugo Rowland Barnwell Newton, RN, was part of a prisoner of war exchange with the Italians on 28 March 1943. His submarine, HMS *Cachalot*, was the fourth *Porpoise*-class to be lost in the war. A fifth, HMS *Porpoise*, would also fail to return from a war patrol. She is believed to have been sunk on 11 January 1945 by Japanese aircraft in the Malacca Strait, to the east-northeast of Pulo Perak (Perak Island).[48]

HMS *PORPOISE* ARRIVES IN THE MEDITERRANEAN

On 26 October 1941, HMS *Porpoise* (Lt. Comdr. Edward Fowle Pizey, RN) arrived at Alexandria from Holy Loch, as the relief for HMS *Rorqual*. *Rorqual* ended a war patrol at Gibraltar that same day, and two days later left for Holy Loch and a refit in the UK. While in passage to Alexandria, *Porpoise* had stopped at Gibraltar and Malta—delivering to the latter British Colony: passengers (two officers and twelve ratings), spare gear, six 18-inch torpedoes, and bombs for use against the Corinth Canal.[49]

Porpoise made a storage trip to Malta from 3-18 November, and then took up a war patrol at month's end. She departed Alexandria on 28 November with orders to patrol off the southwest approaches to the Antikithera Channel—the southern and widest of three channels leading from the Greek island of Antikithera to the Aegean. On 7 December, Pizey shifted the patrol to the Kithera Channel (the middle channel), and a day later to the Navarino area in southwest Greece. On 9 December, *Porpoise* torpedoed the Italian passenger ship *Sebastiano Venier* about five miles south of the port town of Navarino (Pylos). Heavily damaged, the ship grounded and wrecked off Cape Methoni.[50]

HMS *Porpoise* ended her first war patrol in the Mediterranean on the 19th at Alexandria.[51]

SUMMATION OF EFFORT IN THE MEDITERRANEAN

Mining by British submarine and surface minelayers in the Med in 1941 was limited to ten operations by the *Rorqual*, in widely dispersed sites between the Aegean and Sardinia; one by the *Abdiel* off Cephalonia, Greece; and one small lay each by the gunboat *Aphis* and by two MTBs (motor torpedo boats), in the Gulf of Bomba—on the northern coast of Libya—and off Bardia, a seaport in eastern Libya. *Manxman* laid a field to the south of Leghorn, Italy, while operating from the UK. (Chapter 1 is devoted to this operation, code named MINCEMEAT.)[52]

Swordfish of 830 Squadron, later assisted by 828 Squadron, shouldered the main aero-duties; mining Tripoli almost continuously, only occasionally visiting other targets, including ports in Sicily. Wellingtons carried out a few operations mining North African ports, and one to the Corinth Canal. Another Swordfish squadron (No. 815) mined Brindisi in April, from an airfield in Greece; and planes from HMS *Ark Royal* lightly mined Spezia in February.[53]

9

The "Rats of Tobruk"

Eight months ago, the enemy isolated the fortress of Tobruk and laid siege to it. Today Tobruk is no longer besieged and her garrison is pursuing the retreating enemy to the westward.

During those fateful eight months the task of maintaining the garrison with all its bodily needs and war supplies has fallen on the Navy and units of the Merchant Navy. Most of the work devolved on destroyers and small ships.

Units from the Royal Navy, the Royal Australian Navy, and Indian Navies and the naval forces of the Union of South Africa all took their part whilst amongst the crews of the merchant ships were officers and men of the Allied Nations.

When the tale of the siege of Tobruk comes to be written, the part played by these craft will provide a story worthy of the highest traditions of our naval history.

I have watched with admiration the work of the "little ships". They have borne the burden of the day but neither fatigue nor the assaults of the enemy have deterred them. Their achievement is one of which they may all be proud.

—Special Message issued by Adm. Sir Andrew Cunningham, Commander-in-Chief, Mediterranean, on 19 December 1941, a few days after the siege of Tobruk was lifted[1]

The battle for North Africa was a struggle for control of the Suez Canal and associated access to oil from the Middle East and raw materiel from Asia. Britain, the first major nation to field a completely mechanised army, was particularly dependent on oil, and the canal also provided her a valuable link to her overseas dominions–part of a lifeline that ran through the Mediterranean Sea. Thus, the North African campaign and the naval campaign for the Mediterranean were intertwined. The struggle for control of North Africa had begun as early as October 1935, when Italy invaded Ethiopia from its colony Italian Somaliland (present-day northeastern, central and southern Somalia), extending south from Cape Asir to the boundary of Kenya. That move made Egypt very wary of Italy's imperialistic aspirations, and she granted Britain permission to station large forces in her territory. Britain and France also agreed to divide the responsibility for maintaining naval control of the Mediterranean, with the main British base located at Alexandria, Egypt.[2]

Italy had been a wild card in the Mediterranean strategic equation at the outset of war. If the Italians remained neutral, British access to the vital sea lanes would remain almost certain. However, if Italy sided with Germany, the powerful Italian navy had the capability to close the Mediterranean. Its main base was at Taranto, a port city on Italy's southeast coast, and operations from there would be supported by air force units flying from bases in Sicily and Sardinia. Italy did remain neutral when Germany invaded Poland in September 1939, but Benito Mussolini could not resist the opportunity to grab his share of the spoils when Germany invaded France in June 1940. On 11 June, six days after the British evacuation at Dunkirk, France, Italy declared war on Britain and France. Britain and Italy were now at war in the Mediterranean, and the French surrender on 25 June, placed the entire burden of controlling the Mediterranean sea lanes on the Royal Navy.[3]

In North Africa, Italian Marshal Rodolfo Graziani had some 250,000 troops in Libya, while Gen. Lord Archibald Percival Wavell, British Commander-in-Chief of the Middle East, had only 100,000 troops to defend Egypt, Sudan and Palestine. The British ground forces, however, were better organised, trained, and equipped and had superior leadership. The British and Italian armies faced each other across the Libyan-Egyptian border, an inhospitable area with no vegetation and virtually no water, known as the Western Desert.[4]

By October 1940, the British began to reinforce Wavell. Through December, an additional 126,000 troops arrived in Egypt from Britain, Australia, New Zealand, and India. On the night of 11-12 November, twenty-one canvas-winged Fairey Swordfish aircraft took off from HMS *Illustrious* to carry out one of the most pivotal aerial attacks of the war. With unprecedented determination and bravery, the pilots of the torpedo planes pressed home a surprise attack on Taranto; crippling the Italian fleet and rendering the Italian Navy ineffective for the duration of hostilities.[5]

It is likely the Taranto raid influenced the Imperial Japanese Navy staff's planning for the attack on Pearl Harbor, as successful attacks on both of these shallow harbours required finding ways to prevent air-dropped torpedoes from striking the bottom and being rendered ineffective. Lt. Comdr. Takeshi Naito, the assistant naval attaché to Berlin, flew to Taranto to personally investigate the attack, and discussed his observations at length with Comdr. Mitsuo Fuchida. Fuchida led the first wave of Japanese air attacks on 7 December 1941.[6]

On 9 December 1940, the British Western Desert Force, under Lt. Gen. Sir Richard O'Connor, attacked the Italian Tenth Army at Sidi Barrani, a town in Egypt, near the Mediterranean Sea, about fifty-nine miles east of the border with Libya, and around 150 miles from Tobruk, Libya. Pushing the Italian forces out of Egypt on 3 January 1941, the British scored a major victory at Bardia, just inside Libya. Continuing with the drive into eastern Libya, the British took the vital port of Tobruk on 22 January. O'Connor continued to pursue the Italians, and after trapping them at Beda Fomm on 7 February, they collapsed. The impressive British and Commonwealth successes against the Italians in Libya were short-lived, as they were answered by a formidable German counter-offensive. The early victories were reversed and all that stopped Generalleutnant Erwin Rommel's Afrika Korps march on Egypt was the defiant garrison at Tobruk.[7]

Map 9-1

Eastern Mediterranean and surrounding areas

It was vital for the Allies' defence of Egypt and the Suez Canal to hold Tobruk with its harbour, as this forced the enemy to bring most of their supplies overland from the port of Tripoli, across 930 miles of desert. Additionally, expenditure of resources and time for the supply effort diverted enemy troops from their advance. Tobruk was subject to repeated ground assaults and almost constant shelling and bombing by aircraft. Nazi propaganda called the tenacious defenders "rats," a term the Australian soldiers embraced as an ironic compliment. Tobruk offered the only significant port on the North African coast between Tripoli and Alexandria. Further, Rommel's Afrika Korps was in no position to attack across the Egyptian border toward Cairo and Alexandria while the Tobruk garrison threatened the resupply of his front-line units. Its capture would dramatically shorten the German and Italian supply lines.[8]

Surrounded by German and Italian forces, the Tobruk garrison, mostly men of the 9th Australian Division, withstood tank attacks, artillery barrages, and daily bombings for eight long months. Maj. Gen. Leslie Morshead's aggressive policy of patrolling outside the garrison to "make the besiegers the besieged" kept the enemy at arm's length, enabling the Australians to dominate "no man's land" and in doing so, stave off the German advance toward Egypt. While pursuing this strategy, the troops lived in dug-outs, caves, and crevasses while enduring the desert's searing heat, bitterly cold nights, and hellish dust storms.[9]

Photo 9-1

Members of the 2/10th Australian Infantry Battalion standing by in a "hot section of the front," 400 yards from the enemy. Australian War Memorial photograph 009514

THE "TOBRUK FERRY SERVICE"

> *In all, the Australian destroyers made a total of 139 runs in and out of Tobruk during the period of the regular 'Ferry'. Vendetta held the record with 39 individual passages into Tobruk, 11 from Alexandria and 9 from Mersa Matruh; and from Tobruk 8 to Alexandria and 11 to Mersa Matruh. From the end of May until the first week in August [1941] she was without intermission on the Tobruk shuttle service and carried 1,532 troops to Tobruk; brought 2,951 away, including wounded and prisoners of war; and transported 616 tons of supplies into the port.*
>
> —Reference made in the official history of the Royal Australia Navy to the Tobruk Run and the destroyer HMAS *Vendetta*.[10]

The "Tobruk Ferry Service" was the name given to the force of Royal Navy and Royal Australian Navy ships involved in the supply of Allied forces during the Siege of Tobruk. Its mission was to keep the besieged Allied forces supplied with badly needed stores such as ammunition, spare gun barrels, medical supplies and mail, and bring out the wounded. For a long period, the ships of the 10th Destroyer Flotilla, of which the Australian destroyers were a part, did the run "solo" (unaccompanied by any other vessel). The practice was to leave Alexandria early in the morning after loading the night before and make the 350-mile transit at high speed so as to arrive at Tobruk about midnight, unload stores and embark the wounded and depart a couple of hours later.[11]

Decades later, Commodore Rodney Rhoades, DSC RAN (Retired), who had commanded HMAS *Vendetta*, described the lengthy work still remaining before completing a typical run:

> We then sped back at full speed to Mersa Matruh [Egypt] halfway along the coast towards Alexandria, put the wounded ashore there and sailed again in the afternoon with fresh stores for Tobruk, where we unloaded, embarked wounded and then sailed for Alexandria about 0200.
>
> It was no picnic as you could imagine, as we were the target for bombers and submarines, to say nothing of mines and we had many a narrow escape.[12]

The danger posed by attack from air and sea eventually resulted in a pair of destroyers making each run. By this means, they could help protect each other, and if one were damaged or sunk, the other could provide assistance or recover survivors.[13]

SOUTH AFRICAN NAVY SHIPS ARRIVE AT TOBRUK

At the outbreak of World War II, there was no South African Navy, only men of the Royal Naval Volunteer Reserve (SA) and former cadets of the South African training ship *General Botha*. This vessel, the former British cruiser HMS *Thames* commissioned in 1888, was used exclusively for the nautical training of British and South African boys, so that they could subsequently serve in ships of the British Empire. The prime minister of the Union of South Africa, Gen. Jan Christiaan Smuts, asked Rear Adm. Guy Waterhouse Hallifax to form a South African naval service. Hallifax, who was retired from the Royal Navy and living in South Africa, agreed, and on 15 January 1940, a new naval unit titled the Seaward Defence Force (SDF) was formed.[14]

Hallifax took charge of the seaward defences at Cape Town and arranged for the purchase/chartering of suitable whalers and trawlers from South African harbours, for employment as minesweepers and anti-submarine patrol vessels. These ships were armed with available guns of the arsenal at Simon's Town. Four of the vessels were former whalers (of approximately 344 tons each) of the Southern Whaling & Sealing Co. Ltd., Durban. After being taken over by the South African Government, they were each fitted with one 3-lb (4-inch) gun forward, 20mm and machine-guns, and sonar and depth charges. Ship's complement was 20-25 officers and ratings, all South Africans.[15]

On 15 December 1940, the four vessels comprising the 22nd Anti-Submarine Group, South African Seaward Defence Force—HMSAS *Southern Floe* (Lt. John E. J. Lewis), *Southern Isle* (Sub Lt. Louis Botha Ribbink), *Southern Maid* (Lt. David Alfred Hall), and *Southern Sea* (Lt. John Charles Netterburg)—departed Durban, South Africa, for Alexandria, Egypt. The Senior Officer, Lt. Comdr. A. F. Trew, was aboard *Southern Maid*. The commanding officers were all SARNVR (South African Division of the Royal Naval Volunteers Reserve) officers. Active operations in the Mediterranean began on 20 January 1941 off Alexandria, when the *Southern Isles* dropped depth charges on a sonar contact believed to be an enemy submarine.[16]

GERMAN AIRCRAFT MINE TOBRUK HARBOUR

Events in early February 1941, precipitated an immediate need for minesweeping for magnetic mines at Tobruk and surrounding waters. At noon on 4 February, the British troopship *Ulster Prince* and merchant motor ship *Devis* left Alexandria with troops for Tobruk. Upon learning of aerial mining with magnetic mines of Tobruk Harbour, the ships were recalled to Alexandria. As there were no minesweepers available to counter this type of mine at Tobruk, four 205-foot Flower-class corvettes—HMS *Peony*, *Gloxinia*, *Salvia*, and *Hyacinth*—were recalled to Alexandria from the Kithera Patrol off the southernmost part of mainland Greece, for refitting with 'LL' electric cables to enable clearing of Tobruk Harbour.[17]

Photo 9-2

Painting by Edward Tufnell, RN (Rtd.) of HMS *Hydrangea*, a Flower-class corvette. Naval History and Heritage Command photograph #NH 86394-KN

Magnetic mines resting on the bottom of a waterway are triggered upon sensing a disturbance of the Earth's magnetic field, created by a passing vessel. LL (double L) minesweeping involves towing a pair of electric cables parallel to each other on floats behind a minesweeper. The cables are rigged to emit a strong electric pulse generating a magnetic field which detonates the magnetic mines.[18]

On 5 February a motor schooner was mined at Tobruk, killing the Assistant King's Harbour Master, Lt. Comdr. James Cochrane. Three days later, the Dutch tanker *Adida* was damaged by two mines, and the former Italian steamer *Rodi* by one exploding close aboard. On 9 February, the British steamer *Crista* was damaged by a mine at Tobruk. Despite the toll mines were taking on shipping, the destroyer HMS *Decoy* left Tobruk with the minesweeping trawlers *Arthur Cavanagh* and *Milford Countess* for Benghazi on 11 February. Presumably, there was a more urgent need for minesweepers at Benghazi. The following day, the armed boarding vessel *Chakla* (with Senior Naval Officer, Inshore Squadron, Capt. Albert Lawrence Poland, aboard) accompanied by the destroyers *Stuart*, *Voyager* and *Vampire*, minesweeper *Fareham*, and minesweeping corvettes *Peony* and *Hyacinth* left Tobruk at 0730 to arrive at Benghazi early on the 13th.[19]

LOSS OF HMSAS *SOUTHERN FLOE* TO A MINE

On 11 February 1941, HMSAS *Southern Floe* was sunk by a mine off Tobruk—the first of four units of the South African Navy that would be lost during the war. The *Southern Floe* and *Southern Sea* had arrived at Tobruk via the Suez Canal on 31 January to take over patrol duties from their sister ships which had arrived earlier. Initially tasked with helping protect the Tobruk Harbour from attack by enemy submarines, South African ships later also began to escort resupply ships to the besieged seaport. Although enemy submarines were not a threat in the first six months of the Western Desert campaign, the discovery of numerous floating mines suggested the existence of extensive moored mine fields. Except for sweeping the narrow coastal traffic route and harbour entrances, there had been little minesweeping accomplished. At the time, there had been insufficient time to accurately locate these fields, let alone clear them.[20]

When the ships arrived, their main duty was to patrol the nearest section of the swept channel and to escort shipping along it. The port was subject to air raids, and as a consequence, it was littered with sunken wrecks and possibly contained bottom-mines. Patrol duties were complicated by sandstorms that strong off-shore winds extended many miles out to sea, resulting in low visibility, heavy cross-seas, and much discomfort to personnel. To these challenges was added an ill-defined and unlighted coast opposite the harbour.[21]

On the morning of 11 February, *Southern Sea* arrived at a designated patrol rendezvous point, two miles east of Tobruk. She found no sign of *Southern Floe* her sister ship. That evening, a passing Allied destroyer picked up a man clinging to some wreckage—all that remained of the *Southern Floe* and her crew. The single survivor was Leading Stoker C. J. Jones, RNVR (SA), lent from HMS *Gloucester* to fill a vacancy just before *Southern Floe* sailed from Alexandria. Jones had been in the stokehold when, at about 0400, there had been a heavy explosion and water rapidly filled the ship. In darkness, he made his way into the flooded engine room and escaped through the skylight as *Southern Floe* sank. There had been a few others in the water, and Jones had done his best, to support a wounded man until he was unable to continue.[22]

It was believed that a mine, either floating or moored, had been responsible for sending *Southern Floe* to the bottom. Her loss was a grievous blow to the flotilla and to the South African Seaward Defence Force. The ships had been on station for barely a month. At their home in South Africa, few were aware that they had arrived and had been in action. The casualties were the first naval losses suffered by the SDF and the sense of loss was profound.[23]

A week later, the Naval trawler *Ouse* (Sub Lt. W. V. Fitzmaurice, RNR) was sunk by a mine at Tobruk. Fitzmaurice and eight ratings were wounded—but survived. A/Sub Lt. E. P. Ede RNR, Gunner J. Edwards, and ten ratings were lost. In addition to the threat posed by aero-laid mines, the Germans soon began sending bombers to attack the army garrison, shipping, and shore installations at Tobruk.[24]

Photo 9-3

The effect of a thousand-pound bomb on a small ship lying in Tobruk Harbour. Australian War Memorial photograph 007580

AWARDS FOR VALOUR

The three remaining little ships, joined by HMSAS *Protea* (Lt. R. P. D. Dymond RNVRSA) as a replacement for *Southern Floe*, performed magnificently in the following months. Twenty-three individuals aboard *Southern Maid*, *Southern Isles*, and *Southern Sea*, received the Distinguished Service Cross, Distinguished Service Medal, Conspicuous

Gallantry Medal, or Mention in Despatches for heroic actions during the period 15 January 1941 to 31 July 1942:

HMSAS *Southern Maid*

DSC	DSM	CGM	MID
Lt. David Alfred Hall	Petty Officer T. S. Hupton		
Sub Lt. C. B. B. Watson	Stoker Petty Officer C. H. Arnold		
Sub Lt. (E) C. L. Evans	Leading Seaman N. M. Hardwich		
	Leading Steward B. D. Tarling.		

HMSAS *Southern Isles*

DSC	DSM	CGM	MID
Sub Lt. Louis Botha Ribbink	Petty Officer W. H. Dean	Leading Stoker Rene Sethren	Lt. A. C. Matson
	Leading Seaman H. Offer		AB Seaman T. E. E. Overton
	Stoker (1st Class) A. Mooney		Signalman P. D. Kockett
	Leading Stoker H. M. Jewell		

HMSAS *Southern Sea*

DSC	DSM	CGM	MID
Lt. John Charles Netterburg	AB Seaman P. L. Lewis		Sub Lt. J. P. Baker
Lt. Allan Thomas	AB Seaman J. Riley		Leading Cook R. C. E. Miles
	Stoker (1st Class) R. H. T. Keppler[25]		

DSC: Distinguished Service Cross DSM: Distinguished Service Medal
CGM: Conspicuous Gallantry Medal MID: Mention in Despatches

Rene Sethren, Leading Stoker aboard HMSAS *Southern Isles*, was the only one of these men awarded the Conspicuous Gallantry Medal, and the only South African to earn one in the war. On 30 June 1941, the *Southern Isles*, sloop HMS *Flamingo*, and gunboat HMS *Cricket* were escorting a Tobruk-bound convoy comprised of the Greek steam ships *Miranda* and *Antiklia*, which had left Mersa Matruh, a seaport in Egypt, a day earlier. In early afternoon the convoy came under attack by a large German bombing force of Stukas and Junkers 88s. Sethren replaced a wounded man on a twin-Lewis and was himself seriously wounded. Despite his injuries he shot down an enemy aircraft. The citation for his CGM states that he was hit by eleven bullets; medics later found twenty-seven wounds.[26]

CANADIANS AND BRITISH CREW MINESWEEPERS

In late March 1941, the 109th Minesweeping Flotilla—HMS *Swona*, *Skudd III*, *Skudd IV* and *Skudd V*—arrived at Alexandria. The four small British ships, like the South African ones that preceded them at Tobruk, were former whalers. *Swona* was a 313-ton whaler of South Georgia Co. Ltd., completed in August 1925 and taken over by the Admiralty in April 1940. The three *Skudd*s were former Norwegian whalers requisitioned by the Admiralty, as was *Skudd VI*, included in the table. HMS *Sotra* (which arrived at a different time), like *Swona*, had been a whaler of South Georgia Co. Ltd. The data column in the table lists each ship's completion date, date it was commissioned into the Royal Navy, and displacement.[27]

Ship	Data	Commanding Officer	Disposition
HMS *Skudd III*	27 Jun 29 3 Nov 40 245 tons	Lt. Robert Cunningham MacMillan, RCNVR	Sunk by German dive bombers at Tobruk, Libya, on 27 August 1941
HMS *Skudd IV*	21 Jun 29 4 Nov 40 245 tons	T/Lt. Kjell Tholfson, RNR	Renamed *Spate* on 3 April 1944, returned to owner in October 1945
HMS *Skudd V*	28 Aug 30 5 Nov 40 265 tons	Sub Lt. Robert Arthur Neville Cox, RNR	Renamed *Surge* on 3 April 1944, returned to owner in March 1945
HMS *Skudd VI*	25 Sep 30 6 Jul 40 323 tons	T/Lt. William Jackson Wolfe, RNVR	Renamed *Sleet* on 3 April 1944, returned to owner in November 1945
HMS *Sotra*	Aug 25 Sep 39 313 tons	T/Lt. John Mortimer Davies, RCNVR	Sunk by *U-431* on 29 January 1942 about 80 nautical miles east of Tobruk, Libya
HMS *Swona*	Aug 25 Apr 40 313 tons	T/Lt. Richard Leigh Smith, RCNVR[28]	Became a danlayer in 1944

The commanding officers of three of the six whaler minesweepers were Canadians. The Royal Canadian Navy's greatest contribution to the war would be the role it played in the Battle of the Atlantic—the grim and unrelenting struggle against German U-boats. However, RCN personnel would also serve aboard a variety of Royal Navy vessels, from aircraft carriers, to light cruisers and landing craft, in European and Pacific waters. Two branches of the Royal Navy in which Canadians formed a substantial presence were coastal forces and naval aviation—

largely because Britain was permitted to recruit in Canada for these specialties. Canada's last Victoria Cross recipient, Lt. Robert Hampton Gray, was a Corsair pilot aboard HMS *Formidable*. He received the award (posthumously), killed in the final few days of the war when his plane was shot down as he was attacking a Japanese warship in Japanese home waters.[29]

The executive officer of HMS *Skudd V*, John McDonald Ruttan, had begun his Naval service with the Port Arthur Half Division of the RCNVR (Royal Canadian Navy Volunteer Reserve) in Spring 1937. He reported aboard the Canadian three-masted training schooner HMCS *Venture* in late December 1937 and was assigned to the Winnipeg Division in the rank of sub-lieutenant in May 1938. Following the outbreak of war, Ruttan travelled to Britain in June 1940 for a six-week basic officers course at HMS King Alfred and then the minesweeping course at Lochinvar, Scotland. He was appointed as executive officer of *Skudd V* in September 1940. The whaler minesweeper sailed for Alexandria in October 1940 and operated in the eastern Mediterranean, including Tobruk and Greece in 1941. In 1941, Ruttan was serving in HMS *Svana* as commanding officer.[30]

LOSS OF HMS *SKUDD III* & OTHER DAMAGE AND LOSS OF PERSONNEL DUE TO AIRCRAFT ATTACKS

Photo 9-4

German Air Force Stuka Ju87 aircraft in flight; Europe, circa 1941.
Australian War Memorial photograph 106483

On 13 April 1941, HMS *Skudd IV* was damaged by German dive bombing at Tobruk. Ordinary Telegraphist Maurice O. Fagg died two days later after being wounded during this attack. On 22 August, Seaman James Gay, a crewman aboard HMS *Skudd V*, was killed. This may or may not have been associated with an aircraft attack on her, but several individuals received medals for valour as a result of this action. HMS *Skudd III* was sunk at Tobruk on 27 August during an air raid by forty dive-bombers.

HMS *Skudd III* Casualties

T/Midshipman John T. Bloxham, RNR	Killed 27 August 1941
Seaman Augustus A. Oldford, Seaman, RNPS	Killed 27 August 1941
Engineman William H. Thomsen, RNPS	Killed 27 August 1941
Leading Wireman Paul A. Lawler	Missing, Presumed Killed
Ordinary Seaman Alfred Oram, RNPS	Missing, Presumed Killed
T/Sub-Lt. Edward R. Swift, RNVR[31]	Died of Wounds 28 August 1941

RNR: Royal Naval Reserve
RNPS: Royal Naval Patrol Service
RNVR: Royal Naval Volunteer Reserve

Photo 9-5

The upper part of a minesweeper (possibly *Skudd III*) bombed and sunk while berthed alongside a wrecked Italian ship; photograph taken 5 October 1941.
Australian War Memorial photograph 021069

AWARDS FOR VALOUR

The below listed officers and men of HMS *Skudd III* and *Skudd V* received awards for valour for "Bravery and endurance while minesweeping, and when attacked by enemy aircraft." MacMillan's award was a bar to a DSC he had previously received (signifying a second act of gallantry meriting the award). Oldford, Swift, and Gay were awarded a MID posthumously.

M/S Whaler HMS *Skudd III*

Awardee	Date "Gazetted"	Award
Sub Lt. Robert Cunningham MacMillan, RCNVR	1 Jan 42	DSC
Lt. Robert Cunningham MacMillan, RCNVR	6 Feb 42	Bar to DSC
Seaman Thomas Harry Jessop, RNPS	27 Jan 42	DSM
A/Leading Seaman Augustus Albert Oldford, RNPS	23 Jan 42	MID (PH)
Ord Signalman Edward McCarroll	27 Jan 42	MID
Ord Seaman Thomas Morrison	23 Jan 42	MID
T/Lt. Edward Rhodes Swift, RNVR	27 Jan 42	MID (PH)

M/S Whaler HMS *Skudd V*

Awardee	Date "Gazetted"	Award
Lt. Robert Arthur Neville Cox, RNR	23 Jan 42	DSC
Lt. John MacDonald ("Mac") Ruttan, RCNVR	23 Jan 42	DSC
Wireman Charles Robert Ladds	27 Jan 42	DSM
Seaman James Gay	27 Jan 42	MID (PH)[32]

BATTLE HONOURS

"Battle Honour - LIBYA 1940-42" was awarded to all ships of the Inshore Squadron (Force W). Fleet Air Arm squadrons, both carrier and shore based, were also included. The operations covered were those in which the Royal Navy and Dominions supported the Army in the Western Desert between Port Said and Benghazi.

In the following tables, all of the South African Naval Forces (minesweeping whalers and anti-submarine whalers) are listed, as are the Royal Australian Navy destroyers and sloops that participated in the "Tobruk Ferry Service." Extensive numbers of Royal Navy ships were involved; only those cited in this chapter are listed. Two of the four ships that South Africa lost during the war were sunk at or near Tobruk by German forces, and the remaining two by mines in Greek waters. Four of the Australian warships which earned Battle Honour - LIBYA 1940-42 were also lost in the war, as were two of the Royal Navy *Skudd*s.

South African Naval Forces (complete list of ships)

M/S Whaler	Year(s)	A/S Whaler	Year(s)
HMSAS *Bever* (3)	42	HMSAS *Southern Floe* (1)	41
HMSAS *Boksburg*	42	HMSAS *Southern Isles*	41-42
HMSAS *Gribb*	42	HMSAS *Southern Maid*	41-42
HMSAS *Imhoff*	42	HMSAS *Southern Sea*	41-42
HMSAS *Langlaate*	42	HMSAS *Treern* (4)	42
HMSAS *Parktown* (2)	42		
HMSAS *Protea*	41-42		
HMSAS *Seksern*	42		

1. Sunk by a mine off Tobruk, Libya, 11 February 1941.
2. Sunk by German E-boat off Tobruk, Libya, 21 June 1942.
3. Sunk by a mine off Pireaus, Greece, 30 November 1944.
4. Sunk by a mine off the east coast of Greece, 12 January 1945.

Royal Australian Navy (complete list of ships)

Destroyer HMAS *Nizam*	41	Destroyer HMAS *Vendetta*	40-41
Sloop HMAS *Parramatta* (2)	41	Destroyer HMAS *Voyager* (4)	40-41
Destroyer HMAS *Stuart*	40-41	Destroyer HMAS *Waterhen* (1)	40-41
Destroyer HMAS *Vampire* (3)	40-41	Sloop HMAS *Yarra*	41

1. Heavily damaged by Axis aircraft on 29 June 1941, while operating with the Tobruk Ferry Service; sank the following day; the first RAN ship lost to combat in the war.
2. Torpedoed and sunk on 27 November 1941 by *U-559*.
3. Sunk by Japanese aircraft in the Bay of Bengal on 9 April 1942.
4. Ran hard aground on 23 September 1942 at Betano Bay, Timor, during troop disembarkation operations. Japanese air raids hampered efforts to offload salvageable materiel, and ship was destroyed by detonation charges and abandoned.

Royal Navy (partial list of ships)

Auxiliary Minesweeper HMS *Arthur Cavanagh*	41	M/S Corvette *Salvia*	41
Minesweeper HMS *Fareham*	41	M/S Trawler HMS *Sotra* (2)	41-42
M/S Corvette *Gloxinia*	41-42	HMS *Skudd III* (1)	41
M/S Corvette *Hyacinth*	41-42	HMS *Skudd IV*	41
M/S Trawler HMS *Milford Countess*	41	HMS *Skudd V*	41-42
M/S Corvette *Peony*	41-42		

1. Dive bombed and sunk by German Ju87 aircraft at Tobruk on 27 August 1941.
2. Torpedoed and sunk by submarine *U-431* off Bardia, Libya, on 29 January 1942.

AFTERMATH

The defenders of Tobruk—principally the Australian 9th Infantry—did not surrender and did not retreat. Half the Australian garrison was relieved in August 1941, the rest in September-October. However, 2/13 Battalion could not be evacuated and was still there when the siege ended on 10 December 1941, the only unit present for the entire siege. Total losses of the 9th Division and attached troops from 1 March-15 December were 832 killed, 2,177 wounded and 941 prisoners. The determination, bravery, and humour of the soldiers of the garrison who held the port against the Afrika Korps, during the Siege of Tobruk, resulted in lasting fame for the "Rats of Tobruk." They were later immortalised in the 1953 American war film "The Desert Rats" from 20th Century Fox.[33]

Six months later, Tobruk—now garrisoned by the British South African Division, which included the Eleventh Indian Brigade—fell to Rommel. With support from artillery and dive-bombers, his panzer division occupied the port on 21 June 1942 and, unable to resist any longer, South African Maj. Gen. Henrik Balzazar Klopper ordered his officers to surrender early that morning.[34]

BRITISH AND NORWEGIAN WHALE CATCHERS/ FACTORY SHIPS TAKEN FOR MILITARY SERVICE

By 1942, whaling by British and Norwegian vessels, with very few exceptions, had ceased. The whale catchers were requisitioned for employment as minesweepers or anti-submarine ships, and the factory ships, which could carry oil and cargo, put into the Atlantic convoys. The latter ships, large, and slow, were easy targets for the enemy. All the British and most of the Norwegian factories were lost in the war. In November 1942, the Hector Whaling Company shareholders were informed that there were no trading results, all vessels were under requisition, and both floating factories had been lost to enemy action.[35]

Photo 9-6

Norwegian whaler *Hector VI* in 1929; *Hector X* became the South African minesweeper whaler HMSAS *Bever*.
Ansgar Theodor Larsen (1875-1932)

10

RN and RNN Coastal Forces

Although the situation was then in hand, any increase in British mining would cause difficulties.

—1940 War Diary entry made by Kommodore Friedrich Ruge, Senior Officer Minesweeping Forces West, at year's end.[1]

Photo 10-1

British trawler HMS *Vulcan* with 1st MTB Flotilla alongside circa 1937. *Vulcan* was purchased by the Admiralty on 11 July 1936 and used as a Coastal Forces depot ship at HMS VERNON before deploying to the Mediterranean in 1939.
Collection of Rob Hoole

Coastal Forces, a division of the Royal Navy, was established in World War I and later disbanded. On 27 April 1937, the 1st Motor Torpedo Boat Flotilla was formed with six Scott-Paine boats powered by three Napier Sea Lion W12 petrol engines, giving a top speed of 38 knots.[2]

The 60-foot, hard chine boats were each armed with two 18-inch torpedoes (originally designed for employment by aircraft) and .303-

inch machine guns fore and aft. The torpedoes were loaded through hatches in the deck above the engine room, and stowed facing forward on overhead rails. Hinged girders provided extensions to the discharge rails abaft the boat's transom, and could be folded up when not in use. When firing a torpedo, doors fitted over the stern apertures were opened, and the girders were hinged down to form a continuation of the overhead rails. The torpedoes were launched by accelerating the boat, which allowed the torpedoes to slide backwards down the rails into the sea. The torpedoes could not be launched with the boat stationary, and the rails were flimsy and liable to buckle. Finally, it took great skill in aiming the bow of the boat at the target, firing, then turning sharply away to avoid being struck by the torpedo, which entered the water nearly stopped before beginning its run. On later torpedo boats this bizarre and unsafe launching arrangement was replaced by torpedo tubes enabling the weapons to be fired ahead of the boat.[3]

Ordered to the Mediterranean for evaluation, the 1st MTB Flotilla, under Lt. Comdr. Guy Bourchier Sayer, sailed on 27 June 1937, calling at Brest, Corunna, and Lisbon before arriving at Malta on 17 July. The flotilla was later allocated the boats of the 3rd Flotilla, which were originally bound for Singapore but then retained at Malta. The boats worked on occasion with the British Mediterranean Fleet, but their true value was not appreciated by the fleet's senior officers who remained convinced that future naval requirements lay in its heavily armed battleships, battle cruisers and destroyers. In November 1939, ten weeks after the outbreak of war, the flotilla received orders to return to England, and sailed at midnight on 11 November.[4]

The return was a particularly treacherous passage in bad weather through the Med to Marseille. One boat was lost. The most seaworthy of the boats then made the remaining transit through the French canals, with nine boats arriving back in the UK shortly before Christmas 1939. The recall of the 1st Flotilla signaled the beginning of the expansion of Coastal Forces, as greater numbers and more capable motor torpedo boats, motor launches, and motor gunboats were added to its resources.[5]

FOUNDING OF COASTAL FORCES

Following its return, the 1st MTB Flotilla became operational from Felixstowe (a seaside town on England's east coast, sited on an estuary opposite Harwich) in January 1940. In October 1940, five boats—*MTB 14, 15, 16, 17,* and *18*—comprised the flotilla. Its readiness marked the defacto establishment of a unit of the Royal Navy that became known as Coastal Forces. With expansion came the need for more senior leadership. In November, Rear Adm. Piers Kekewich was appointed Flag Officer Coastal Forces to coordinate design and production of boats and to act as a liaison between the Admiralty and the various commands. His headquarters was originally at HMS Vernon and later at HMS Hornet, at Gosport across the harbour from Portsmouth, on the south coast of England.[6]

Over the course of the war, Coastal Forces craft operated mainly in the English Channel and in North Sea waters, but were also employed in the Mediterranean, off the Norwegian coastline, and for raids on St. Nazaire and Dieppe in France. The craft were manned by British personnel and those of other Allied nations, including the Netherlands, Norway, Canada, Australia, and New Zealand. Coastal Forces operations included attacks against German convoys and their "E-Boat" escorts, and clandestine raids and landings, including recovering secret agents in Norway, and Brittany, France. As the size of the organisation increased, the number of bases in support of small boat operations grew in consort.[7]

Map 10-1

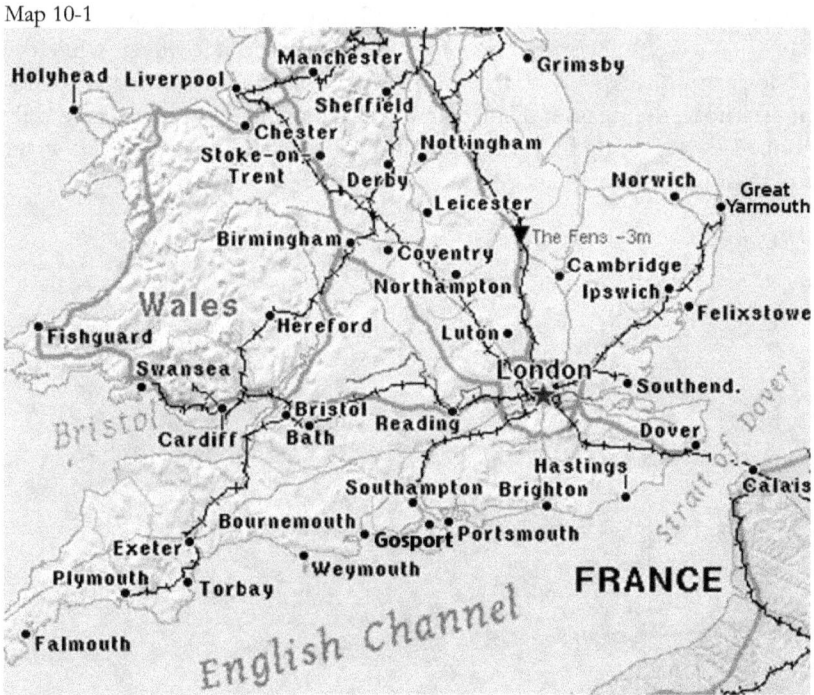

England's south and southeast coasts

Chapter 10

A partial list of Coastal Forces bases follows. Only relatively small numbers of torpedo boats and launches were employed for minelaying. Of the many bases that came into existence, these were the ones most involved in those duties.

Coastal Forces Headquarters and Partial List of Bases

Bases	Location	Parent Command
HMS Midge	Great Yarmouth (east coast)	Nore Command
HMS Beehive	Felixstowe (east coast)	Nore Command
HMS Wasp	Dover (south coast)	Dover Command
HMS Hornet (HQ)	Gosport (south coast)	Portsmouth Command
HMS Grasshopper	Weymouth (south coast)	Portsmouth Command
HMS Black Bat	Plymouth (south coast)	Plymouth Command[8]

COASTAL FORCES CRAFT USED FOR MINELAYING

From the end of 1940, as a result of an ambitious construction programme, motor torpedo boats (MTBs) and Fairmile motor launches (MLs) became available in increasing numbers. These craft were modified to lay ground and moored mines, by fitting chutes in the former and traps in the latter. The 112-foot, 85-ton Fairmile B motor launches were the workhorses of the mining work. However, while the 72-foot motor torpedo boats with planing hulls could operate at speeds of 40 knots, the heavier displacement hull Fairmiles could manage only about 17. A planing hull is designed to glide on the surface of the water as the boat gains speed.[9]

Photo 10-2

British Fairmile B-class motor launch *ML 187;* MLs configured for minelaying would carry mines aft in lieu of depth charges for anti-submarine duties. Courtesy of naval-history.net

Commencement of operations using the motor boats was delayed to January 1941 by a need to modify the A Mk 1 mine—a British ground magnetic mine designed to be laid by aircraft—to ensure that it would operate satisfactorily under conditions where it might be subjected to severe washing by waves, sea spray, or rain while aboard craft. The first MTB minelaying operation was carried out on 7 January 1941, by the 4th Motor Torpedo Boat Flotilla operating from HMS Beehive at Felixstowe, Suffolk.[10]

Beehive, a Royal Naval Coastal Forces Base newly commissioned on 1 July 1940, was headquartered at Felixstowe Dock in the Little Ships Hotel (the former Pier Hotel). The base (which included the hotel, the surrounding dock buildings and adjacent basin, along with old RAF seaplane hangars) would host motor torpedo boats, motor gun boats, and motor launches providing living accommodations for crews, administrative offices, maintenance facilities for the boats, and necessary storage facilities.[11]

4TH MTB FLOTILLA MINELAYING OPERATIONS

The operation on 7 January 1941 involved the nighttime mining of the West Deep entrance to the Zuydcote Pass located east of Dunkirk. For the mission, the lead boat *MTB 31* carried torpedoes for defence of the group. The other two boats, MTBs *32* and *34*, were loaded with four mines each—A Mk 1 magnetic types, with period-delay mechanisms set for two, three, four, and five days. This feature enabled the MTBs to get safely away after the lay, and potentially extended the life of the field. By staggering the intervals at which the mines armed themselves, a ship(s) entering the field on, say day two, would at most, set off two of the eight mines. If the field was first entered on day three, four of the eight mines would be active, and so forth. The lay was completed with eight mines positioned in a line between, and just to the east of, Buoys 1 and 2 in the Zuydcote Passage.[12]

Following this initial operation, 4th Flotilla motor torpedo boats carried out six more mining excursions by the end of April 1941. Zuydcote Pass was mined a second time, and five locations off the northern French coast, once each. These were the first of a series of minelaying operations ("PW") in Nore Command which would be replaced in September by two new identities (QK and QL). During the second operation at Zuydcote Pass on 14 February, sailors aboard the two participating MTBs sighted the wreckage of a ship (the German torpedo boat *Wolf*, mined on 8 January) one mile to the south of the previously laid field. This early success was of great encouragement to the flotilla.[13]

Photo 10-3

British Motor Torpedo Boat HMS *MTB-263* (ex-USS *PT-14*) ready for delivery to the Royal Navy, circa mid-1941. She has been modified to British specifications, with Type 21-inch torpedo tubes, a 20mm machine cannon, and other changes. Naval History and Heritage Command photograph #NH 100911

Map 10-2

The West Deep, a coastal area off Nieuwpoort (Nieuport), lies northeast of Dunkirk. Farther up the coast is the Hook of Holland (Hoek van Holland).

50TH MOTOR LAUNCH FLOTILLA ESTABLISHED

In May, as the 4th Flotilla carried out additional "PW" series minelaying operations, four motor launches—MLs *101*, *103*, *104*, and *220*—were formed into the 50th Motor Launch Flotilla under Commodore Robert Cunliffe, Commander-in-Chief, Dover. Following necessary stiffening and fitting out of the craft, followed by workup practices, MLs *103* and *220*—based at HMS Wasp in Dover—carried out the first in a series of "NL" mining operations on the night of 22 June. It was planned to use magnetic ground mines to start with, and later to add Mk XIX moored contact mines when they became available.[14]

A second operation was carried out on 25 May, after which a waxing gibbous moon, transitioning to a full moon, offered too little concealment for small craft on such dangerous missions. From experience gained in these two missions, it was agreed that an extra officer needed to be carried in the lead boat to perform navigational duties. Also, as a result of lessons learned, arrangements were made to fit a number of boats in each flotilla with taut wire measuring gear and echo sounding; the latter being essential for the laying of moored mines. A set of criteria was also developed for helping to choose suitable nights for mining missions:

- The night must be moonless for a sufficient period
- The state of the sea must not be greater than slight (sea state 3)

- The direction of the wind must be such that engine noises were unlikely to be heard on the enemy coast
- Visibility must be neither extreme nor less than two miles[15]

Provided that the weather was suitable, orders for each operation were issued before 1800 on the night that it was to be carried out. The senior officer had the authority to postpone an operation in the event of any unforeseen worsening of the conditions. During the dark period of the lunar cycle in July, MLs carried out four operations off Calais and adjacent Cape Blanc Nez, in northern France.[16]

NORE COMMAND

In August, 4th MTB Flotilla boats (MTB *30, 32, 44, 45, 47*) operating from HMS Beehive at Felixstowe, carried out three operations of the "PW" series at Zuydcote Pass off the Dutch coast, and farther to the south off northern France between Cape Gris Nez and Dunkirk. Two MTBs were employed for each of the missions to collectively mine waters at Zuydcote Pass, and between the Dyck and Inner Ratel Banks, and at Inner Ratel.[17]

Map 10-3

Northern France. Dunkerque is also spelled Dunkirk. The port city of Boulogne is located about midway between Etaples and Cape Gris Nez.

The newly formed 51st Motor Launch Flotilla was collocated at HMS Beehive with the 4th Motor Torpedo Boat Flotilla. MLs *106* and *110* conducted the flotilla's first mission on the 28th of September, sowing the West Deep with twelve A Mk I-IV magnetic mines.[18]

50TH ML FLOTILLA AT DOVER COMMENDED
September marked a considerable increase in Coastal Forces minelaying, due to herculean efforts by three motor launches of the 50th Flotilla. Twelve of eighteen total operations were carried out by *ML 101, 103*, and *104* over fourteen consecutive nights (15-29 September 1941); earning a commendatory signal from the First Sea Lord, Adm. Sir. Dudley Pound, which expressed his appreciation for their daring, good seamanship and good maintenance.[19]

During this month, the waters of two new areas were visited—the Schoonveld off the Netherlands, and the Hook in the southwestern corner of Holland. Type A Mk I-IV magnetic mines were used, some of a new variety fitted with sterilisers to restrict the life of the field, and a more sensitive double-acting relay. The A Mk I-IV influence mines laid by aircraft and Coastal Forces became more sophisticated over the course of the war. These type mines are triggered by the influence of a ship or submarine, rather than direct contact. The A Mk I-IV evolved from magnetic to acoustic to combination magnetic-acoustic. The Mk XIX moored contact mine was also laid for the first time during a mission on 24 September.[20]

BAD WEATHER IMPACTS LATE-YEAR OPERATIONS
October brought bad weather, reducing Coastal Forces operations to a single mine lay, mid-month, between the coastal cities of Etaples and Boulogne in northern France. Poor weather persisting through the remainder of 1941 resulted in only four operations in November and another four in December; with the 50th ML Flotilla responsible for seven of the total eight lays.[21]

MINE DAMAGE TO GERMAN SPERRBRECHER
In total, over the course of the year, 589 mines were laid in enemy coastal channels, from the Hook of Holland to Boulogne. The early success in sinking the *Wolf* had not been repeated for several months, but in June, the first German Sperrbrecher minesweeper was damaged. Sperrbrecher, which literally means "barrier breaker" referred to ships used as blockade runners or, in this case, to sweep mines. To counter the effectiveness of these ships against their magnetic ground mines, the British laid in each field a number of mines having such a coarse relay setting (to complete the detonating circuit) that only the powerful magnetic field generated by these type vessels could activate them, and only when directly above them.[22]

Photo 10-4

German Sperrbrecher, location and date unknown.
www.german-navy.de/pics/postcards/saar_001.jpg

Through the remainder of 1941, the laying of magnetic mines in narrow and well-defined swept channels leading into enemy ports and harbours resulted in the damage of another eight sperrbrechers as well as the sinking of the one in June. It is likely that duty aboard these German mine clearance ships was not highly sought after. By 1940, magnetic mines laid by British aircraft during the night, posed the biggest threat for enemy ships entering or leaving their bases. To offset this threat once it was recognised, sperrbrechers were equipped with a VES-System, basically a huge magnetic field generator that could explode magnetic mines at a safe distance—this worked until the British adjusted the sensitivity settings on their mines.[23]

Coastal Forces magnetic mines, combined with those laid by aircraft, were responsible for 50-percent of the Sperrbrecher barricade-breaking force being out of action during the last months of 1941. Compounding the mine clearance challenges of magnetic mines facing the Germans, were those of moored mines laid by motor launches. The enemy used R-boats to undertake this work; unfortunately, they were not available in sufficient numbers, owing to heavy casualties from air attacks. The R-boats were often used for a variety of tasks other than minesweeping, including minelaying, but they were designed as minesweepers.[24]

Photo 10-5

German R-Boats operating off Norway during World War II.
Naval History and Heritage Command photograph NH 61828

About half of the over 100 sperrbrecher (mainly former merchant ships of about 5,000 tons) were lost in their dangerous role in the war. Following the end of hostilities, some of the surviving ships returned to service as merchant vessels.[25]

ROYAL NORWEGIAN NAVY EXILED IN ENGLAND

When the war in Norway came to an end in June 1940, most of the small Norwegian Navy had been sunk or put out of action. King Haakon VII (born Christian Frederik Carl Georg Valdemar Axel) and his government left Norway aboard the heavy cruiser HMS *Devonshire* on 7 June 1940 for Great Britain. What was left of their navy sailed for England to take part in the war against Germany. All the units flew the Norwegian naval ensign but were under British operational control.[26]

Thirteen naval ships, and about 500 officers and men left for Britain with the king. Of these ships only one, a small destroyer, was relatively modern having been launched in 1936. A second destroyer was hopelessly old (launched in 1908). Both destroyers were quickly pressed into service, while the remaining eleven, mostly old ships—patrol vessels and fishery protection vessels—were re-equipped to function as a minesweeping flotilla ordered to Dundee in Scotland. This flotilla, later reinforced by other Norwegian minesweepers, was tasked with keeping coastal convoy routes swept clear of mines near this location.[27]

After establishing its headquarters in London, the Norwegian Navy was able to assist British naval authorities in addressing a general shortage of minesweepers, patrol vessels, and auxiliaries beyond those previously described. At the time of Germany's invasion of Norway, a large Norwegian whaling fleet was in the Antarctic, and could not return home following the whaling season. The whalers were sturdy, seaworthy ships of 300-500 tons, well suited for naval service. As such, large numbers were requisitioned by the Norwegian government. Sixteen were converted to minesweepers and patrol vessels manned by the Norwegian Navy. Another forty-nine were transferred to the Royal Navy (some of these were later reinstated in the Norwegian Navy), and others joined the Royal Canadian Navy.[28]

Photo 10-6

Norwegian sailors in Nova Scotia, Jorgen Steiger Kristensen is in the back row. Courtesy of his daughter, Freda Martin

When Norway fell it had a major merchant navy with over 1,000 ships at sea. Their government in exile ordered them to go to a British or Allied port. During the spring and summer of 1940, seven factory ships and 22-23 whale catchers with upwards of 2,000 men arrived in

Halifax. The whalers were converted into patrol vessels and minesweepers for the Royal Norwegian and Royal Canadian navies. While the ships in their new roles required fewer men for their new duties, the extra ones couldn't go home and weren't needed at sea. Accordingly, Norwegian authorities paid for and built a camp in Lunenburg, Nova Scotia, about sixty miles southwest of Halifax, where the unemployed whalers would live under Norwegian military control.[29]

"CAMP NORWAY" IN CANADA

The facility, aptly called "Camp Norway," was opened on 29 November 1940. It consisted of a barracks to house about 800 men. Later, a mess hall, two storage buildings, a garage, and a carpentry shop were added. Much of the work was done by Norwegians since there were carpenters, pipefitters, and other tradesmen in the whaling crews. The camp served as a Royal Norwegian Navy training depot for seamen and whalers who were being taken into the navy. It trained about 450 men as crewmen for the converted whale catchers and other vessels, and 635 gunners for armed merchant ships.[30]

Map 10-4

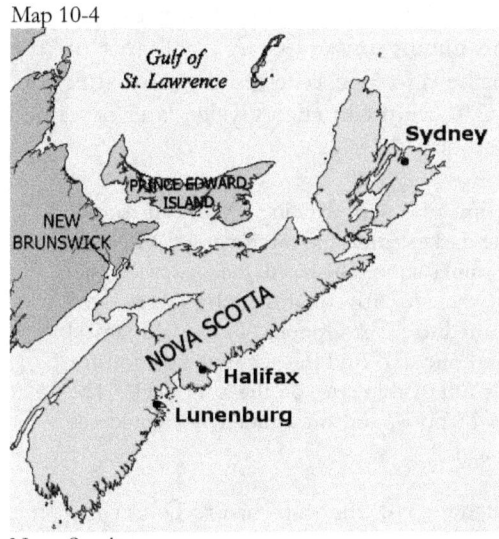

Nova Scotia

OPERATIONS BY THE 52ND ML FLOTILLA IN 1942

During the first half of 1942, as Coastal Forces continued to sow mines in enemy waters, a new flotilla was established for duties with the Portsmouth Command. Units of the 52nd Motor Launch Flotilla—MLs *123*, *125*, *128*, and *210*—were manned entirely by officers and men of the Royal Norwegian Navy. The 52nd Flotilla was tasked with mining waters along the coastal route between Cap de la Hague, the tip of the Cotentin Peninsula in Normandy, and Point de Barfleur in the lower Normandy region of France. However, after the flotilla performed its first mission on 6 June (KB1), the series was discontinued, and it was transferred to the Dover Command to operate with the 50th Motor Launch Flotilla.[31]

Photo 10-7

Officers and men of the Norwegian 52nd Motor Launch Flotilla at Dover. Petty Officer Telegraphist Finn Christian Gysler is at the top left.
Courtesy of Colin Bernard Torlief Stromsoy

Over the remainder of June, 52nd Flotilla motor launches operating from Dover carried out nine mining missions. A small area off the Outer Ruytingen Bank, in northern France, received the most attention. The telegraphist aboard *ML 210*, Finn Christian Gysler, later described such operations in French waters:

> The minelaying could at times be very exciting. We went out at dusk so we could have the darkest time for minelaying. When we got to the point where the mines should be layed, the Coxswain [sic] had to remove all the safetypins, to prove that the mines were layed alive. They had to be returned to [T/A Gunner (T) Donald James] Pangborne on return to harbour. To find the accurate position for the minefield, we had reels full of thin wire on the after deck. The wire was dropped at "No 1" buoy, and on a meter, we could see how many yards we had sailed.
>
> When approaching the enemy coast, the ships had to be in close contact. Signal lamps and radio could not be used, we actually had to talk to each other ship to ship. At this point, we were going at slow speed, "Dumbflows" on, hardly a sound could be heard from our ships.
>
> The order could be, either to lay mines in enemy convoy routes, or to block enemy [harbour] entrances. The Germans had no radar, or we could not have done what we did. But they had hydrophones

[to listen] placed in certain positions. When we came inside the range of these, searchlights played up, and they opened fire at us. We were never hit from shore batteries. All the shells seemed to go well above us. At this moment we knew that the E-boats would be alerted, and ordered to intercept, so many times we laid a smokescreen, and disappeared.

As the wireless operator on the Senior ship, I often received a coded signal from VAD (Vice Admiral Dover) to steer courses clear of the enemy track. What we did not know, at times, was the position of enemy minefields.[32]

Photo 10-8

HMS Vernon's MX Wrens (Women's Royal Naval Service members) assembling special anti-minesweeper devices for use with British mines. M stood for Mining; X for anything militarily secret, special or experimental. Collection of Rob Hoole

Over the course of seven ("NP" series) operations, forty-six moored mines were sown off the Outer Ruytingen Bank, with the field heavily guarded by "sweep-obstructors" emplaced at the same time. These devices were mixed in with ordinary mines to make sweeping more hazardous and were specifically intended to target Sperrbrecher minesweepers. These ships often had their cargo holds filled with buoyant materiel to aid in flotation should they hit a mine.[33]

ML 103 LOST WITH KILLED AND INJURED

On many nights during July, August, and September, when there was intense E- and R-boat activity in the Dover Command area of responsibility, the minelaying MLs were directed to back up area patrols.

Just before dawn on 24 August, motor gunboats engaged a large force of German light craft, some of which were laying mines in the swept channel off Dungeness.[34]

While en route to the scene to assist in the action, MLs *210* and *103* unwittingly crossed the laying track of the enemy, and *ML 103* (of the 50th Flotilla) struck a mine. The detonation set the boat aflame, and an explosion ensued. Fortunately, the explosion extinguished the flames, enabling HM trawler *Fyldea* to take the motor launch in tow. Nevertheless, she sank at 0730. The Norwegian *ML 210* had previously taken off the survivors, four badly burned. Two members of the crew—Stoker 1st Class Robert B. McKinna, and Ordinary Seaman Eric F. Glover—were logged as Missing, Presumed Killed.[35]

From July through September, enemy action and bad weather both played a restrictive role in the activities of Coastal Forces minelayers. Only eleven missions were completed. *ML 108* reported to the 50th Flotilla as a welcome reinforcement in September. The last quarter of 1942 was a busy one for minelayers—though bad weather precluded any operations in December. Operations were marred by the death of and injuries to crewmen aboard the Norwegian *ML 125* (52nd Flotilla) and significant damage to the boat from a mine strike. See below for details. On a separate occasion, enemy shore batteries opened fire on motor launches operating near Calais; which experienced several near misses but they fortunately emerged unscathed.[36]

ML 125 DAMAGED WITH MEN KILLED AND INJURED

On 2 November, motor launches *ML 101*, *102*, *104*, *125*, *210*, and *213* left Dover by dusk, fully loaded with mines to carry out an operation off the Outer Ruytingen. The weather was calm and clear and as the group approached the position where the mines were to be laid, speed was reduced and all hands took up their stations. Aboard *ML 210* the coxswain had removed all the safetypins, in preparation for minelaying. Suddenly at 2035, there was a terrific explosion, and the whole ship shuddered. *ML 125* had struck a drifting mine south of Sandettie Bank, which blew off her bow and destroyed a part of the wheelhouse.[37]

Telegraphist Gysler, aboard *ML 210*, later described efforts taken by the Norwegian and British to rescue survivors in the water:

> The stern was floating intact, with all the mines, which now had to be thrown overboard. A dinghy was lowered from *ML 213* and manned by S. Lt. Rolf Berntzen and [C]oxswain Knut Johannessen. Four men were floating around in the water. One of them was dead.

As they picked up the survivors, the dinghy capsized, and now they were all in the water.

One of the British ML's put a searchlight on, and picked up all the Norwegians. One of the British sailors, and the cook on *ML 210* [Olav Martin Stromsoy] jumped into the water, to help Rolf Berntzen secure lines around the most wounded.... *ML 213* took what was left of *ML 125* in tow, and we all went back to Dover. Here ambulances and doctors were on the jetty.[38]

Three men—the first lieutenant, telegraphist, and the British base echo-sounding officer—were killed. Four ratings were injured; one died later of his wounds. A list of the casualties follows:

Name	Status
Ty/Electrical Lieutenant Ivor L. Linay, RNVR	Missing, presumed killed
U/f Harold August Johannessen	Killed
U/dm Magne Sigvart Mortensen	Killed
U/dm Haakon Olav Austrheim[39]	Died of wounds on 6 November

ML 213 took what was left of *ML 125* in tow for Dover. Her stern was taken to Wellington Dock, and the motor launch was later rebuilt and returned to service. Debris from the exploded mine found aboard her revealed it was of German origin, thought to have broken adrift from a newly laid E-boat field.[40]

Photo 10-9

Photo of the *ML 125* incident from Finn Christian's papers. Courtesy of Colin Bernard Torlief Stromsoy

Photo 10-10

Norwegian sailor Olav Martin Stromsoy.
Courtesy of Colin Bernard Torlief Stromsoy

TALLY FOR 1942

During the whole of 1942, motor launches and motor torpedo boats of the Coastal Forces laid a total of 1,164 moored and ground mines, together with 186 sweep obstructors. Twenty-seven enemy ships, totaling 13,219 tons were sunk, and twelve (19,836 tons) damaged. Of these thirty-nine casualties, twenty-one were German minesweepers: seven sperrbrechers sunk and another eight damaged, and five R-boats sunk and one damaged.[41]

11

British Offensive Mining in Northwest Europe in 1942

During 1942, offensive surface ship and submarine mining reached its peak in February—primarily owing to concentrated efforts by the fast minelayers HMS *Manxman* and *Welshman*. At the end of January, the Admiralty received intelligence suggesting that the Kriegsmarine Heavy Squadron would shortly be leaving Brest, France, to return to Germany. Ordered to mine the squadron's expected route, the two minelayers laid down six fields between the 3rd and 9th of February. RN destroyer minelayers had sown two fields off Brest the preceding year, as a threat to the German battleships *Scharnhorst* and *Gneisenau*, which had arrived there at the end of March.[1]

Photo 11-1

German battleships *Scharnhorst* (left) and *Gneisenau*, circa spring or early summer 1939. Naval History and Heritage Command photograph #NH 97537

This tasking completed, *Manxman* and *Welshman* continued work previously begun to sow fields in the Bay of Biscay (intended to target U-boats operating from French ports); and laid another five by 21 February. After this date, *Welshman* began a refit on the Tyne while *Manxman* sailed to the Mediterranean to relieve the damaged *Abdiel* in the Eastern Fleet. (Earlier in February, *Abdiel* had grounded on departure from Port Anson, damaged her port shaft, and returned to Colombo on one engine.) On completion of her refit, *Welshman* laid three additional fields in the Bay of Biscay, extending the mined area southward to coincide with enemy shipping routes. *Welshman* was then transferred to the Mediterranean as well, essentially bringing offensive mining by surface ship minelayers in northwest Europe to an end until the spring of 1944.[2]

Following repair and refit at a commercial shipyard at Tyne in northeast England, *Abdiel* laid a single field, in December 1942, off Les Heaux de Brehat (an island and site of a lighthouse), as part of her post-refit workup before returning to the Mediterranean.[3]

Map 11-1

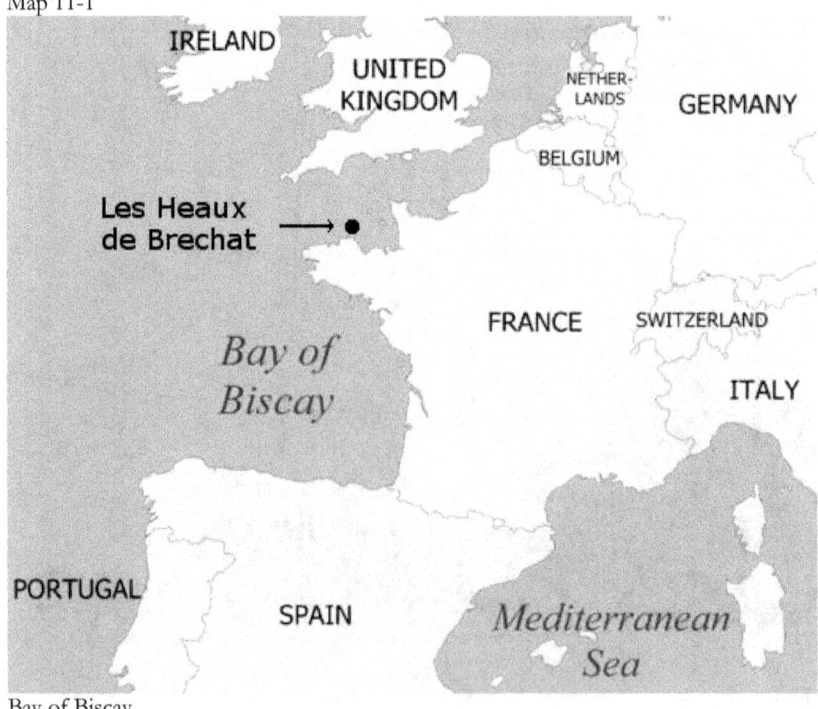

Bay of Biscay
https://commons.wikimedia.org/wiki/File:Bay_of_Biscay_map.png

SINGLE ROYAL NAVY MINELAYING SUBMARINE
The Free French submarine *Rubis* was the only minelaying submarine available for operations in Home Waters in 1942. She laid four fields in the southernmost areas of the Bay of Biscay, against blockade runners and the Spanish iron ore trade (supplying Germany with wolfram ore, a vital material necessary for production of armaments); one in the German Bight; and two off the west coast of Norway.[4]

The latter operations, and five visits to the Karmsund (a strait separating the island of Karmøy and mainland Norway) by aircraft of Bomber Command, comprised the general mining plan attention paid to that area during the year.[5]

RAF TASKED WITH AERIAL MINING IN NW EUROPE
The Royal Air Force's Bomber Command, in March 1942, assumed responsibility for all air minelaying in northwest Europe. This initiation followed the previous month's introduction into service of new long-range multi-mine carrying aircraft—increasing the capability of Bomber Command. RAF Coastal Command, which had previously acted as the coordinating agent for all sea mining operations, took no part in these, except for a small contribution between August and December, by Swordfish aircraft and crews on loan to them by the RN.[6]

Because bombing tasks took priority over mining, mining was generally restricted to nights when the weather was unsuitable for bombing on the continent. Nevertheless, from May onward, an average monthly laying rate of over 1,000 mines was achieved, largely carried out by RAF bombers, with the number of sorties rising almost four-fold over what had been previously achieved. About seventy percent of the mine loads were deposited in the Bay of Biscay and the North Sea. The overall increase in operations, and availability of long-range aircraft, also facilitated mining in the Baltic—which previously had been immune from attack because of the range of the aircraft employed.[7]

Important German U-boat trials and training areas in the Gulf of Danzig were first mined in May. In February, in anticipation of a breakout by German warships from the Royal Navy-blockaded French port of Brest, RAF aircraft laid mines near Terschelling (an island in the northern Netherlands) to cover the expected route of the Squadron.[8]

BREAKOUT FROM BREST BY GERMAN "HEAVIES"
The German battleships *Scharnhorst* and *Gneisenau* had entered Brest in March 1941, following a number of successful attacks on British Atlantic convoys. They were joined there by the heavy cruiser *Prinz*

Eugen. During their time at Brest, the ships were under constant attack by Bomber and Coastal Command aircraft. Furthermore, Hitler had demanded additional naval protection for occupied Norway, as he feared a possible British invasion. For these reasons, the German Naval Command ordered that all three heavy warships return to Germany.[9]

A plan was devised, code named Operation CERBERUS, for the principal ships and escorting destroyers to break out of Brest and proceed up the heavily guarded English Channel and through the Dover Strait to Germany. In anticipation of such a move, the British had a counterplan, Operation FULLER, by which the Royal Navy and Royal Air Force would block the threat of such a breakout. The British believed the German ships would sortie from Brest by day and make the passage through Dover Strait by night; and as a consequence, the RAF flew constant patrols to monitor the ships at Brest and provide early warning of their sailing.[10]

Photo 11-2

USS *DD-939*, the former German destroyer *Z-39*, under way off Boston, Massachusetts, on 22 August 1945. She, like *Z-25* and *Z-29*, was a *Z23*-class destroyer.
Naval History and Heritage Command photograph #NH 75374

This expected action occurred, but not as anticipated by the British. VAdm. Otto Ciliax, commander, Battleships (Befehlshaber der Schlachtschiffe) initiated the breakout from Brest at night, at 2230 on 11 February. A six-ship destroyer screen took up position while torpedo boats swept mines blocking the port, and protection from over 200 Luftwaffe day and night fighters was readied to supply air-cover. By 2345, the three major ships were under way, and by 0130 the entire group was outbound in the channel making 30 knots. The destroyer *Z-29* led, with the *Scharnhorst* next in line, followed by *Gneisenau* and *Prinz Eugen*. Two destroyers were positioned along both flanks with destroyer *Z-25* covering the rear.[11]

NO. 825 SQUADRON SENT TO ATTACK WARSHIPS

The navy will attack the enemy whenever and wherever he is found.

—Response by the British First Sea Lord Dudley Pound to insistence by Lt. Comdr. Eugene Esmonde that sending his Swordfish crews to attack the heavily armed ships and their fighter cover in broad daylight would be tantamount to a suicide mission.[12]

Luck was with the Germans, for the breakout, in darkness and bad weather, proceeded as planned. The inclement weather and faulty British radar allowed the enemy ships to transit 300 miles up the English Channel undetected. Their good luck continued when these conditions hindered operations by Bristol Beaufort bomber squadrons at airfields along the northern coast.[13]

As the German ships passed Dover, shore battery fire opened, but was ineffective. As torpedo bombers approached, intense anti-aircraft fire from the destroyer screen forced the aircraft to release their ordnance two miles distant of their targets, with no direct hits made. Six Swordfish torpedo planes were lost in the attack to German land-based fighter support. The British also sent out 242 bombers to attack *Scharnhorst* and the other ships. Of these, at least 39 were able to drop their bombs, but none found their target.[14]

Lt. Comdr. Eugene Esmonde, in command of No. 825 Squadron of the Fleet Air Arm, had learned the morning of 12 February that *Scharnhorst*, *Gneisenau* and *Prinz Eugen*, escorted by some thirty surface-craft (E-boats had joined the group) and fighter cover, were entering the Straits of Dover. His orders were to attack the ships before they reached the sandbanks northeast of Calais, France.[15]

As part of Operation FULLER, the six Swordfish of the squadron had flown from their base at Royal Naval Air Station Lee-on-the-Solvent to RAF Manston (located in Kent on the Isle of Thanet) on 4

February. The squadron had been on five-minute standby on the 11th to react to any enemy attempt to break out of Brest—but was stood down as there was no perceived threat that such would occur.[16]

In early afternoon on the 12th, in response to the news of the breakout, Esmonde and his six Swordfish took off at 1225 from Manston. The squadron's two flights of torpedo bombers were to rendezvous with three RAF Fighter squadrons, but only ten Spitfires arrived. The British single-seat fighter aircraft were soon engaged with the German fighter escort, leaving the Swordfish vulnerable and open to attack. Within ten minutes of taking flight, the squadron was set upon by Luftwaffe fighters. As Esmonde led his squadron on, in steady flight toward the target, all six planes were shot down by Messerschmitt Bf 109 fighters, with thirteen of eighteen aircrew members lost including Esmonde. (A pilot, an observer, and a radio operator/gunner comprised each Swordfish aircrew.)[17]

At dawn on 13 February, the German force entered Brunsbuttel in northern Germany (sited on the mouth of the Elbe River, near the North Sea) almost intact. *Scharnhorst* and *Gneisenau* had both been damaged by mines, and one of the E-boat escorts was lost, along with seventeen covering fighter planes. The Germans had successfully moved the squadron from Brest to the German port, though now the *Scharnhorst*, *Gneisenau*, and *Prinz Eugen* were bottled up northeast of Britain, limiting their immediate role in the Battle of the Atlantic.[18]

AWARDS FOR VALOUR

Lt. Comdr. Eugene Kingsmill Esmonde, RN, was awarded the Victoria Cross (posthumously) for this heroic attack. The other aircrewmen received either Gallantry Medals or Mention in Despatches.

Flight 1	Flight 2
Swordfish Aircrew	**Swordfish Aircrew**
Lt. Comdr. Eugene Esmonde, RN (killed)	A/Sub. Lt. John Chute Thompson, RN (killed)
Lt. Williams, RN (killed)	Sub Lt. Robert Laurens Parkinson, RN (killed)
Petty Officer Airman William Johnson Clinton, RN (killed)	Leading Airman Tapping (killed)
Swordfish Aircrew	**Swordfish Aircrew**
A/Sub. Lt. Kingsmill, RNVR (wounded)	Sub. Lt. Wood, RN (killed)
Sub. Lt. Samples (wounded)	A/Sub. Lt. Fuller-Wright, RNVR (killed)
Petty Officer Airman Bunce (wounded)	Leading Airman Henry Thomas Albert Wheeler, RN (killed)

Swordfish Aircrew	Swordfish Aircrew
A/Sub. Lt. Rose, RNVR (wounded)	Sub. Lt. Bligh, RNVR (killed)
A/Sub. Lt. Lee (wounded)	Sub. Lt. Beynon, RNVR (killed)
Petty Officer Airman Ambrose Lawrence Johnson, RN (killed)	Leading Airman Granville-Smith (killed)[19]

LOSS OF SWORDFISH EXPOSES WEAKNESSES

The entire combat action by the squadron from take-off to the downing of the last Swordfish lasted just twenty minutes, and none of the torpedoes dropped found their target. Five of the nine crewmembers in the first flight of three Swordfish were pulled from the water and survived. The second flight of Swordfish following it had no chance, and all nine crewmen were killed.

Photo 11-3

A Fairey Swordfish biplane being readied with a mine for a nighttime mission. This type aircraft could carry one 1,670lb torpedo or one 1,500lb mine mounted centreline, or eight 60lb RP-3 rockets (beginning with the Mk II series).
Australian War Memorial photograph 005693

After this tragedy, Swordfish were never again used as torpedo bombers. They were easy targets for fighter aircraft, and particularly vulnerable while setting up for that type of attack. The aircraft were relegated to mining missions, usually of enemy shipping channels. Overall losses were relatively light, because the Swordfish were primarily employed where they would not be opposed by land-based fighters.[20]

SUMMARY OF MINING IN NW EUROPE IN 1942

Following departure of the Heavy Squadron from Brest, air mining off the Biscay ports was directed mainly against U-boats. These mines sank two and damaged a third submarine. Separately, one U-boat was sunk and one damaged by air-laid mines in the western Baltic Sea. In total, 14,041 mines were laid offensively in northwest European waters in 1942. Of this number: aircraft accounted for 9,711, fast minelayers for 2,644, Coastal Forces craft for 1,164, submarines for 223, and HMS *Plover* for 299 mines.[21]

As a result of this greatly increased mining, casualties to enemy-controlled shipping rose steeply in 1942 to 203 vessels (70 sunk and 124 damaged)—mostly due to aircraft. A minefield laid by *Plover* in the English Channel sank a destroyer in January; the efforts of the fast minelayers claimed four small vessels damaged, and one sunk; and submarine-laid mines were responsible for the destruction of eight small vessels, one of them off the Norwegian coast. In the Channel and off the Netherlands coast, Coastal Forces garnered twenty-seven vessels sunk and twelve damaged.[22]

In areas under British air-mining attack, German commanders were put under intense pressure from February onward, particularly in the Kattegat and Baltic. Casualties rose by over 300 percent in these areas from the previous year, and an acute shortage of minesweepers led to frequent interruption of ship movements. Sperrbrechers were pressed into clearance and escort duties, being able to operate in adverse weather conditions which had precluded alternative forms of sweeping. Loss and damage of these vessels in the German Northern and Western Commands, to RN magnetic ground mines was particularly severe. The former command suffered seven sunk and twenty-two damaged, and the latter, seven sunk and twenty damaged.[23]

12

French Minelaying Submarine *Rubis*

> *I am offered candidates who, although very worthy and valiant, do not meet the very exceptional conditions which justify accession to the Order. That's why only 1036 people, 5 communes and 18 fighting units were awarded this prestigious decoration between January 1941 and January 1946.*

—Handwritten note from Gen. Charles de Gaulle to the Council of the Order dated 3 December 1945, attesting to the exceptional character of the the Cross of the Liberation. The Free French submarine *Rubis* was one of only eighteen Army, Air Force, or Navy units to earn the Croix de la Libération, second only to the Légion d'Honneur.[1]

Photo 12-1

Photo of Free French submarine *Rubis* in difficulty off Norway, taken by an unknown crewman aboard a Coastal Command aircraft.
Coastal Command, Ministry of Information, 1942

Following the breakout of the German heavy ships from Brest in February 1942, submarine patrols in the Bay of Biscay were effectively discontinued until mid-September. The Free French submarine *Rubis*, being unable to safely operate in northern waters during the period of short nights, carried out three minelays in the Bayonne-Bordeaux area. The *Rubis* was a *Saphir*-class submarine built by Arsenal de Toulon (Toulon, France), and commissioned on 4 April 1933. After serving at Toulon with the 7th and later 5th Submarine Squadrons, she had been transferred in 1937 to Cherbourg. At war's commencement, the modest sized, 216-foot submarine was at Bizerte, Tunisia.[2]

The profile of *Rubis* was characterised by a boat-like bow projection, tapered stern, a flat surface deck and centrally-located conning tower, and dive planes forward and low along her sides. Propelled by two diesel engines, she had a top speed of 12 knots when on the surface. One or more diesels operating as generators, were used to charge the submarine's batteries as necessary. Submerged, two battery-powered electric motors producing a combined 1,100 horsepower, enabled a top speed of 9 knots for a short period.[3]

Her armament consisted of a standard 75mm deck gun ahead of the conning tower, a twin-barrel 13mm gun aft, and two 8mm machine guns. She was also fitted with five torpedo tubes (three 21.7-inch tubes inside the boat and two external 15.7-inch tubes) and could carry thirty-two mines. Ship's complement was forty-two officers and men.[4]

PREVIOUS SERVICE IN NORWEGIAN WATERS

In April of 1940, *Rubis* operated with the British Home Fleet and laid mines off the Norwegian coast during the Norway Campaign (9 April-10 June 1940). Six Norwegian ships were sunk and one damaged by mines she sowed on 10 May, 27 May, and 9 June.

Norwegian Vessels Sunk or Damaged by *Rubis*-laid Mines

Date	Ship	Fate
26 May 40	Norwegian vessel *Vanso*	Sunk by a mine laid by *Rubis* on 10 May 1940 near Egersund, Norway
28 May 40	Norwegian (sailing) vessel *Blaamannen*	Sunk by a mine laid by *Rubis* on 27 May 1940 near Haugesund, Norway
31 May 40	Norwegian merchant vessel *Jadarland*	Sunk by a mine laid by *Rubis* on 27 May 1940 near Haugesund, Norway
10 Jun 40	Norwegian merchant vessel *Sverre Sigurdsson*	Sunk by a mine laid by *Rubis* on 9 June 1940 near Herdla, Norway
7 Jul 40	Norwegian merchant vessel *Almora*	Damaged by a mine laid by *Rubis* on 10 May 1940 near Egersund, Norway
24 Jul 40	Norwegian merchant vessel *Kem*	Sunk by a mine laid by *Rubis* on 10 May 1940 near Egersund, Norway
28 Jul 40	Norwegian merchant vessel *Argo*	Sunk by a mine laid by *Rubis* on 10 May 1940 near Egersund, Norway[5]

At the time of the French surrender on 22 June 1940, *Rubis* was at Dundee, Scotland, from which she had been operating. The officers and men serving in the French Navy faced a difficult choice. The armistice with Germany demanded the return of all naval units to their home country. French commanders could comply and face possible integration into the Kriegsmarine, wait out the war interned in port, or continue the fight as British auxiliaries. Lt. Georges E. J. Cabanier and most of his crew elected to join the Free French Forces and continue operations in Norwegian coastal waters and the Bay of Biscay.[6]

RUBIS SEVERELY DAMAGED, NEARLY SUNK

In August 1941, *Rubis* was ordered to mine several shipping channels about two miles off the Norwegian coast. The submarine's former first lieutenant, Henri L. G. Rousselot, was in command following Georges Cabanier accepting a new assignment in May. On 21 August, Rousselot carried out a minelay off Egerøy, Norway, and later sighted while gazing through the periscope, a lone tanker proceeding in the opposite direction. He ordered the number 3 external stern tube fired. There was a whoosh, and a countdown to expected target impact began, but seconds passed without an explosion.[7]

When queried about whether the torpedo had been fired, the crewman responsible insisted it had. Pressure gauges indicated the tube was open; but use of the periscope revealed no sign of a wake of a running torpedo. Then crewmembers detected a humming sound. The torpedo was stuck in the tube, running and armed—meaning that it would explode on contact. As long as *Rubis* didn't strike anything with her stern the submarine and her crew were safe—or so the crew hoped. Meanwhile, the lucky tanker disappeared over the horizon.[8]

Since nothing could be done about the torpedo, *Rubis* moved to her next area to mine and proceeded to do so. Immediately following the release of the final mine, a four-ship convoy emerged from the channel, steaming toward the submerged submarine. A pair of escort vessels were screening two merchant ships flying the swastika. Rousselot ordered action stations remanned and fired both bow tubes. After a short interval, a tremendous explosion rocked the submarine, causing a loss of lighting inside the boat, and *Rubis* angled upward steeply, threatening to breech uncontrolled. The Finnish merchant vessel SS *Hogland* had been torpedoed, but one or more warhead detonations so close nearby had damaged the submarine.[9]

Rousselot ordered the crew to move forward inside the submarine to level the boat. He then took the *Rubis* down, resting her on the bottom to await expected counterattacks. None developed, but the submarine remained on the sea floor for a lengthy period until any lurking hunters and/or rescue ships left the area. Nearing eighteen hours on the bottom, men started to feel lethargic, signifying that carbon dioxide was building to dangerous levels inside the submarine.[10]

Rousselot ordered ballast tanks blown to bring *Rubis* to the surface, but she merely scraped along the sea floor, then settled again. The use of a second set of ballast tanks produced the same result; normally, one set got the sub to surface. The explosion had pierced the outer hull, and as high-pressure air pushed water out of the tanks, sea water from outside the submarine flowed in. In desperation, Rousselot ordered all remaining ballast tanks blown at once, and this last-ditch attempt succeeded. *Rubis* surfaced in darkness just after 2100.[11]

Unable to dive, she had to escape out to sea as quickly as possible. This meant traversing a charted German minefield protecting the Norwegian coast. Just as *Rubis* had moved out of sight of land, heading away from the coast, her engines died, setting her adrift in the field. The batteries had started leaking acid and shorted out, and, as the acid combined with water in the bilges, chlorine fumes filled the interior, forcing the crew to take refuge on deck.[12]

The Admiralty response to Rousselot's distress report brought the news that British ships would not meet the *Rubis* unless she cleared the mined area. If she was unable to do so, Catalina flying boats would take off the crew, and the submarine would be scuttled. Fortunately, this proved unnecessary. Engineers in gas masks were able to get a few of the batteries to work and the engines sputtered to life. Slowly, the sub made it to open water, met by a cruiser, four destroyers, and two tugs. Refusing a tow, *Rubis* returned to Dundee under her own power.[13]

Several months later, the German *U-702* (Kptlt. Wolf-Rüdiger von Rabenau) fell victim to a Rubis mine, with the loss of all hands.

Finish Merchant Vessel and German U-boat Victims

Date	Ship	Fate
21 Aug 41	Finish merchant vessel *Hogland*	Torpedoed and sunk by *Rubis* off Egerøy, Norway
31 Mar 42	German submarine *U-702*	Sunk west of Denmark, by a mine laid by Rubis on 21 Mar 1942[14]

OPERATIONS OFF SOUTHWEST FRANCE

Rubis left Dundee on 27 May for the first of three lays in the Bayonne-Bordeaux area (designed to threaten ships running iron ore cargos between Spanish and French ports) under escort by the Dutch minesweeper HNLMS *Jan van Gelder*. The two parted company off Wolf Rock, southwest of Land's End, in Cornwall, and the sub proceeded alone to her operational area, diving by day. On the afternoon of 5 June, *Rubis* laid two groups of 16 Vickers T III mines each, within a radius of one mile about positions 43°37'N, 01°35'W and 43°46'N, 01°33'W. The moored mines were set at a depth of seven feet below the surface, with flooders to operate on 9 June.[15]

At completion, *Rubis* patrolled in the vicinity of Arcachon Point, off southwest France, for three days before returning to Portsmouth.

She sailed from there on 30 June to carry out a second mining operation between Arcachon Point and the mouth of the Gironde. She arrived off the Point on 7 July and laid thirty-one mines in three groups that day. *Rubis* then set a course for the Lizard, a peninsula in southern Cornwall, where she was to rendezvous with an escort. After waiting three hours, *Rubis* made Falmouth alone and was later escorted to Dartmouth, which lay on the English Channel in southwest England, and finally to Portsmouth. (The French sub-chaser *Chasseur 8* had been bombed and sunk while in passage.)[16]

On 8 August 1942, *Rubis* sailed from Portsmouth in the evening bound for a third lay off Arcachon Point. She deposited thirty-two Vickers T III units in eight groups of four in early morning on the 14th. On completion, she returned to the UK, meeting her escort off the Lizard and entering Falmouth. She then proceeded to Dundee and continued operations off the Norwegian coast.[17]

Victims of *Rubis*-laid Mines off Southwest France

Date	Ship	Fate
12 Jun 42	German auxiliary minesweeper *Marie Frans* (M 4212)	Sunk south of Vieux-Boucau, France, by a mine laid on 5 June by the *Rubis*
26 Jun 42	French tug *Quand Meme*	Sunk south of Vieux-Boucau by a mine laid on 5 June by the *Rubis*
10 Jul 42	German auxiliary minesweeper *Imbrim* (M 4401)	Sunk northwest of Arcachon, France, by a mine laid by the *Rubis*
18 Aug 42	German armed trawler *Hans Loh* (V 406)	Sunk southwest of the Gironde Estuary, France, by a mine laid on 14 August by the *Rubis*
20 Sep 42	German auxiliary minesweeper *Antoine Henriette* (M 4448)	Sank off Bayonne, France, after hitting a mine possibly laid by the *Rubis* on 5 June
10 Jul 43	German auxiliary minesweeper *Gauleiter A. Meyer* (M 4451)	Sunk off Arcachon, France, by a mine laid by the *Rubis*[18]

RUBIS SINKS TWENTY-THREE SHIPS IN THE WAR

Rubis continued to achieve great success fighting alongside the British. By war's end, she had laid 684 mines over 23 operations, and was responsible for sinking 21,000 gross register tons of shipping—more than the rest of the Free French Navy combined. Of the twenty-three vessels she sent to the bottom (twenty-two with her Vickers T III mines, and one by use of torpedoes), fourteen were German combatants.

Her most fruitful month was in December 1944 when a German merchant vessel, three submarine chasers, and a minesweeper all sank after striking mines laid by the *Rubis*.

Ships Sunk or Damaged off Norway by *Rubis*-laid Mines

Date	Ship	Fate
26 Sep 44	German auxiliary submarine chaser *Grönland* (UJ-1106)	Sunk 16 nautical miles southwest of Stavanger, Norway by a mine laid by *Rubis* on 24 Sep 1944
27 Sep 44	German auxiliary submarine chaser *Lesum* (UJ-1715); German merchant vessel *Cläre Hugo Stinnes*; Norwegian merchant vessel *Knute Nelson*	All three vessels sunk 16 nautical miles southwest of Stavanger, Norway by mines laid by *Rubis* on 24 Sept 1944
27 Oct 44	German patrol vessel *Seehund* (V 5304)	Heavily damaged after hitting a mine laid by *Rubis* on 18 Oct 1944
24 Nov 44	Norwegian merchant vessel *Castor*	Damaged after hitting a mine off Egersund, Norway, laid earlier that day by the *Rubis*
21 Dec 44	German merchant vessel *Weichselland*; German auxiliary submarine chasers *UJ-1113*, *UJ-1116*, *UJ-1702*; and the small German minesweeper *R-402*[19]	All five vessels sunk off Feiestein-Rinne, Norway, after hitting mines laid by the *Rubis* on 19 Dec 1944

HONOURS

As a result of *Rubis'* exemplary service with the Free French Naval Forces, she was made a companion of the Ordre de la Libération by a decree issued by Gen. Charles de Gaulle on 14 October 1941.

Medal of the Order of Liberation, called the Croix de la Libération ("Cross of Liberation"), established by Gen. Charles de Gaulle on 16 November 1940. Second only to the Légion d'Honneur (Legion of Honour), the Order of Liberation was awarded to heroes of the Liberation of France during World War II. *Rubis* was one of three Naval units to receive it, the others being the corvette *Aconit*, and the 1er Régiment de Fusiliers Marins.

Following service in the French submarine pavilion, then the Free French Naval Forces during the Second World War, and back with the French Navy, *Rubis* was struck on 4 October 1949. She was sunk off the French coast on 31 January 1958 for use as a sonar target.[20]

13

U.S. Army and Navy Minelaying Forces

Torpedoes [mines] are not so agreeable when used by both sides; therefore, I have reluctantly brought myself to it. I have always deemed it unworthy of a chivalrous nation, but it does not do to give your enemy such a decided superiority over you.

—Rear Adm. David G. Farragut, USN, March 1864

In autumn 1942, the U.S. Navy despatched Atlantic Fleet minesweepers and minelayers to the African Theatre for the Allied invasion of French North Africa. Thereafter, mine forces supported amphibious landings in the Mediterranean and European theatres. Navy tugs and salvage ships accompanied invasion forces to save vessels damaged in combat. They also worked with minesweepers and clearance divers to open harbours critical to sea-supplied support of Allied troops ashore. (Comdr. Bruhn's book, *We Are Sinking, Send Help!* describes efforts of tugs, salvage ships, USN divers, and British Port Clearance "P" Parties to open harbours.)

This chapter introduces the U.S. Navy's surface minelaying force, which consisted of 7 minelayers (CM), 20 light minelayers (DM), 3 coastal minelayers (CMc), and 10 auxiliary minelayers (ACM). Although the combined numbers total 40, only 37 ships were involved as there is some double-counting because of vessel reclassifications. Coastal minelayers *Monadnock* and *Miantonomah* were reclassified as minelayers, and later, *Monadnock* a second time to auxiliary minelayer. Of this force, only about a dozen minelayers operated in Europe, Africa and the Med. Seven, identified in the table, earned battle stars: two ex-cargo ships, two ex-ferries, and three smaller former U.S. Army mine planters.

Battle stars earned by U.S. Navy ships were similar to Royal Navy battle honours. The criteria which allowed crewmembers to affix a star to the campaign ribbon on their uniform blouses was defined in the U.S. Navy and Marine Corps Awards Manual:

The prerequisite to the wearing of a star on an area service ribbon shall be honorable service in a ship, aircraft unit or shore-based force at the time it participated in actual combat with the enemy. In instances in which the duty performed did not result in actual combat with the enemy but is considered equally hazardous, the Chief of Naval Operations may award an operation of engagement star to the units concerned.

U.S. Navy vessels could receive only a single battle star for any particular operation, no matter how many individual combat actions. Royal Navy battle honours were (and are today) awarded to the ship's name rather than to the hull itself, so that the honour lives on in future ships of the same name, long after the one that earned it rests in the depths—or has met its demise in a breaker's yard.[1]

U.S. Navy Minelayers Awarded Battle Stars

North African Occupation: Algeria-Morocco Landings

Ship	Period	Commanding Officer (s)
Miantonomah (CM-10)	8-11 Nov 42	Lt. Comdr. Raymond Dorsey Edwards, USN
Monadnock (CM-9)	8-11 Nov 42	Lt. Comdr. Frederick O. Goldsmith, USNR

Sicilian Occupation

Ship	Period	Commanding Officer (s)
Salem (CM-11)	6-15 Jul 43	Comdr. Henry Goodman Williams, USN
Weehawken (CM-12)	9-15 Jul 43	Lt. Comdr. Robert Edwin Mills, USNR

Invasion of Normandy

Ship	Period	Commanding Officer (s)
Chimo (ACM-1)	6-25 Jun 44	Lt. Comdr. John Winston Gross, USNR
Miantonomah (CM-10)	6-25 Jun 44	Comdr. Austin Edward Rowe, USNR

Invasion of Southern France

Ship	Period	Commanding Officer (s)
Barricade (ACM-3)	15 Aug- 25 Sep 44	Lt. Charles Percy Haber, USN
Planter (ACM-2)	15 Aug- 25 Sep 44	Lt. Theodore Thomas Scudder Jr., USNR; Lt. Richard Albert Knapp, USN

U.S. ARMY MINE PLANTER SERVICE

The service of auxiliary minelayers *Chimo*, *Planter*, and *Barricade* began with the U.S. Army Mine Planter Service, which was established near the end of the First World War. Founded on 22 July 1918 and placed under the U.S. Army Coast Artillery Corps, the Mine Planter Service was responsible for installing and maintaining underwater minefields as part of the armament of U.S. coastal fortifications, including those at the approaches to the Panama Canal, and associated with the defence of Manila Bay in the Philippines.[2]

Prior to 1918, the Coast Artillery Corps had been responsible for mine defences in major ports. The vessels it employed to lay and maintain minefields ranged in size from small motorboats to 1,000-ton ships. The ships were operated day-to-day by civilian mariners, under the command of a Coast Artillery officer, with embarked enlisted mine specialists carrying out the technical aspects of mining operations. However, conflict between the soldiers and the civilians revealed the need for military personnel to operate the vessels. Friction developed in particular over civilian officers and crews leaving during operations to take other employment.[3]

In addition to establishing the U.S. Army Mine Planter Service, the act of July 1918 provided for army personnel operation of vessels. It introduced the rank and grade of warrant officer, and directed that warrant officers serve as masters, mates, chief engineers, and assistant engineers of each mine planter. In preparation for its personnel manning the vessels, the Army opened a school to train "soldier mariners" at Fort Monroe, Virginia, commanded by an officer who was a Naval Academy graduate. (The Warrant Officer Personnel Act of 1954 eliminated the U.S. Mine Planter Service.)[4]

Warrant officers in the Mine Planter Service wore simple strips of brown cloth on their uniform sleeves as their insignia of rank. There were three pay levels authorised. Masters wore four stripes, and those lesser in rank, fewer stripes—with deck officers sporting an embroidered brown fouled anchor above their braid, and engineer officers a brown three-bladed propeller.[5]

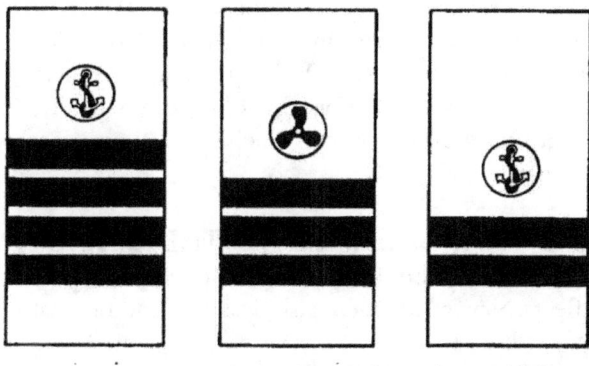

Master   Assistant Engineer   Second Mate

PRE-WORLD WAR II U.S. ARMY PLANTERS

Photo 13-1

Contact mines aboard the Army mine planter *Gen. E. O. C. Ord*, circa World War I. Naval History and Heritage Command photograph #NH 91709

The Mine Planter Service operated ships designated USAMP (U.S. Army Mine Planter) to lay fields of controlled mines to help guard (in conjunction with shore guns) the approaches to coastal fortifications. The controlled mines were designed to be remotely detonated from shore upon the sighting of an enemy ship within lethal range of one or more of the sea-emplaced weapons. In American ports and harbours so protected, mine planters, and smaller vessels assisting them with the work, were assigned to parent mine flotillas.[6]

Of the nineteen mine planters constructed between 1904 and 1937, nine of the ships continued their service into, or beyond World War II; eight for the Army and one, *Dekanawida* (YT-334), as a Navy harbour tug. The other ten ships were scrapped, sold commercially, or transferred to the U.S. Lighthouse Service, which in 1939 was merged into the U.S. Coast Guard.[7]

MINE PLANTERS TRANSFERRED TO THE U.S. NAVY

In 1940, as war loomed closer, the Army ordered sixteen new planters to replace its aging fleet. New mines were also designed and built, along with a few additional facilities to support their use. Following America's entry into the war in December 1941, the Army planted defensive minefields at major harbours in the continental United States. During the war, responsibility for the development and supply of mine materiel was transferred from the Army Coast Artillery Corps to the Ordnance Department, and ground mines (which rest on the seafloor) largely replaced moored mines (which are suspended in the water column).[8]

TEN ACM MINELAYERS SERVED IN THE WAR

Photo 13-2

USS *Chimo* (ACM-1) passing movies to USS *Obstructor* (ACM-7) via light line while en route to Saipan, September 1945.
Naval History and Heritage Command photograph #NH 86268

Sixteen 188-foot, 1,315-ton mine planters were constructed by Marietta Manufacturing, Point Pleasant, West Virginia. Fourteen were eventually transferred to the U.S. Navy—nine late in the war, and five others after the war. Eight of the nine were commissioned and served during the war: *Chimo* (ACM-1), *Planter* (ACM-2), *Barricade* (ACM-3), *Barbican* (ACM-5), *Bastion* (ACM-6), *Obstructor* (ACM-7), *Picket* (ACM-8), and *Trapper* (ACM-9). (*Buttress* ACM-4—the former escort patrol craft USS *PCE-878*—*Monadnock* ACM-10, initially designated CMc-4 and subsequently CM-9, accounted for the Navy's other two ACMs.)

Sixteen U.S. Army Mine Planters Constructed in 1942-1943
(Fourteen were later transferred to the U.S. Navy)

Mine Planter	Built	Disposition
Gen. Henry Knox (MP-1)	1942	To USN 1945 as *Picket* (ACM-8), to USCG 1946 as *Willow* (WLB-332), struck 1969, later crab processing plant *Royal Alaskan*
Col. Henry J. Hunt (MP-2)	1942	To USN 1945 as *Bastion* (ACM-6), to USCG 1946 as *Jonquil* (WLB-330), struck 1969
Col. George Armistead (MP-3)	1942	To USN 1945 as *Barbican* (ACM-5), to USCG 1946 as *Ivy* (WLB-329), later *Agnes Foss*
Gen. Samuel M. Mills (MP-4)	1942	Sold 1960
1Lt. William G. Sylvester (MP-5)	1942	To USN 1945 as *Obstructor* (ACM-7), to USCG 1946 as *Heather* (WLB-331), struck 1967
Brig. Gen. Henry L. Abbott (MP-6)	1942	Later cargo ship 1951, *Neptune* 1969, sank 1975
Maj. Gen. Wallace F. Randolph (MP-7)	1942	To USN 1951 as *Nausett* (ACM-15), sold 1961 as *Sea Searcher*, later *Thunderbolt*, reefed 1986 off Marathon, Florida
Col. John Story (MP-8)	1942	To USN 1944 as *Barricade* (ACM-3), to USCG 1946 as *Magnolia* (WLB-328), sold 1971 as *Galaxy*, burned and sank 2002
Maj. Gen. Arthur Murray (MP-9)	1942	To USN 1945 as *Trapper* (ACM-9), to USCG 1946 as *Yamacraw* (WLB-333), to USN 1959 as ARC-5, scrapped 1967
Maj. Gen. Erasmus Weaver (MP-10)	1942	To USN 1949 as *Canonicus* (ACM-12), sold 1948, later *New Jersey* 1963
Maj. Samuel Ringgold (MP-11)	1942	To USN 1951 as *Monadnock* (ACM-14), sold 1960 as *Tahiti*, later *Amazonia* 1983, *Dear* 1984, *Maxim's Des Mers* 1986
Brig. Gen. Royal T. Frank (MP-12)	1943	To USN 1944 as *Comanche* (ACM-11), sold 1948, later *Pilgrim* 1961, *Cape Cod* (date unknown)
Col. Alfred A. Maybach (MP-13)	1943	To USN 1951 as *Puritan* (ACM-16), sold 1961
Col. Horace F. Spurgin (MP-14)	1943	To USN 1950 as *Miantonomah* (ACM-13), sold 1960 as *Nautilus*, later *Aleutian Mist*, *New Star* 1991, scrapped 2013
Col. Charles W. Bundy (MP-15)	1943	To USN 1944 as *Chimo* (ACM-1), sold 1948, later tuna seiner *Day Island* 1963
Col. George Ricker (MP-16)	1943	To USN 1944 as *Planter* (ACM-2), sold 1948 to Foss, later *San Juan* 1963, *Purple Aster*, *Tiger Fish* 2[9]

SEVEN U.S. NAVY DEEP-WATER MINELAYERS (CM)

With the exception of *Terror*, the U.S. Navy's deep-water minelayers were all converted civilian vessels; one ex-passenger/cargo ship, two ex-cargo ships, and three former ferries. USS *Oglala*, built in 1907 as a passenger/cargo ship, had served in WWI as the minelayer *Shawmut*

(CM-4). She was the fleet's principal minelayer into the early 1940s, despite many deficiencies associated with her civilian origin and advanced age. In recognition of these shortfalls, the Chief of Naval Operations had written to the Secretary of the Navy on 11 September 1936 that the only minelayer then in service was not in a satisfactory condition to continue in commission much longer without a large expenditure of funds and then would not be entirely satisfactory.[10]

Following its consideration of input from Commander-in-Chief, United States Fleet, and those of mine warfare commands and experts, the General Board called for a ship not to exceed 7,000 tons standard displacement, carrying 600 assembled mines on at least four under-cover tracks, plus not less than 300 unassembled mines stowed in holds with assembly space for these. The board further suggested: a speed of at least 18 knots sustained, with an endurance of at least 10,000 miles at 15 knots; and an armament of four 5-inch/38-calibre dual-purpose guns. SecNav approved this recommendation on 17 November 1937.[11]

Photo 13-3

USS *Terror* (CM-5) at sea, 1942-1943; the only U.S. Navy purpose-built minelayer. National Archives photograph 80-G-411681

Terror was laid down at the Philadelphia Navy Yard on 3 September 1940 and commissioned on 15 July 1942. After participating in laying a defensive minefield off the port at Casablanca (as part of the support and reinforcement of Operation TORCH, the Allied invasion of French North Africa), the minelayer returned to America's East Coast. Here she received four 40mm quad mounts at the Navy Yard in Norfolk,

Virginia, in April 1943. These anti-aircraft weapons replaced her original two 1.1-inch quad mounts and augmented her 20mm gun battery (which was also increased and rearranged at the same time, resulting in fourteen 20mm guns), greatly improving her self-defence capabilities.[12]

Terror was sent to the Pacific in October 1943, and used to transport mines, conduct defensive mining, and assist in cargo and personnel movements. She was fitted out as the flagship for commander, Minecraft, Pacific Fleet in December 1944, and during the Iwo Jima and Okinawa invasions in February-May 1945, served as flagship and support ship for all the mine craft in the operations. At Okinawa on 1 May 1945, *Terror* suffered 171 casualties (41 dead, 7 missing, and 123 wounded) while sustaining damage from a kamikaze attack. Following repairs in San Francisco, she returned to duty in the Pacific until relieved in December 1945 by the amphibious command ship *Panamint* (AGC-13) as flagship for commander, Minecraft, Pacific Fleet.[13]

In need of minelayers—in addition to *Oglala* and *Terror*—capable of carrying large numbers of mines to lay deep-water fields, the Navy obtained and converted two ex-cargo ships and three ex-ferries for this purpose. Summary information about all seven minelayers is provided in the table. *Keokuk* was laid down in 1914 and took up minelayer duties on 18 May 42. The letters "LD" in the commissioning date column refer to the year in which she and others were laid down as merchant vessels. With the exception of the *Oglala*, the commissioning date for a particular ship reflects its first service as a minelayer.

Following her minelaying service in World War I, *Shawmut* (CM 4) was converted to an aviation tender. She was reclassified a minelayer (CM-4) in 1920 and renamed *Oglala* on 1 January 1928. *Oglala* was sunk by enemy action on 7 December 1941 at Pearl Harbor. After being raised and repaired, she was put back into service as an internal combustion engine repair ship (ARG-1).

Miantonomah was sunk on 25 September 1944 at Le Havre, France, as a result of mine damage. With the exception of the relatively new *Terror*, the remaining CMs were decommissioned soon after war's end.

U.S. Navy Minelayers (CM) in World War II

Minelayer	Length (ft.)/ Displ. (tons)	Comm. Date	Disposition/Fate
Oglala (CM 4) (formerly *Shawmut*) ex-psgr./cargo ship	386/3,746	1907 built 1920 (CM-4)	Decommissioned 11 July 1946
Terror (CM-5) built as minelayer	454/5,875	15 Jul 42 (CM-5)	Decommissioned 6 August 1956

U.S. Army and Navy Minelaying Forces   161

Keokuk (CM-8) ex-ferry	353/6,150	1914 LD 18 May 42	Decommissioned 5 December 1945
Monadnock (CM-9) ex-cargo ship	292/3,110	1938 LD 2 Dec 41	Decommissioned 3 June 1946
Miantonomah (CM-10) ex-cargo ship	292/3,110	1938 LD 13 Nov 41	Struck a mine and sank 25 Sep 1944 At Le Havre, France
Salem (CM 11) ex-ferry	350/3,500	1916 LD 9 Aug 42	Decommissioned 6 December 1945
Weehawken (CM-12) ex-car ferry	350/6,525	1920 LD 30 Sep 42	Decommissioned 11 December 1945

Mine Division 50 was formed on 25 May 1942 with five of these seven ships—*Monadnock*, *Keokuk*, and *Miantonomah*—with the remaining two, *Weehawken* and *Salem*, to join upon commissioning. *Keokuk* served as a minelayer between 18 May 1942 and November 1943. During this period, she mined waters off the Atlantic Coast and, as the war in Europe intensified, off Casablanca and later Gela, Sicily, prior to the landings there. During the latter operations, *Keokuk* was attacked on 11 July 1943 by six enemy planes, which anti-aircraft fire drove off. All of the Navy's CMs, save *Oglala*, participated in minelaying at Casablanca.[14]

Photo 13-4

USS *Salem* (CM-11) under way off Norfolk, Virginia, on 29 April 1944.
Navy Photograph #80-G-229558, in the collections of the National Archives

Monadnock was reassigned to the Pacific Fleet in late autumn 1943. She reached Pearl Harbor on 7 January 1944 and reported for duty to commander, Minecraft, Pacific Fleet. *Miantonomah* remained a unit of the Atlantic Fleet and took part in the landings at Normandy in June 1944. On 21 September 1944, she delivered port clearance materials (in support of salvage and clearing operations) to Le Havre in northwestern France on the English Channel. The port city had been liberated by sea and land less than two weeks before. After exiting the harbour on the 25th, *Miantonomah* struck an enemy mine at 1415 and sank about twenty minutes later with the loss of some fifty-eight officers and men.[15]

Salem laid 202 mines off Casablanca on 27 and 28 December and helped fight off an air raid there on 31 December 1942. On 20 January 1943, she sailed from Casablanca and arrived at Norfolk on 9 February. She left the United States again on 13 June and arrived at Oran, Algeria, on 5 July 1943. *Salem* got under way the next day as part of the Sicily invasion force. On 11 July, she laid 390 mines off Gela, on the southern coast of Sicily, in company with *Weehawken* and *Keokuk*. Returning to Norfolk in autumn, she carried out local operations along the Atlantic coast until 11 May 1944, when she stood out of Hampton Roads for duty with Service Squadron 6 in the Pacific.[16]

Weehawken participated with other minelayers in sowing a defensive minefield off the harbour at Casablanca in late December 1942. Seven months later, *Weehawken* joined *Keokuk* and *Salem* in mid-July 1943 in laying defensive minefields around the invasion beaches at Gela, on the southern coast of Sicily. During the operation, *Weehawken*'s group underwent a series of heavy attacks by the Luftwaffe on 11 July, but she came through unscathed save for some fragments from a stick of bombs which exploded just off her starboard bow.[17]

On 20 April 1944, Mine Division 50—*Miantonomah*, *Salem*, and *Weehawken*—was dissolved and *Weehawken* was assigned to the Pacific Fleet. *Weehawken* cleared Hampton Roads on 11 May in company with *Salem*, bound for the Pacific. The two minelayers reached Pearl Harbor on 14 June for duty with Service Squadron 6.[18]

14

American Forces Land in French North Africa

In order to forestall an invasion of Africa by Germany and Italy, which, if successful, would constitute a direct threat to America across the comparatively narrow sea from western Africa, a powerful American force equipped with adequate weapons of modern warfare and under American command is today landing on the Mediterranean and Atlantic coasts of the French colonies in Africa.

The landing of this American army is being assisted by the British Navy and Air Force, and it will in the immediate future be reinforced by a considerable number of divisions of the British Army.

This combined Allied force, under American command, in conjunction with the British campaign in Egypt, is designed to prevent an occupation by the Axis armies of any part of northern or western Africa and to deny to the aggressor nations a starting point from which to launch an attack against the Atlantic coast of the Americas.

In addition, it provides an effective second front assistance to our heroic allies in Russia.

—With these words,
President Franklin D. Roosevelt announced the
landing of American troops on African soil on
Sunday, 8 November 1942.[1]

Following a meeting between the Russian Minister of Foreign Affairs and Franklin D. Roosevelt in Washington, D.C. in June 1942, a press release was issued stating that the American president agreed on the "urgent task of creating a second front" that year. Inaction by the Allies in Europe had enabled Germany to concentrate her army on the eastern front, and it was questionable whether Russia could hold out unless something was done quickly to divert German forces elsewhere, via an operation in Europe or Africa. On the heels of this announcement came news of the fall of Tobruk—a port city on Libya's eastern Mediterranean coast, near the border with Egypt—with the advance of German general Erwin Rommel's panzer division into Egypt. Rommel, known as the Desert Fox because of his cunning tactics, was poised to take Alexandria, gain control of the Suez Canal, and push the British out of Egypt. The Allies were thus threatened with both the defeat of Russia and the cutting of the Suez Canal lifeline. Discussions by the British about opening a front in Africa, which had preceded the entry of the United States into the war, had envisioned a landing of about 55,000 men in the vicinity of Casablanca, a large port city in western "Vichy" controlled French Morocco on the Atlantic. After America's entry, the plan was enlarged to include landings not only near Casablanca but also in the Mehdia-Port Lyautey area—a beach village and port on the Sebou River (known today as Kenitra) to the north of Casablanca—and the port city of Safi to the south. Planners then expanded the operation to include occupation of the entire North African coast as far east as Tripolitania, the coastal region of what is today Libya. Occupation by Allied forces of Morocco, Algeria, and Tunisia would help safeguard Mediterranean convoys, thus dramatically shortening the route to the Middle East around the Cape of Good Hope.[2]

Map 14-1

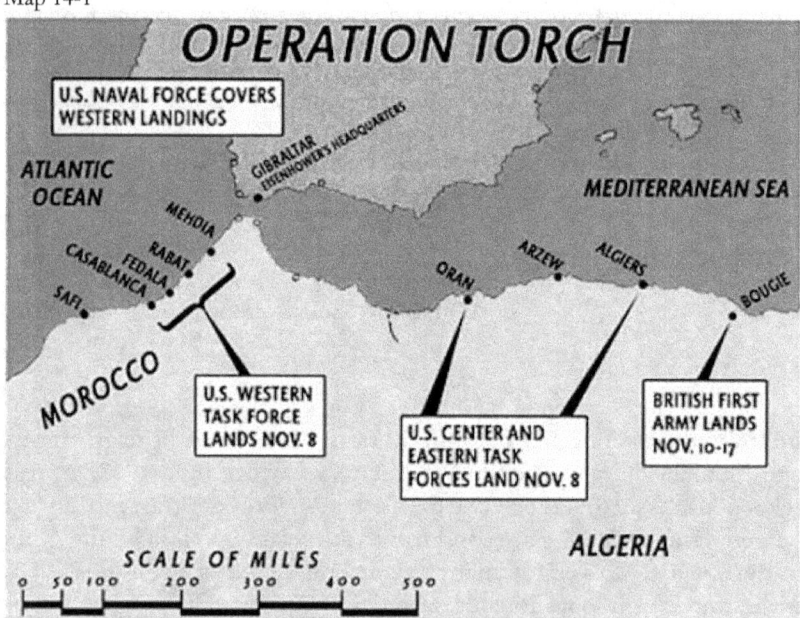

As part of Operation TORCH, assault troops of the U.S. Western Task Force landed on the west coast of French Morocco on 8 November 1942.

The United States was to have responsibility for the military and naval operations on the Atlantic coast of Morocco. Oran and Algiers, cities on the Mediterranean on the northern coast of Algeria, were to be captured by two joint British and American forces. For the joint landings, the British were to supply all the naval service except for a few transports while the landing forces were to be partly American and partly British. Allied occupation of French North Africa was to be achieved through simultaneous assaults by three attacking forces on Casablanca, Oran, and Algiers. Lt. Gen. Dwight D. Eisenhower was given command over the forces, with the exception that British naval units permanently assigned to the Mediterranean would remain under the control of the British Admiralty.[3]

THE MOROCCAN EXPEDITION

The naval component of the Moroccan expedition under the command of Rear Adm. Henry K. Hewitt, U.S. Navy, was designated the Western Naval Task Force. The Army component under Maj. Gen. George S. Patton, U.S. Army, was titled the Western Task Force. The mission assigned to the Naval Task Force was:

> To establish the Western Task Force on beachheads near Mehdia, Fedala and Safi, and support the subsequent coastal military operations in order to capture Casablanca as a base for further military and naval operations.

The objective of the landings at Fedala and Safi was to enable the capture of Casablanca from the land side. Mehdia was to be occupied as a prelude to taking the adjoining airfield at Port Lyautey. At a conference of about 150 naval and army officers convened by Admiral Hewitt at Norfolk, Virginia on 23 October 1942, the day before the task force sailed, General Patton predicted that all the elaborate landing plans would break down in five minutes, then the Army would take over and win through. He stated in part:

> Never in history has the Navy landed an army at the planned time and place. If you land us anywhere within fifty miles of Fedala and within one week of D-day, I'll go ahead and win.... We shall attack for sixty days, and then, if we have to, for sixty more.[4]

ATLANTIC CROSSING

Hewitt placed the Western Naval Task Force (Task Force 34) organisation in effect at Norfolk, Virginia, at 0400 on 23 October. Prior to departure, in order to avoid too great a concentration of ships in the Hampton Roads area (where the James, Nansemond and Elizabeth

Rivers pour into the mouth of the Chesapeake Bay), he moved the Covering Group north to Casco Bay, an inlet of the Gulf of Maine. The Air Group plus three old destroyers being fitted out for special service, were staged at Bermuda. The remaining units, which consisted primarily of combat loaded transports and cargo vessels, with destroyer, cruiser, and battleship escorts sailed from Hampton Roads in two groups on 23 and 24 October 1942.[5]

Photo 14-1

Adm. Royal Eason Ingersoll, USN, Commander-in-Chief, Atlantic Fleet, watches as the Operation Torch task force stands out from Hampton Roads, Virginia, on 24 October 1942, en route to North West Africa.
Naval History and Heritage Command photograph #NH 90944

Comprising part of the task force were units of Mine Divisions 19, 21, and 50. Five destroyer minesweepers—*Hamilton* (DMS-18), *Hogan* (DMS-6), *Howard* (DMS-7), *Palmer* (DMS-5), and *Stansbury* (DMS-8)— were assigned to the convoy screen along with the ships of Destroyer Squadrons 13 and 15. Minesweepers *Auk* (AM-57), *Osprey* (AM-56) and *Raven* (AM-55), along with minelayers *Miantonomah* (CM-10) and *Monadnock* (CM-9), sailed in the body of the nine-column convoy. Upon arrival off the assault beaches, the minesweepers were to sweep ahead of the invasion forces to clear any enemy mines that would impede putting troops ashore. The minelayers were to lay defensive minefields as necessary to protect Allied vessels in transport areas from attack by German U-boats, and later captured ports and harbours.[6]

Once joined, Task Force 34 (also designated Task Force HOW) followed a route to a point south of Newfoundland to give the impression of a regular troop convoy to the United Kingdom. From there to a point southeast of the Azores, the advance was more or less toward Dakar, the capital of Senegal and an Atlantic port in West Africa. The Western Naval Task Force then proceeded toward the Strait of Gibraltar to avoid potentially unfriendly air search from the Azores and the Canary Islands (should either Portugal or Spain be providing vessel locating information to Germany), until the force split up into three attack groups on 7 November to proceed to final destinations.[7]

Map 14-2

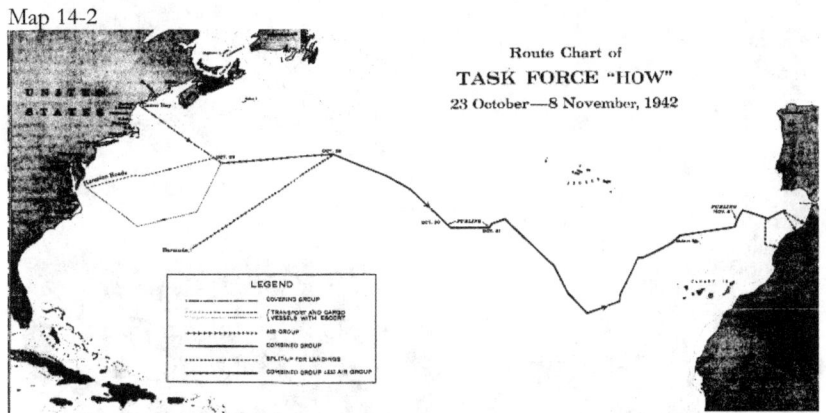

Navigation track of the Western Naval Task Force (Task Force 34/Task Force HOW) en route to Operation TORCH, the Allied invasion of French Morocco.
ONI Combat Narratives, The Landings in North Africa, 1944

In execution of Operation TORCH, the Covering Group was to contain the Vichy French Fleet at Casablanca and, if necessary, the French units at Dakar, while the attack groups effected landings at Safi just south of Casablanca; at Fedala, just north of Casablanca; and at Port Lyautey, further to the north of Casablanca. Nearing French Moroccan waters, the Southern Attack Group left the convoy at daylight on 7 November to proceed to Safi, 125 miles south of Casablanca. With it went the minesweepers *Hamilton* and *Howard*, and minelayer *Monadnock*. In late afternoon, the Northern Attack Group veered off, accompanied by the minesweepers *Osprey* and *Raven* headed eastward toward Port Lyautey, 65 miles north of Casablanca. The Centre Attack Group continued on to Casablanca, with minesweepers *Auk*, *Hogan*, *Palmer*, and *Stansbury*, and minelayer *Miantonomah*. As darkness fell, the men aboard the ships became silent and tense. There was nothing ahead of them, then, but unknown North West Africa waters.[8]

Photo 14-2

A U.S. destroyer passes astern of the aircraft carrier USS *Ranger* (CV-4) at sunset on 8 November 1942, the first day of landings on North Africa.
U.S. Navy photograph 80-G-30232, now in the collections of the National Archives

ASSAULT ON PORT LYAUTEY

> *We are about to embark on a difficult and historical task—the opening of a second front. In America our people back home and the entire United Nations will watch us with consuming interest. Don't let them down. To successfully carry out our task we must live up to the glorious traditions of the Navy which will require the utmost from all and duty beyond the usual call. This call is your opportunity to strike a blow for America and her allies. Let us be ready to give the enemy hell when and where he shows himself. Make every shot count.*
>
> —Rear Adm. Monroe Kelly, USN, commander, Battleships, Atlantic Fleet and commander, Task Group 34.8—the Northern Assault Group—embarked in the battleship *Texas* (BB-35)[9]

The 9,078 troops of Maj. Gen. Lucian K. Truscott's Sub-Task Force GOALPOST were made up of the 60th Infantry Division; 9th Infantry Division; 1st Battalion, 66th Armored Regiment, 2nd Armored

Division; elements of the 70th Tank Battalion; and seven coast artillery batteries. They were all carried aboard Northern Assault Group transports. Their mission was to simultaneously land on five beaches on either side of the village of Mehdia—where the Sebou River emptied into the Atlantic—in order to capture Port Lyautey and its airfield for use by Army and Navy aircraft. From this location, these aircraft could then attack Casablanca to the south. A follow-on objective after gaining control of the Sebou was to seize the Rabat-Sale airfield, southwest of Port Lyautey on the coast.[10]

Osprey and *Raven* led the Northern Attack Group approaching Mehdia, streaming sweep gear about dusk on 7 November. But the formation soon increased speed to 14 knots, thereby running up on the minesweepers ahead and forcing them to make full speed. With the increased speed, *Raven*'s sweep wires were carried away and *Osprey* lost her gear as well. About midnight the ships slowed to 5 knots and the minesweepers were able to again stream gear, but the transports they were supposed to lead in abruptly made a turn toward their assigned anchorage, leaving the minesweepers with no vessels following. While the troops aboard the transports were embarking in landing craft, the minesweepers took in their gear and assumed their new duties as control ships for the initial assault wave. *Osprey* guided the first wave from *Susan B. Anthony* into Beach Red north of Wadi Sebou, while *Raven*, farther south, led *Henry T. Allen*'s first wave into Blue and Yellow Beaches.[11]

By 0600 on 8 November, the beaches opposite *Raven* and *Osprey* had been taken with little opposition. Between these points, at Green Beach off Mehdia, the fighting became intense. Vichy French shore batteries opened fire at 0600. This was answered six minutes later by counterbattery from gunfire support vessels. Soon a new threat developed as Vichy fighters and two-engine bombers from the airfield at Rabat-Sale began to strafe and bomb the landing craft. Assault troops making the beach were met ashore with bullets, bayonets, and 75mm fire from the 1st and 7th Regiments Moroccan Tirailleurs (French Army designation for infantry recruited in colonial territories), the Foreign Legion, and naval ground units. During the assault, *Osprey* helped drive off two attacking French planes.[12]

The next few days brought new tasking for *Raven* and *Osprey*, including a run on 13 November up the Sebou River to Port Lyautey, to bring desperately needed gasoline, ammunition, and supplies to American troops there. They were used as lighters because the river passage was so narrow. They found the winding Wadi Sebou littered with ships, which the Vichy French had scuttled and beached in an unsuccessful effort to block the river passage, thereby preventing capture of the riverport and airfield by Allied forces.[13]

Previously on 10 November, the destroyer *Dallas* (DD-199) had landed a party of Army raiders that had seized the airfield. During the transit up river at speeds up to 25 knots, her screws had at times churned the mud bottom dragging her down to 5 knots. The first chapter of my book, *We Are Sinking, Send Help! The U.S. Navy's Tugs and Salvage Ships in the African, European, and Mediterranean Theaters in World War II*, details how a demolition party comprised of one officer and ten men from the fleet tug *Cherokee* (AT-66) and a second officer and five men from the salvage ship *Brant* (ARS-32) breached a steel cable anti-ship boom stretched across the Sebou to allow the assault to commence. Protected by shore guns at Fort Kasbah on the heights above, this boom had prevented farther movement upriver to the airfield adjacent to Port Lyautey which was itself guarded by a number of anti-aircraft guns.

NAVAL BATTLE OF CASABLANCA

As the Centre Attack Group, charged with capturing Casablanca, closed the coast in preparation for putting troops ashore over beaches at Fedala (located about twelve miles to the northeast of Casablanca), the Covering Group took up position between the Centre Attack Group and Casablanca. By this means, the battleship *Massachusetts*, heavy cruisers *Wichita* and *Tuscaloosa*, and destroyers *Jenkins*, *Mayrant*, *Rhind*, and *Wainwright* could protect the Centre Attack Group during debarkation of troops into landing craft and also be in position to open fire on Casablanca at sunrise, should this action be necessary.[14]

The destruction of the French fleet at Casablanca was the main objective of the Covering Group. In effecting its elimination, the group came into contact with shore defences. The series of naval engagements fought between American ships covering the invasion of North West Africa and Vichy French ships defending French Morocco would be termed the Naval Battle of Casablanca. The final stages involved operations by German U-boats which reached the waters off Fedala the same day that the French troops ashore surrendered.[15]

The performance of French naval forces and shore batteries throughout the Battle of Casablanca was excellent. Their gunnery was accurate—first salvos frequently straddling their targets. That more hits were not scored on U.S. Navy units was due to the skill with which the American ships manoeuvred and to the high angle of fall of French projectiles, necessitated when firing long ranges at distant targets.[16]

PRELUDE TO THE LANDINGS AT FEDALA

Map 14-3

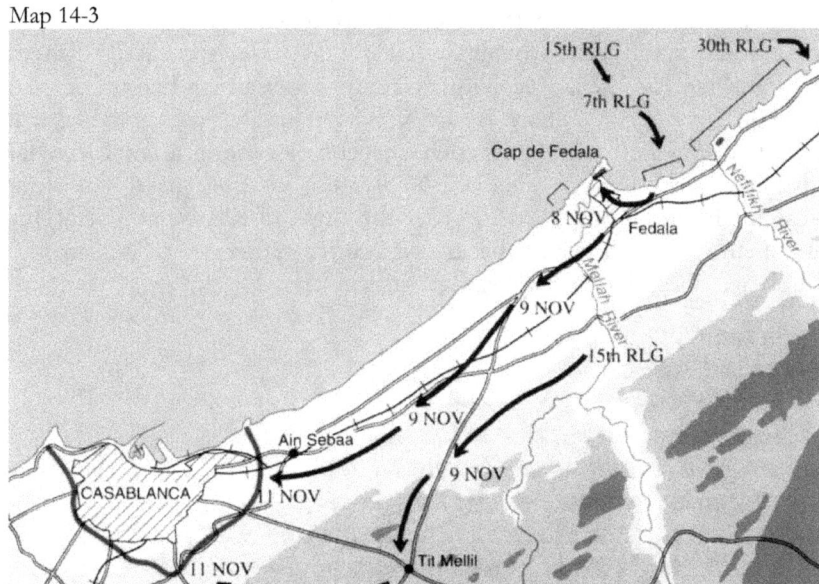

Maj. Gen. Jonathan W. Anderson's Sub-Task Force BRUSHWOOD, consisting of troops of the 3rd Infantry Division (three regimental landing groups) and of the 2nd Armored Division and the 756th Tank Battalion, came ashore at Fedala and rapidly advanced to Casablanca.
Algeria-French Morocco, US Army Campaigns in World War II, CMH Pub 72-11

The attack on Fedala was the principal one for the invasion of French Morocco. Its first object was the landing of Sub-Task Force BRUSHWOOD. The force was comprised of U.S. Army 3rd Infantry Division troops (the 7th, 15th, and 30th Infantry Regiments), the 1st Battalion, 67th Armored Regiment, and the 82nd Reconnaissance Battalion (both of the 2nd Armored Division), and the 756th Tank Battalion. This force (19,364 officers and men) was larger than the other two combined. Once ashore, it was to advance rapidly to Casablanca for junction with assault forces converging from the north (Mehdia) and from the south (Safi).[17]

HOGAN CAPTURES FRENCH CORVETTE *VICTORIA*

On 8 November, prior to arrival of the task force transports off Fedala, the minesweepers *Palmer, Stansbury, Hogan* and *Auk* made an exploratory sweep through the area for moored, magnetic, and acoustic mines. No mines were found. Later the next day, *Hogan* took up duties at 0215 as a unit of the anti-submarine screen patrolling an east-west line, north of the planned transport anchorage. Shortly before dawn, she left her station at 0515 to investigate two vessels approaching Fedala on a course that would take them through the anchorage. *Hogan* challenged them by signal light at 0525. Receiving no reply, her commanding officer, Lt. Comdr. Ulysses S. G. Sharp Jr., USN, placed his ship across the course of what appeared to be a French corvette (escorting a small coastal steamer). From a distance of 500 yards, he then used both the searchlight and a bull horn to order the ships to reverse course. This command was conveyed by the executive officer, Lt. William H. Sublette, USN, first in English and then in French.[18]

Photo 14-3

Destroyer USS *Hogan* (DD-178) before conversion to a destroyer-minesweeper. Courtesy of Dwight Messimer

It became obvious a few minutes later that the corvette intended to ram his ship, since she was now close aboard, headed for *Hogan*, and her speed had not slowed. Sharp backed both engines full and by this avoidance action, the French warship passed close under the destroyer-minesweeper's bow. Stopping his engines, he then ordered warning bursts from a .30-calibre machine gun fired across the bow of the corvette. After the French ship replied with machine gun fire, *Hogan* opened fire at a range of 150 yards with her No. 1, 3, and 5 20mm guns. Many of the 150 rounds expended hit the corvette's hull and superstructure (most exploding on contact) killing the French captain and nine of his crew. Sharp did not employ *Hogan*'s larger, longer-range 3-inch guns because the firing was in the direction of the transport anchorage.[19]

Badly damaged, and with personnel casualties, the corvette stopped and blinked her running lights, apparently as a sign of surrender. As the French ship, and small vessel she was escorting, continued to lay to, *Hogan* resumed her anti-submarine patrol nearby. At 0626, *Auk* put a prize crew aboard the corvette. She was found to be the *Victoria* (W-62). After removing the breech blocks from the guns aboard the French ship to *Auk* (rendering the guns inoperable) the prize crew sailed the corvette to the transport area. Upon arrival, they anchored *Victoria* near the transport *Ancon* (AP-66) for treatment of her wounded. Rear Admiral Hewitt directed that the body of the French naval officer who had been killed in action be landed with all due ceremony and that secret and confidential files aboard the corvette be delivered to the force commander.[20]

LANDINGS AT FEDALA

0641-42: Enemy splashes were again falling very close – Over, Short and close astern. Morning twilight showed splashes to seem unreasonably small and of a blackish-green color. It was not understood why no sign of fragments from near misses appeared to come on board. Later explained by knowledge these were A.P. projectiles. Did not detonate on water impact.

—USS *Murphy* (DD-603) deck log entry. The destroyer was hit at 0643 on 8 November 1942, by a 138mm armor-piercing round from a shore gun, which penetrated her after engine room, putting it out of action, killing and wounding personnel, and setting the space and a berthing compartment on fire.

While *Hogan* was investigating the *Victoria* and escorted coastal steamer, the first wave of assault craft landed at Fedala. The port city (today called Mohammédia) lay on a shallow bay between the rugged projection of Cap de Fedala to the southwest and the bold headland of Cherqui, three miles to the northeast. The Vichy French defences included several gun batteries. Two 77mm guns on the top of Cap de Fedala could, with a range of 9,000 yards, fire on the beaches on which major landings were planned. Near the foot of the cape, four 100mm guns (the Batterie de Fedala, or Batterie du Port) could fire salvos out to 15,400 yards. The most powerful battery (the Batterie du Pont Blondin) was on the Cherqui headland. Its four 138mm guns were capable of firing on targets 20,000 yards distant. This shore battery opened at 0608, and task force gunfire support ships returned fire four minutes later. The light cruiser *Brooklyn* (CL-40) reported three guns silenced at Pont Blondin by 0655 and the fourth and last one at 0725. Firing on the beach then generally ceased.[21]

The destroyers *Ludlow* (DD-438), *Murphy* (DD-603), *Swanson* (DD-443) and *Wilkes* (DD-441) were operating as control vessels, leading landing boat assault waves from the transport area to their designated lines of departure and sending these waves at predetermined times with proper navigational information to enable them to arrive at their landing beaches on schedule. The first four boat waves were despatched between 0525 and 0540. At about 0525, when the searchlights at Cherqui found the boats en route to the central beaches, the support ships opened fire on the lights. At 0604, the shore batteries at Cherqui and Fedala opened fire on the control vessels and U.S. troops ashore. The destroyers' guns then opened, with *Wilkes* and *Swanson* firing on Fedala, and *Murphy* and *Ludlow* on Cherqui. The guns were silenced and the landing operation went on to a successful conclusion.[22]

AUK AND *TILLMAN* DESTROY ENEMY CORVETTE

Photo 14-4

The minesweeper *Auk* (AM-57) off the Norfolk Navy Yard, circa May 1942. Naval History and Heritage Command photograph #NH 84027

During the opening hours of the invasion on 8 November, *Miantonomah* (CMc-5), loaded with mines, had orders to stay away from the transport area. Shortly before noon, she began preparing to lay buoys for use in navigation, as a prelude to planting a defensive minefield near the transport area as protection from enemy submarines. In preparation for laying, her crew had removed shoring, wedges, cable, etc., securing the deadly mines in place two hours earlier. Following reports of sightings of enemy mines in the area that she was to mine, *Auk* received orders to sweep ahead of her. However, *Miantonomah* proceeded alone after *Auk* informed her that it would take forty-five minutes to rig gear. Fortunately, reports of the presence of mines proved to be false.[23]

Miantonomah began laying mines at 1245, at the relative breakneck speed of twelve knots, and ceased at 1458 after her full complement of 380 mines had been sown. Every man, from the commanding officer down to the lowest seaman, was happy to get the mines off the ship. During the Atlantic crossing, her crew had experienced anxious moments in rough weather as the minelayer and its deadly cargo took heavy rolls. This unease persisted in spite of the fact that it appeared the ordnance was unaffected by the violent shaking. *Miantonomah*'s movements during the operation were tracked by radar aboard the transport *Leonard Wood* (AP-25) in order to plot the precise location of the minefield.[24]

Meanwhile, *Auk* had received orders from the destroyer *Tillman* (DD-641) to investigate a ship shoreward, which had pointed its guns at *Miantonomah* and nearby merchant vessels. The offender proved to be a Vichy French corvette, the *Estefette* (W-43), escorting a small coastal convoy of six ships. The commander aboard the corvette refused to stop or comply with *Auk*'s orders to have the corvette and the other ships follow the *Auk* and *Tillman* back to the transport area. Shortly afterward as the corvette continued to proceed, she was directed to stand out to deeper water, where the *Auk* and *Tillman* were located, or they would fire. The corvette had led its brood into shallow water in which the U.S. Navy ships could not approach, and three of the escorted ships appeared to have beached and the others anchored to avoid grounding.[25]

At 1235, after repeated warnings, *Tillman* fired one 5-inch round across the bow of the *Estefette*. The corvette returned fire and an exchange of fire amongst the three ships continued for five minutes until 1240, when the burning and sinking French ship went aground. Apparently, she was not under command at this time, as her crew had leapt overboard. With the action over, *Tillman* returned to her station in the anti-submarine screen. She patrolled westward of the transport area until the following morning, 9 November, when she received orders to return to the site of the coastal convoy. Arriving there at 1345,

Tillman sighted the wreckage of the *Esteffette* on the beach with, as before, three convoy vessels beached and the other three at anchor.²⁶

Tillman lowered a motor whaleboat with a boarding party to visit and search the six ships. They proved to be the SS *Foudrayant*, SS *Loup de Mer*, SS *Simon Duhamel*, SS *Dahomey*, SS *Strasbourgois*, and SS *Lorrain*, all of French registry. Prize crews from the destroyer were sent aboard the anchored *Simon Duhamel*, *Foudrayant*, and *Loup de Mer* to sail them to the transport area. The captured ships arrived off Fedala about 2300. (*Auk* had earlier delivered the convoy commodore, Lt. Michel Lacozi, French Navy, to the transport *Leonard Wood*, AP-25, as a prisoner of war.) At 2320, Colonel E. S. Johnson, U.S. Army, reported that Fedala had been occupied by American forces and was quiet except for a small group of enemy forces resisting in the rocks on the Point, which would soon be "reduced"—a euphemism for "destroyed."²⁷

LANDINGS AT SAFI, SOUTH OF CASABLANCA

The objective of the landings at Safi was to secure that port as an unloading point for the tanks carried aboard the transport and aircraft ferry *Lakehurst* (APV-3). The intention was to use them in the subsequent attacks on Casablanca, 135 miles to the north. D-Day was on 8 November 1942. The naval convoy transporting Maj. Gen. Ernest N. Harmon's Sub-Task Force BLACKSTONE to Safi arrived eight miles offshore Safi shortly before midnight on 7 November. Aboard the transports were the 47th Infantry, 9th Infantry Division; two reinforced battalions of the 67th Armored Regiment, 2nd Armored Division; elements of the 70th Tank Battalion; and several artillery batteries. As the landing was not preceded by a softening-up shore bombardment, the debarkation of troops and equipment was attempted in silence.²⁸

But as landing craft made their way toward shore, the French fired on the transports. The U.S. Navy ships immediately returned fire. As naval gunfire pounded French batteries, the first American troops to land in French Morocco—Company K, 47th Infantry—came ashore at 0445. As waves of boats continued to land, the troops coming ashore extended their beachhead inland. By daybreak, the Americans controlled the port facilities and other installations at Safi, and all roads leading into town. Daylight made possible more accurate naval gunfire, and by 1045 all French batteries were out of action. The French garrison commander, clearly understanding that his relatively small force was outnumbered and outgunned, surrendered at 1530—only eleven hours after the first American troops had come ashore.²⁹

During the initial approach to Safi, sweepers *Howard* and *Hamilton* had led the Southern Attack Group in, but swept no mines, and *Monadnock* trailing astern had laid no mines. In late morning on that day, *Howard* and *Hamilton* swept the approach channel to the entrance of the harbour. After recovering gear, they took up patrol stations in the anti-submarine screen, until ordered to serve as escorts for the light cruiser *Philadelphia* (CL-41) off Safi. The following afternoon, *Howard* and *Hamilton* sailed from Safi with the battleship *New York* (BB-34), bound for Fedala, arriving there on 10 November.[30]

That same day, the American tanks at Safi started rumbling along the road to Casablanca, but before they got very far, word was received that hostilities were over. General Patton had planned an all-out attack on Casablanca on the morning of 11 November, but at 0700, fifteen minutes before this attack was to commence, the French army commander sent Patton a flag of truce. There was no surrender, only a cease-fire that developed into full cooperation between Allied and French authorities, both military and civil.[31]

U-BOATS ATTACK SEVEN SHIPS OFF FEDALA

French capitulation at Casablanca did not, however, end hostilities in the area. Between 11 and 15 November, two German submarines, *U-130* and *U-173*, carried out torpedo attacks on six transports and one destroyer off Fedala. Four of the seven ships sank. The remaining three, although damaged, remained afloat.

Ship	Date/Time	U-boat	Result
Joseph Hewes (AP-50)	11 Nov/1953	*U-173*	Sank at 2045 on 11 Nov
Winooski (AO-38)	11 Nov/1954	*U-173*	Damaged, remained afloat
Hambleton (DD-455)	11 Nov/2118	*U-173*	Damaged, remained afloat
Hugh L. Scott (AP-43)	12 Nov/1730	*U-130*	Abandoned and sank that night
Edward Rutledge (AP-52)	12 Nov/1731	*U-130*	Abandoned and sank about 1848
Tasker H. Bliss (AP-42)	12 Nov/1737	*U-130*	Sank at 0230 on 13 Nov
Electra (AK-21)	15 Nov	*U-173*	Damaged, made port[32]

AK (cargo ship)   AO (fleet oiler)   AP (transport)   DD (destroyer)

CONCLUSION OF OPERATION TORCH

By mid-November, Patton's Western Task Force had taken French Morocco, and Oran and Algiers had fallen to the joint British and American Centre and Eastern Task Forces, giving the Allies control of French Algeria as well as Morocco. Patton expressed his appreciation of the assistance provided by the Navy in a despatch to Hewitt, which the admiral in turn transmitted on Sunday, 15 November 1942, to every ship of Task Force 34 remaining in African waters:

> It is my firm conviction that the great success attending the hazardous operations carried out on sea and on land by the Western Task Force could only have been possible through the intervention of Divine Providence manifested in many ways. Therefore, I shall be pleased if in so far as circumstances and conditions permit, our grateful thanks be expressed today in appropriate religious services.[33]

U.S. DEFENSIVE MINEFIELD OFF CASABLANCA

> *Laying of mine field was attended with success. Only one premature [explosion of a mine] occurred during actual laying operations. Adequate and efficient screening [of vessels] kept contacts from occurring near the field of operations. Explosions were reported heard in the west and east sections of the fields during the night of November sixteenth. It is hoped that these were due to [enemy] submarine activity. All of the mines should have ripened [been operational] before that time.*
>
> —Report by Comdr. Howard W. Fitch, USN, commanding officer, USS *Terror* (CM-5) regarding the laying of a defensive minefield off Casablanca, Morocco, on 16 November 1942.[34]

The minelayer *Terror* (CM-5) arrived at Casablanca in mid-November to join *Miantonomah* and *Monadnock* in laying a defensive minefield to enclose the anchorage area between Fedala and Casablanca. This action was being undertaken to protect Allied ships at anchor from enemy submarine attack. Since subs were known to be operating in the area to be mined, a request had been made to commander, Task Force 34, Rear Adm. Henry Kent Hewitt, USN, to furnish protection for the minelayers during the operation.[35]

Under command of Comdr. Howard Fitch, *Terror* had made the Atlantic crossing with Convoy UGF 2. After clearing the swept channel leading out of the naval anchorage at New York City Harbour on 2 November, she had joined the convoy and taken her assigned station. "UG" was the designation of U.S. military convoys sailing from America to North Africa for the delivery of troops allocated to Operation TORCH and then subsequent operations in North Africa and the western part of the Mediterranean. The letter F designated a convoy of faster ships, and the number 2 denoted the second such convoy.[36]

During several days of particularly foul weather (characterised by heavy rain squalls, strong winds and heavy seas off the port beam of the ships), the convoy course was shifted to right of track to bring the seas on the quarter after several ships reported cargo in danger of shifting. Fortunately, *Terror* experienced no trouble with her cargo of high explosive mines and depth charges, while rolling up to 32 degrees to starboard. Two life float nets were, however, carried away by seas, and a third was salvaged just as it was about to break adrift.[37]

Photo 14-5

USS *Terror* (CM-5) loading Mk 6 mines at Yorktown, Virginia, for the Casablanca operation, circa October 1942.
Naval History and Heritage Command photograph NR&L(M) 35192

At dawn on 14 November, upon orders from commander, Task Force 38, the minelayer left the convoy with one destroyer for escort. The two ships proceeded at 20 knots toward Casablanca. When met by a patrol destroyer off the harbour entrance, *Terror* released the destroyer to rejoin the task force, and proceeded into the inner harbour. With the aid of a pilot and tugs, she moored in the Delpit Basin, her stern with quarter lines to Delure Jetty, and bow, with both anchors veered to sixty fathoms of chain, pointed toward the fairway. Such an arrangement is termed a "Mediterranean Moor."[38]

Photo 14-6

Casablanca Harbour on 16 November 1942.
Navy photograph #80-G-1003967, now in the collections of the National Archives

Terror had orders to load the *Miantonomah* with mines and top-off the *Monadnock* to capacity. Spanning 454 feet, with a 60-foot beam, and 5,875-ton displacement, she was able to carry a much larger quantity of mines than were the other two ships. The *Miantonomah* and *Monadnock*, at 292 feet in length, with 48-foot beams and 3,110-ton displacements, were relatively svelte in comparison to her. Upon reporting to commander, Task Force 34 for duty, Fitch was informed that *Miantonomah* would reach port the following day, and that *Monadnock*, which had not laid a field, was full to capacity. Meanwhile, various destroyers entering port to receive fuel also required depth charges. In preparations for their transfer, crewmen shifted some mines up to the main deck to gain access to unload the depth charges.[39]

Miantonomah arrived in port around noon on 15 November and moored alongside *Terror* to receive mines. The transfer of mines to her continued until well into the night, concurrently with the moving of depth charges up to the main deck.[40]

DEFENSIVE MINEFIELD OFF CASABLANCA

Map 14-4

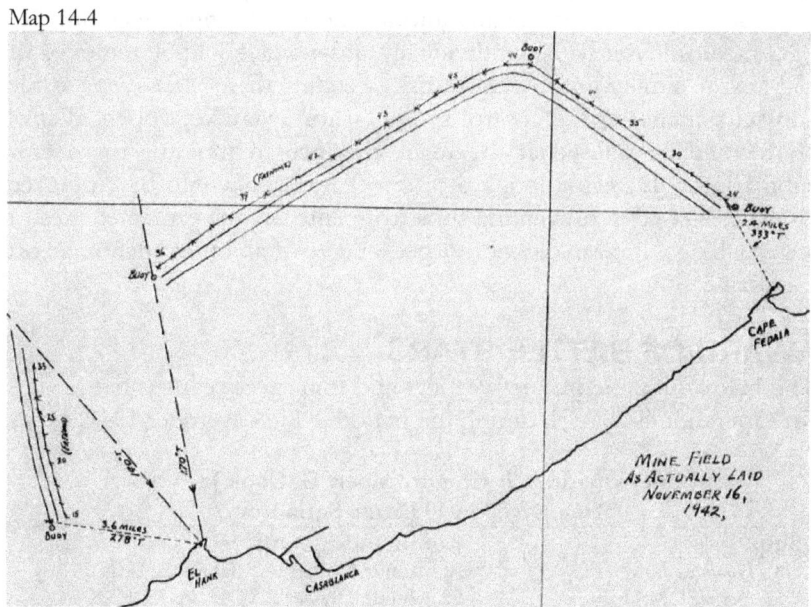

Minefield laid by *Terror, Miantonomah*, and *Monadnock* on 16 November 1942, to enclose the Allied anchorage area between Fedala and Casablanca.
U.S. Atlantic Fleet Task Unit USS *Terror*, Flagship, Operation Order No. 1-42

Minelaying commenced in mid-afternoon the following day. The short western field was laid first, with the ships starting from an accurate fix from the south. Upon a signal from *Terror, Monadnock* dropped a marker buoy 300 yards from the starting point of the minefield. "Cease mining" was executed two degrees before the west limiting bearing for the entrance was reached, providing 300-400 yards safe area outside of the bearing in case the current should set ships slightly over this line.[41]

Upon completion of the western field, the minelayers moved to the Fedala area to sow the main field, which took the shape of a dog leg. The last mines were dropped just short of the east entrance limiting bearing, 350° from El Hank light. Additionally, a marker buoy was dropped at the outer end of the first leg, and another at the end of the second leg of this field.[42]

The fields, as laid, provided a main entrance to Casablanca between bearings 320° and 350° from El Hank light. Two smaller entrances also existed—one between the shore and the south end of the western field, and the other between Cape Fedala and the south end of the eastern leg of the main field. The commanding officer of *Terror* recommended that these emergency entrances (or at least the western one) be sealed with controlled minefields, similar to the British unit type.[43]

Such units consisted of an indicator loop enclosing a given number of controllable mines, laid individually and which could be removed or laid next to similar units, without risk of setting them off, as long as the contact switches at the control station ashore were kept open. Patrol craft could be utilised at the main entrance to monitor for enemy submarines. Detector loops on the ocean floor would be employed elsewhere to aid in preventing subs from entering the protected area or approaching within maximum torpedo range of an outer anchorage off Casablanca.[44]

AWARD OF BATTLE STARS

The below identified minesweepers and minelayers earned battle stars for Operation TORCH, during the period 8-11 November 1942.

Mine Squadron 7: Comdr. Albert G. Cook Jr., USN

Ship	Commanding Officer
Mine Division 19 (Mine Squadron 7)	
USS *Hamilton* (DMS-18)	Lt. Comdr. Robert Ray Sampson, USN
USS *Hogan* (DMS-6)	Lt. Comdr. Ulysses S. G. Sharp Jr., USN
USS *Howard* (DMS-7)	Lt. Comdr. Charles Joseph Zondorak, USN
USS *Palmer* (DMS-5)	Lt. Comdr. Joshua Winfred Cooper, USN
USS *Stansbury* (DMS-8)	Lt. Comdr. Joseph Benedict Maher, USN
Mine Division 21 (Mine Squadron 7)	
USS *Auk* (AM-57)	Lt. Comdr. William David Ryan, USN
USS *Osprey* (AM-56)	Lt. Comdr. Cecil Llewellyn Blackwell, USN
USS *Raven* (AM-55)	Lt. Comdr. Colby G. Rucker, USN
Mine Division 50	
USS *Miantonomah* (CM-10)	Lt. Comdr. Raymond D. Edwards, USN
USS *Monadnock* (CM-9)	Lt. Comdr. Frederick O. Goldsmith, USNR

AM (Minesweeper)   CM (Minelayer)   DMS (Destroyer-minesweeper)

15

Invasion of Sicily – Operation HUSKY

The amphibious assaults were uniformly successful. The only serious threat was an enemy counterattack on D plus one day against the 1st Infantry Division when a German tank force drove across the Gela plain to within one thousand yards of the DIME beaches. The destruction of this armored force by naval gunfire delivered by U.S. cruisers and destroyers, and the recovery of the situation through naval support, was one of the most noteworthy events of the operations.

—Vice Adm. Henry K. Hewitt, USN,
commander, Western Naval Task Force[1]

Following the surrender of the Axis forces in North Africa, the Americans were eager make a cross-Channel invasion, with the purpose of drawing the Germans into a decisive battle that would end the war. The British, however, felt that the combined Allied forces were not yet ready to face the Germans in France, arguing instead for an assault on the Continent out of the Mediterranean into what Churchill called the "soft underbelly." The British argued that the Germans were weaker there and attacking this soft underbelly in either Italy or the Balkans would relieve pressure on Britain as well as give the Americans more combat experience before they confronted the main body of the German Army. It would also draw German divisions from the Eastern Front, which would serve to fulfil Stalin's demand for an Allied second front in Europe.[2]

Another argument that favored an invasion from the Mediterranean was that Italian resolve was wavering and if they could be taken out of the war it would ease Allied efforts considerably. There was also an active Italian partisan movement that would harass the German rear echelons. The decision was made to invade Sicily since seizure of the island would provide a steppingstone to the Italian mainland.[3]

There were about 250,000 Axis troops defending Sicily, most of them Italians with a stiffening of German troops and a large number of

Luftwaffe personnel. The coast defence formations were intended to act as a delaying force, giving time for the regular Italian divisions and the Germans to react decisively. The largest German units present for the landings were the Hermann Göring Panzer Division and the 15th Panzergrenadier Division. Between them they had about 150 tanks, including 17 heavy Tiger tanks together with infantry and supporting artillery.[4]

On the night of 9-10 July 1943, Allied forces commenced Operation HUSKY—the code word for the invasion of Sicily—which launched the Italian Campaign. This large scale amphibious and airborne operation, followed by six weeks of ground combat that ended on 17 August, drove Axis (Italian and German) air, land, and naval forces from the island. It also opened sea lanes in the Mediterranean for use by the Allies and helped topple Italian dictator Benito Mussolini from power. On 24 July 1943, the Italian Grand Council of Fascism voted a motion of no confidence against Mussolini. That same day, King Victor Emmanuel III replaced him with Marshal Pietro Badoglio and had him arrested.[5]

Mussolini did not remain in captivity for long. Less than seven weeks later, a group of German paratroopers freed him from the Campo Imperatore Hotel where he was being held high in the Apennine Mountains. The commandos landed nearby from DFS 230 gliders and overwhelmed Mussolini's captors. Otto Skorzeny, a German Waffen-SS officer whom Hitler had personally selected as the field commander for the mission, greeted Mussolini with "Duce [his nickname meaning "the leader"], the Führer has sent me to set you free!" Mussolini replied, "I knew that my friend would not forsake me!" Mussolini was made leader of the Italian Social Republic (a German puppet state in northern Italy). Near the end of the war in late April 1945, with total defeat looming, he fled Milan, where he had been based, and tried to escape to the Swiss border. He was captured and executed near Lake Como by Italian partisans. Afterward, his body was taken to Milan, where it was hung upside down at a service station for public viewing and to provide confirmation of his demise.[6]

The plan for Operation HUSKY called for the amphibious assault of Sicily by two armies concentrated at its southeastern end, with all landings scheduled for D-day at the same H-hour (exact time when the assault is to take place). British forces including the 1st Canadian Infantry Division were to assault the southern Sicilian coast to the east of Pozzalo, and the east coast, south of Syracuse; while American forces came ashore on the south coast in the Gulf of Gela. The landings would be supported by naval gunfire, tactical bombing, and close air support.

General Dwight D. Eisenhower, Commander-in-Chief, Allied Forces North Africa, was the overall commander. The British general Sir Harold Alexander was his second in command and the Land Forces/Army Group commander. British admiral Andrew Cunningham was the Allied Naval Force commander. He had under him Vice Adm. Bertram Ramsay, RN, commander, Eastern Naval Task Force, and Vice Adm. Henry K. Hewitt, USN, commander, Western Naval Task Force of the U.S. Eighth Fleet.[7]

Map 15-1

The American Western Naval Task Force assaulted beaches in the Gulf of Gela, and the British Eastern Naval Task Force south of Syracuse.

OPERATION HUSKY FORCES

Over 3,200 ships, craft, and boats made up the Allied naval forces assembled to launch the invasion of Sicily. The Western Naval Task Force—of more than 1,700 vessels—was charged with landing the American invasion troops on the southwest coast of Sicily at Licata, Gela, and Scoglitti. The soldiers carried by the British Eastern Naval Task Force were to land on beaches on the southeastern side of the island. Lt. Gen. George S. Patton and his Seventh Army would push across the island to secure Palermo, and then swing east to Messina, while British general Sir Bernard Montgomery would drive his Eighth Army north to Syracuse and Catania, to meet up with Patton at Messina.

The U.S. Army Air Force and British Royal Air Force were tasked to provide air support.[8]

WESTERN NAVAL TASK FORCE

The ships of Admiral Hewitt's Western Task Force sailed from or staged through six different Algerian or Tunisian harbours. The task force was divided into three component attack forces charged with landing the American troops in the Gulf of Gela, with the 3rd Infantry Division coming ashore on the westernmost beaches, the 45th Infantry on the easternmost beaches, and the 1st Infantry on beaches at Gela, which lay between the other two areas.

Attack Force Commander	Task Force	Embarked Troops/Objective
Rear Adm. Richard L. Conolly, USN	JOSS (TF 86)	3rd Infantry Division: capture the town of Licata and its port
Rear Adm. John L. Hall Jr., USN	DIME (TF 81)	1st Infantry Division: capture Ponte Olivo Airfield near the town of Gela
Rear Adm. Alan G. Kirk, USN	CENT (TF 82)	45th Infantry Division: capture the airfields at Comiso and Biscari near the fishing village of Scoglitti[9]

Supporting the attack forces were the Control Force (TF 80)—made up of the Force flagship, (80.1), Escort Group (80.2), Screening Group (80.3), Demonstration Group (80.4), Minelaying Group (80.5), and Reserve Group (80.6)—and the Train (TF 86). The term "Train" was short for Fleet Train (today, known as the Service Force). The fleet train included tenders, oilers, stores ships, and ammunition ships.[10]

ASSAULT LANDINGS AT GELA, SICILY

As Rear Admiral Hall's DIME Force approached Gela, flares were dropped by Allied aircraft revealing the entire area in yellow light. Large fires from associated aerial-bombing to the east and west of Gela, and in the town itself, contributed to the illumination. The additional light aided the amphibious force in finding its way, but also helped expose ships and craft to the enemy. The DIME Force would encounter the toughest opposition during the American landings. Two tank assaults were repelled by naval gunfire.[11]

The opposition encountered for both armies would likely have been more formidable, if a supporting counter-intelligence operation carried out in advance of the landings had not been so successful. Code named MINCEMEAT, it involved purposely allowing a corpse, dressed as a British officer, to wash up on a beach at Punta Umbría in southern Spain on the Atlantic. Attached to the body was a briefcase containing fake top-secret documents which revealed the Allies were planning to invade Greece and Sardinia and had no plans to invade Sicily. German intelligence accepted the documents as genuine, with the result that the General Staff diverted much of its defensive effort from Sicily to Greece.[12]

Assault sweeping began at 2200 on 9 July, as the steel-hulled minesweepers *Steady* and *Sustain* (sweeping for moored mines) and wooden YMSs *207* and *227* (for magnetic mines), cleared the approach channel, and transport and fire support areas. YMSs *62*, *69*, *208*, and *226* expanded the swept area and after daylight, shoreward in landing boat lanes. No mines were found at Gela. While working in the boat lanes, YMSs came under fire several times, but escaped damage.[13]

Photo 15-1

Minelayer USS *Keokuk* near the Norfolk Navy Yard on 1 November 1942. During minelaying operations at Gela on 11 July 1943, she was attacked by six enemy planes, but anti-aircraft fire drove the aircraft off.
National Archives photo 19-N-36498

Most of the assault troops arrived in transports and made the shore in landing craft. The first boats hit the beach exactly at H-hour, at 0245 on 10 July. Everything went smoothly until dawn when enemy aircraft attacks commenced, and reports at 0830 of two columns of about twenty-five Italian tanks each, approaching the beach were received. (These were actually French Renaults captured during the German invasion of France and subsequently passed on to the Italians to bolster their comparatively weak armored forces.) Minelayers *Keokuk*, *Salem*, and *Weehawken* arrived at Gela the morning of 11 July as German aircraft were attacking tank landing ships unloading on the beaches, and destroyers were firing at German tanks, which had arrived earlier that morning, that were within 1,000 yards of the beach.[14]

During this melee, with the entire area under intense bombing and strafing attack, the minelayers went to work, sowing a defensive field of 1,170 contact mines around the anchorage area. The ships streamed abreast one another, dropping their mines every 10 seconds on signal. For their work, they received a message of congratulations from General Eisenhower, and a "Well Done" from Vice Admiral Hewitt, who didn't routinely send such messages. A German shore battery fired one salvo at the loaded minelayers during the operations; fortunately for everyone for miles around, the shots missed.[15]

For reasons of space, details are omitted about the other two American landings; and those of Vice Adm. Sir Bertram Ramsay's Eastern Task Force, between Formiche, on the west side of Cape Passero, and Cassibile, not far south of Syracuse. Although the British had easier meteorological conditions for their landings, as their beaches were sheltered from the wind, they suffered more from enemy air attack. The Americans initially met stiffer opposition, but after 15 July this situation changed.[16]

CONCLUSION OF OPERATION HUSKY

The capture of Sicily was an undertaking of the first magnitude.

—British Prime Minister Winston Churchill

The brilliant achievements of the Allied forces in this conquest, launched on a magnitude which heretofore had never been attempted, were due principally to the singleness of purpose which all forces demonstrated. The appreciation of each other's problems produced an inter-service spirit of co-operation and common endeavor which welded the naval and military forces into a single team possessed with the resolute will to win.

—Vice Adm. Henry K. Hewitt, USN, commander, Western Naval Task Force[17]

General Patton and his staff left Admiral Hewitt's fleet flagship *Monrovia* (APA-31) at 1700 on 12 July, and soon after leaped from a landing barge and waded ashore to the beachhead at Gela. His troops—consolidated from the three landing positions—advanced across the island to Palermo, then fought their way along the north coast, aided by the Navy. On 27 July, when Palermo Harbour was first opened to Allied shipping, Vice Adm. Henry K. Hewitt organised Motor Torpedo Squadron 15, Destroyer Squadron 8, several minesweepers, and a few other warships left in Sicilian waters, into "General Patton's Navy," to support the Seventh Army.[18]

Photo 15-2

British Gen. Bernard L. Montgomery (centre) and Lt. Gen. George S. Patton Jr. look over a map of Sicily, circa July-August 1943.
Naval History and Heritage Command photograph NH-95561

This force was commanded by Rear Adm. Lyal A. Davidson, embarked in the cruiser *Philadelphia* (CL-41). He arrived at Palermo on 30 July with the cruiser *Savannah* (CL-42) and destroyers *Butler* (DD-636), *Cowie* (DD-632), *Glennon* (DD-620), *Herndon* (DD-638), and *Shubrick* (DD-639). After anchoring outside the breakwater, Davidson met with Patton to discuss operations along the north coast of Sicily. These would include gunfire support for the Seventh Army as it advanced, providing craft to Patton for shore-to-shore amphibious operations to "leapfrog" enemy strongholds, and ferrying Army heavy artillery, supplies, and vehicles.[19]

On the night of 16 August 1943, beating the British, the leading elements of the 3rd Infantry Division entered Messina, and the city fell the next day. In just thirty-eight days, the Seventh Army, under Patton's leadership, and the British Eighth Army, under Montgomery, conquered all of Sicily. Thus ended Operation HUSKY, which Patton observed "out-blitzed the inventors of Blitzkrieg."[20]

AWARD OF BATTLE STARS TO MINELAYERS

The U.S. Navy's mine forces in World War II, like those of the Royal Navy, included hundreds of minesweepers—which collectively earned hundreds of battle stars. Citing even those associated with the European Theatre would consume much space in the remaining text. Accordingly, only those awarded to USN minelayers will be identified.

Sicilian Occupation

Ship	Period	Commanding Officer
Keokuk (CM-8)	9-15 Jul 43	Comdr. Leo Brennan, USNR
Salem (CM-11)	6-15 Jul 43	Comdr. Henry Goodman Williams, USN
Weehawken (CM-12)	9-15 Jul 43	Lt. Comdr. Robert Edwin Mills, USNR

16

Coastal Forces Operations in 1943 and early-1944

I have noted with admiration the work of the light coastal forces in the North Sea, in the Channel and more recently in the Mediterranean.

Both in offence and in defence the fighting zeal and the professional skill of officers and men have maintained the great tradition built up by many generations of British seamen.

As our strategy becomes more strongly offensive, the task allotted to the coastal forces will increase in importance, and the area of their operations will widen.

I wish to express my heartfelt congratulations to you all on what you have done in the past, and complete confidence that you will maintain the same high standards until complete victory has been gained over all our enemies.

—Prime Minister Winston Churchill in a letter dated 30 May 1943.[1]

Royal Navy surface minelaying in 1943 was performed almost solely by Coastal Forces craft, whose operations were extended to cover the Brittany coast and the Channel Islands. Off the Netherlands coast, two minefields were completed as far north as Texel Island. There was a marked increase in the number of mines laid by their craft from the previous year. The increase was largely due to the formation of the 10th ML Flotilla in the Plymouth Command for work to the west of the Cherbourg Peninsula, and to the use of two more MTB flotillas in operations off the Netherlands coast, where motor gunboats were also used on occasion. (The Cherbourg Peninsula forms part of the northwest coast of France, and extends northwest into the English Channel toward Britain. To its west lie the Channel Islands and to the southwest, the Brittany Peninsula.)[2]

Four areas were the targets of the units of the Coastal Forces mining programme. The Netherlands coast (which from November

1941 onward was within the German Northern Command) was attended to by the Nore Flotilla. The Belgian and French coasts as far west as Etaples were generally the responsibility of the Dover Command. Portsmouth Command boats visited the waters between Fecamp and Cherbourg, farther down the northern French coast, during Operation MAPLE (discussed in the following chapter). The coast of Brittany and the Channel Islands were allocated to the Plymouth Command. The latter three areas lay within the German Western Command.³

Map 16-1

Northwest coast of France. The port city of Brest is situated near the western tip of the Brittany Peninsula.

BRITISH USE OF SNAGLINE MINES

In 1943, the greater proportion of mining operations were carried out in Netherlands waters with the number of moored mines laid exceeding that of the ground variety. In part, this trend followed the example of the Plymouth boats which, after their first two missions in April, used only moored mines and obstructors. In the Channel, forces made use of a large number of Mk XXVII snagline mines, while also employing acoustic and magnetic types. Buoyed snagline mines were designed to be ensnared by the propeller of a vessel when passing above one of these type mines. The propeller would draw the snagline tight, thereby exploding the ordnance. Ground mines were used almost exclusively off the Dutch coast.⁴

The snagline mines proved troublesome to the enemy and, for a period from March onward, caused the loss of many R-boats—the Germans' only efficient shallow-water, moored-mine sweeper. The absence of any form of self-protection on the boats against the snag mine mandated daylight operations, facilitating the possibility of Allied air attack. In early April, while sweeping by day, seven of the eight boats of the 8th Flotilla were damaged in one attack by fighter aircraft, causing the Commander-in-Chief, Security West (Befehlshaber der Sicherung West) to comment, "If more extensive use is made of these clever mining tactics, we shall be faced with the closure of the Eastern Channel."[5]

With their initial successes, Coastal Force craft continued to lay snag mines, but the Germans used Belgian fishing vessels to find the lines and neutralise the mines. This measure, together with the use of alternate routes (for now very limited shipping traffic), gradually lessened their impact. By September, snag mines no longer posed a significant threat to enemy vessels. Not so for the fishermen engaged by the Germans in disposing of them—as can be gleaned from faint praise in a description of such operations from *British Mining Operations 1939-1945 Vol 1*:

> Sweeping was achieved by the simple means of employing Belgian fishing vessels, working under the guidance of the local [German] minesweeping officer, to fire the mines by pulling on an extended snagline. These latter were 75 feet long and the mine contained only 100 lbs of explosive; but, even so, this method of sweeping must have required considerable sangfroid [the ability to remain composed in a difficult or dangerous situation] on the part of the operators.[6]

JANUARY THROUGH MARCH 1943

In the first quarter of 1943 as weather permitted, the Nore and Dover Coastal Forces continued the same pattern of minelaying as occurred in the previous year, with both motor torpedo boats (MTBs) and motor gunboats (MGBs) being used to back up the Fairmile motor launches (MLs.) The designation "QU" was used for minelaying operations carried out by Nore forces, and "NL" for those of the Dover Command. The January "dark period" (around the New Moon phase, when the moon is darkest) was disappointing, in that only two operations were able to be carried out. Weather conditions improved and between 24 February and 6 March, the Dover boats laid 236 mines and five obstructors. With the observance of enemy activity, escort was provided for the motor launches in most instances, by motor gunboats.[7]

On 6 March while carrying out Operation NL 8, a group of motor launches were attacked in French waters off Gravelines by two darkened vessels, thought to have been torpedo boats. Taking evasive action, including making smoke, the MLs *102, 104, 107, 210,* and *213* escaped.

However, *ML 102* and *ML 104* suffered some damage and casualties. The following listed officers and men of the 50th Flotilla were "gazetted" on 13 July 1943 (when a notice of the DSO, DSMs and MIDs awarded them by the King appeared in *The London Gazette*). Those earned by personnel assigned to MLs *102* and *104* may have received them with respect to the attack on 6 March as well as other operations.

50th Motor Launch Flotilla (based at HMS Wasp in Dover)	
Stoker First Class Andrew Wood (*ML 100*)	MID for bravery, initiative, and devotion to duty in many dangerous minelaying operations
Leading Stoker William Harold Morris (*ML 101*)	DSM for bravery, initiative, and devotion to duty in many dangerous minelaying operations
AB Francis John Churcher (*ML 102*)	MID for bravery, initiative, and devotion to duty in many dangerous minelaying operations
AB Kenneth Lewis Thomas (*ML 102*)	DSM for bravery, initiative, and devotion to duty in many dangerous minelaying operations
A/T/Lt. Comdr. Thomas Aubrey Ashdown, RNR (SO, 50th ML Flotilla, and CO, *ML 104*)	DSO for bravery, initiative, and devotion to duty in many dangerous minelaying operations
AB Gunner James Herbert Barrett (*ML 104*)	DSM for bravery, initiative, and devotion to duty in many dangerous minelaying operations
AB Robert Henry Homewood (*ML 108*)	MID for bravery, initiative, and devotion to duty in many dangerous minelaying operations
Telegraphist John Leslie Sprigg (*ML 108*)	MID for bravery, initiative, and devotion to duty in many dangerous minelaying operations
Telegraphist Ronald Charles Nelmes	DSM for bravery, initiative, and devotion to duty in many dangerous minelaying operations
Ord Signalman Joseph Henry Pead (50th ML Flotilla)[8]	DSM for bravery, initiative, and devotion to duty in many dangerous minelaying operations

MAKE UP OF THE NORWEGIAN 52ND FLOTILLA

In March 1943 *ML 573* joined the Norwegian-manned 52nd Flotilla as a replacement for *ML 125* lost on 2 November 1942. The crewmen were mainly from the Norwegian Merchant Service—two of whom had escaped from Norway by rowing across the North Sea to Scotland in an open boat. Another, Petty Officer Telegraphist Finn Christian Gysler, had a much longer journey in escaping Norway and finding his way to Britain.[9]

When the Germans occupied his country, Gysler was a student at the Kongelige Norske Marine (Norwegian Navy) Navigation School and Radio School. The Germans allowed the schools to continue, because they took all the students to serve in their navy. When Gysler finished he had to get away or they would have taken him for service in the Kriegsmarine, and at that time there was a death penalty for trying to escape Norway.[10]

He first escaped to Sweden on skis, leaving at 2100 one night, and guided by the star Arcturus reached the border to the east. From Stockholm, Gysler flew to Helsinki and then travelled by train to Leningrad and then Odessa. He boarded a cruise ship at Odessa which took him across the Black Sea to Istanbul. From there it was back on a train through Turkey to Alexandretta, where he boarded an English freighter bound for Port Said. Halfway across the Mediterranean, the ship was attacked by a Syrian Vichy plane which dropped a bomb on hatch no. 1. An hour later, an Italian Dornier came in, but its bombs fell 100 yards away.[11]

After reaching Port Said, Gysler and others were sent to Aden and put aboard the troop transport *Empress of Asia*. There he met Prince Philip who was also on his way to England. From Aden, the ship went to Durban, Cape Town, Trinidad and New York. Arriving at New York, he was put up in a hotel near Grand Central Station. One day, he received a telephone call informing him that he was going aboard the tanker *Sevnor* as third mate and telegraphist, and that the ship was to load oil and then proceed to England. Upon arrival in the UK, Gysler was taken into the Norwegian Navy and ended up in Troon, Scotland for duty aboard a motor launch. He was assigned as the wireless operator of *ML 210*, then a part of the 4th Motor Launch Flotilla.[12]

In early 1941, the Royal Norwegian Navy had taken over some British motor launches at Portsmouth (see Chapter 10 for more details about their war involvement.) Their crews were mainly from Norwegian merchant ships and particularly whalers. The launches were sent to Troon, on Scotland's west coast, in late summer for patrol duty, and then to Weymouth on the English Channel coast, where four of the MLs were returned to the Royal Navy. The remaining MLs formed the 52nd Motor Launch Flotilla, and proceeded to Poole to be rebuilt as minelayers. The funnel of each launch was removed, and "Dumbflows" fitted to make the vessels quiet when operating in enemy waters.[13]

The 52nd ML Flotilla arrived at Dover in July 1942, based with other Coastal Forces at the former Lord Warden Hotel in Dover. The Royal Navy had requisitioned the hotel on 2 September 1940 for use as the headquarters of the Coastal Force Base, HMS Wasp. The 52nd Flotilla's tasking was to join the British 50th ML Flotilla at Dover in conducting minelaying along German convoy routes, and blocking enemy port entrances. These operations were carried out during dark, moonless periods.[14]

DOVER FLOTILLA HEADQUARTERS

While under Royal Navy stewardship, the hotel housed Coastal Forces administration, plotting rooms, signals section, maintenance, and billeting for personnel. The boats, for the most part, were berthed in the Camber area of the Eastern Dover Docks. WRNS (members of the Women's Royal Naval Service) assigned duties at HMS Wasp were billeted at Dover College. The college was part of HMS Lynx, the name given to all shore base activities in Dover. When air raids threatened, if time permitted, WRNS would seek shelter in the old Oil Mill caves, Limekiln Street. (The tunnels had been dug in the early- to mid-19th Century to extract chalk for burning and turning into lime.) The hotel was shelled during the war, but survived and HMS Wasp was decommissioned on 14 November 1944.[15]

Photo 16-1

Current view of the Warden Hotel in Dover, which served as HMS Wasp in WWII. A plaque is mounted on the front of the hotel which, unveiled on 10 July 2010, commemorates the men and women of Allied Coastal Forces who served with the Royal Navy at HMS WASP at Dover between 1940 and 1944.
Courtesy of Mark Chapman

The operating area for the Dover-based 50th and 52nd Flotillas was between Zeebrugge, Belgium, and Boulogne, France. Fast German E-boats, which had a top speed of 40 knots and heavy armament, posed the biggest danger to the motor launches, and there was always a threat from shore batteries. The MLs could make 17 knots for a short time, before reducing speed to 15 knots. Possessing little armament, the motor launches had to operate secretly, employing stealth. They would creep up on the French coast, lay their mines, and creep away. The Norwegian boats also took part in other operations, including landing commandos on the French coast.[16]

Under the oversight of T/A Gunner (T) Donald James Pangborne, the mines would come aboard the day of an operation and WRNS would set the depths and install the safety pins. From this point on, there was always a tense atmosphere aboard the launches, as their crews always expected shelling from France. During frequent attacks on Dover, warning could not be given before the first 2,000 lb. gun round would fall. A hit on the mine-loaded MLs would have resulted in severe damage and likely death and injury as well.[17]

Photo 16-2

German heavy naval artillery bombarding Dover in World War II.
Naval History and Heritage Command photograph #NH 59801

APRIL THROUGH JUNE

In April 1943, motor launches from the Plymouth command joined in Coastal Forces minelaying operations. In November 1942, its Commander-in-Chief had proposed employment of a ML flotilla, soon to be allocated to his command, on minelaying operations along the French coast between the Cherbourg peninsula and Ile de Bas. Arrangements were made to fit out eight boats, as they became available. Following a period of practice in minelaying and close station keeping, the 10th Flotilla carried out its first operation (Hostile 3) on the night of 3 April. The lay was one mile offshore in Lannion Bay, on France's Brittany coast. In their second operation on the 11th, owing to enemy movements in waters between the Channel Islands, Guernsey and Herm, only half the boats laid their mines. This operation was designated "Hostile 4," signifying that it also was carried out by forces of the Plymouth Command.[18]

The month of April proved to be a poor one for minelaying; dirty weather and high visibility particularly affected operations from Dover, but the Nore boats managed six operations off the Netherlands coast. One such was abandoned on the night of 29 April, when attack by enemy forces caused mines to be jettisoned. Gales were experienced in the first half of May, but during the latter part of the month seven operations were carried out, three in the "Hostile" series by the Plymouth boats.[19]

ENGLISH CHANNEL AREA

The Channel area, a responsibility of the Dover Flotillas, offered a large expanse of mineable water, the greater part of which lay well within fifty miles of the English coast. A network of narrow channels along the Belgian and French coasts between Ostende and Gravelines offered an abundance of suitable sites. Enemy shipping had the choice of many alternate routes, but it was generally possible to detect any changes in the pattern of movements and to adjust mining operations accordingly. To the west of Gravelines, as far as Cap Gris Nez, the French coast was less cluttered with offshore shoals that might otherwise route traffic, and the enemy had more freedom of action.[20]

In these mining areas frequented by enemy vessels, the likelihood of encountering opposition was increased. An operation by Dover forces comprised of three motor launches, two motor torpedo boats, and four motor gunboats was abandoned on the night of 28 May after two of the motor launches broke down, requiring escort home by the third. The two motor torpedo boats, continuing on to the planned laying position off Dunkirk, were driven off by enemy trawlers. The four motor gunboats in support engaged in combat with enemy patrol craft, resulting in *MGB 110* being sunk with eleven killed, or missing and presumed killed. However, considerable damage was inflicted on the enemy.[21]

Motor Gunboat HMS *MGB 110* (9th MGB Flotilla) Casualties

Name	Rank/Rate	Name	Rank/Rate
Hugh Bailey	Able Seaman	Edward C. Keeping	T/A/Lead Seaman
Stanley R. Burgess	Signalman	George D. K. Richards	Lieutenant.
David J. Crosscurth	Able Seaman	Arthur R. Stagg	Sub. Lt., RNVR
Albert E. Fulton	Able Seaman	Reginald G. Thomas	Ordinary Seaman
Thomas I. Gough	Able Seaman	William Waterman	T/Sub. Lt., RNVR
William Grainger	Able Seaman	[22]	

The weather during the June lunar dark period was not particularly good, some opposition was encountered, and only seven operations were successfully carried out. Motor launches of the 52nd Flotilla were to lay mines off Etaples (Operation NL 86), but were prevented from doing so by enemy interference.[23]

Minelaying off the Cherbourg Peninsula and Brittany to the southwest, had begun by Plymouth Command's 10th Motor Launch Flotilla in April 1943 with a new series of its "Hostile" operations against coastal shipping. During the Hostile 9 operation northeast of Guernsey on 8 June 1943, shore batteries engaged part of the force (MLs *157, 159, 180, 181, 184, 186, 259, 488*). Some superficial damage was suffered, but *ML 159* and *ML 259* laid their mines while under fire.[24]

JULY THROUGH SEPTEMBER

Weather conditions restricted minelaying operations by Coastal Forces craft in July and August, particularly during the dark period at the end of August when strong westerly winds persisted. Things improved in September bringing the total number of weapons laid to 948 mines and ten obstructors. Enemy resistance was less frequent than in the previous quarter. However, on 2 September MTBs *38, 219, 221,* and *240*, having just finished laying sixteen A Mk I-IV ground mines off the western approaches to Dunkirk as part of Operation NL 99 were engaged by enemy trawlers and coastal batteries, fortunately without losses.[24]

An unfortunate incident on 4 September, involving the 50th and 52nd Flotillas in a mining operation off Etaples, resulted in the tragic loss of *ML 108* of the 50th Flotilla. The Norwegian boats of the 52nd Flotilla laid their mines first, and departed. The 50th Flotilla then began to lay their ground mines, in line ahead starting with the rear boat. About ten seconds after the second boat in the line—*ML 108*—had laid her last mine, a heavy explosion astern of her knocked her main engines and steering gear out of action. The flotilla leader, *ML 104*, secured her mines on deck and took the *ML 108* in tow. Despite strenuous efforts to keep her afloat, the damaged motor launch "turned turtle" at 0615 and was lost. All of her crew were picked up.[26]

A Board of Enquiry found that the explosion that sank her was due to a faulty soluble plug in the last mine laid. The Admiralty agreed with this finding, but considered that this defect should have been discovered during the final tests, had they been carried out correctly.[27]

OCTOBER THROUGH DECEMBER

Inclement weather and poor visibility restricted minelaying in the first half of October to a single lay. Thereafter conditions improved and a total of 496 mines and 41 obstructors were laid during the quarter. In November and December, the Plymouth command's "Hostile" operations were extended westward to the coastal route between Ile de Bas and Les Sept Iles, and three fields were laid by their motor launches in that area.[28]

COASTAL FORCES MINING TALLY FOR 1943

Coastal Forces craft laid 2,447 mines (1,310 of them moored) and 83 obstructors in 1943. Twenty-five German controlled vessels of 21,692 tons were sunk, and twenty-one of 14,565 tons damaged. These results were obtained despite steadily decreasing enemy shipping traffic, which averaged only 14,000 gross registered tonnage a month over the course of the year. From July onward, the increased mining by the British helped lower the monthly average to 7,290 GRT. Minesweepers (R-boats suffered most) and fishing vessels constituted the greater part of these casualties. Also, two large merchant vessels were sunk.[29]

GERMAN MINESWEEPING FORCES

Minelaying by Coastal Forces craft and submarines in BSW's (Befehlshaber der Sicherung West) area kept German minesweeping forces fully occupied. All over the Atlantic coast, they were hard-pressed to provide surface escort, and mine-free seaward routes for U-boats. Further, since the Germans were in possession of the Channel Islands, protection of supply shipping to this exposed location had to be provided. The added work associated with mine protection for its shipping was most unwelcome.[30]

Moreover, occasional submarine mining operations by the Allies in the Bay of Biscay added to this burden. The French submarine *Rubis* visited the Bay of Biscay on four occasions between July and October 1943, laying two fields close to Brest and two on the Spanish-iron-ore route. (See Chapter 12 for background information.) The second lay carried out by her, north of Bayonne, preoccupied an entire flotilla of modern minesweepers over a lengthy period, when these ships were urgently required elsewhere. By the end of 1943, only half of BSW's minesweeping forces were operational and Sperrbrecher were used only for inshore work off the Atlantic ports.[31]

BSW's Sweeping Forces at Year's End in 1943

Minesweeper	Operational	Non-operation
M-class	31	25
R-boats	39	7
Auxiliaries	129	59
Sperrbrechers	15	17
Total[32]	214	108

LOSS OF *ML 210* TO GERMAN MINE IN EARLY 1944

In early evening on 15 February 1944, five motor launches—MLs *104, 107, 125, 210,* and *573*—left Dover at 1820 to carry out Operation NL 117. It was a cold night with snowfall. The mission of the combined group of 50th and 52nd Flotilla boats was to lay Mk XXVII mines (snagline version of the Mk XIX in service in 1942 but obsolete by 1944) across the coastal route opposite the Somme Estuary. Motor launches *125* and *573* had completed their lays when at 2232, an explosion blew off the bow of Norwegian *ML 210*, which by then had laid six of her nine mines. Almost simultaneously, *ML 573* was damaged by another explosion close astern, but suffered no casualties and was able to return to harbour under her own power.[33]

The motor launch *ML 210* sank with the loss of two officers and three ratings, with a further ten ratings injured.

Motor Launch *ML 210* Casualties

U/f Alfred Julius Blomberg	U/mm Ivar Martinsen
U/lt Asbjørn Harr Grøneng	U/mm Einar Smith
U/kvm Alf August Little-Kalsøy	

U/lt: lieutenant
U/f: sub-lieutenant
U/kvm: Petty Officer
U/mm: Engine room rating[34]

Surgeon Commander Finsen of the 52nd Flotilla was aboard the motor launch, and provided a graphic account of the traumatic experiences of her crew:

> Just after mine number 6 was laid on the port side, I heard a terrific explosion, and was thrown up in the air and out into the water on the starboard side. When I came to the surface I saw several people floating among the wreckage, about 8 to 10 metres off the starboard quarter of the *ML 210*. She was still afloat, although her bow was down in the water. I heard screams for help and people moaning. The boys helped each other and one or two managed to enter the wreck.

We swam around the stern and saw the British *ML 104* coming slowly towards the wreck's port side. Another ML also closed up, and two men were taken on board. The nine remaining survivors were taken on board *ML 104* by means of lines thrown to us. When we had all been rescued the crew of *ML 104* shouted and then listened, to make sure everyone had been taken on board. No sounds were heard and visibility was good.

We could see all the wreckage on the starboard side of *ML 104*. I was convinced nobody was floating in the water and nobody answered our calls. I reckoned the last man to be taken out of the water had been in the sea for only ten minutes. The crew of ML 104 looked after us in a brilliant way. All of us were taken below, undressed and wrapped in blankets. They gave us rum, warm tea and soup.[35]

FINAL SERVICE OF THE 52ND ML FLOTILLA

Over the next few months, the Norwegian flotilla was assigned mostly daylight duties, which included carrying out patrols and making smoke to cover Allied convoys which passed through the Dover Strait. In early September in 1944, war operations for the 52nd Motor Launch Flotilla ended. Flag Officer Commanding Dover, Admiral Henry Pridham-Wippel, wrote on 7 September 1944:

> Now that the activities of Dover forces are drawing to a close, at any rate so far as enemy seagoing action is concerned, I wish to congratulate all in the Command on their highly successful discharge of the responsibilities assigned to them.[36]

As Norwegian forces sailed for a freed Norway, the last signal they received was from Capt. William Adams, RN, who commanded HMS Wasp and the Coastal Forces Base, Dover: "Thank you. Goodbye and good luck. Your efforts have given your flotilla a reputation, deserved second to none." Records compiled by Captain O'Sullivan, Naval Historical Branch, Ministry of Defence, Empress State Building, London, provide a summary of the results achieved by the 50th and 52nd Motor Launch Flotillas:

- 53 ships sunk, 34 ships damaged
- 2 minesweepers destroyed
- 4,000 to 5,000 mines laid in 170 operations on the French Coast
- 2 British and 2 Norwegian motor launches lost[37]

NORWEGIAN EXPRESSION OF GRATITUDE

Following the war, Finn Christian Gysler, who had journeyed so far to fight the enemy who had occupied his native land, took over the family farm named Stumoen Gard (Gard being Norwegian for farm) on the death of his grandfather. The farm and estate had been in the family since 1760 and today comprises approximately 350 acres of bottom arable land and also about 1,000 acres of pine forest which is managed as a plantation. To be able to take on the trust, Finn had to agree to change his name to Stumoen which he did. For many years as an act of appreciation of the treatment of Norwegian sailors, Finn supplied the town of Dover with its Christmas tree supposedly from his own forest. In actuality it came from a farm near Canterbury. Finn did attempt to bring a tree but the British authorities refused to let him bring it ashore from the ferry.

However, when he was in his 80s, he managed to get the necessary permits and brought a young tree over. It was planted at the junction of Pencester Road and Maison Dieu Road at Dover, one of the principal routes in the centre of the town. There was a planting ceremony and the placing of a plaque, a photograph of which may be found on the following page. This was a gracious act by a grateful Norwegian citizen. In many ways it was not surprising given the tough northern nation's love of freedom and disdain for oppression, and their respect of others.

Photo 16-3

Finn Christian Stumoen, his wife, and the mayor of Dover at the planting of the memorial tree and placing of the plaque.
Courtesy of Colin Bernard Torlief Stromsoy

17

British Fast Minelayer Force Decimated in 1942-1943

Three of the Royal Navy's six fast *Abdiel*-class cruiser minelayers were lost in the war. Others were damaged by enemy action including HMS *Manxman*, whose exploits during a clandestine mission are highlighted in the first chapter of this book. The first of these ships lost was HMS *Latona* (M 76), on 25 October 1941 to an enemy aircraft attack. Details about this incident may be found in Chapter 8. This chapter also describes the important support lent by fast minelayers to the Allied military garrison at Tobruk, to Allied forces on Cyprus, and to the British colony of Malta, in addition to their mining duties.

The losses of, or damage to, the following three fast minelayers all occurred in the Mediterranean, over a twenty-one-month period.

Date	Ship	Disposition
1 Dec 42	HMS *Manxman* (M 70)	Damaged by a torpedo fired by the German submarine *U-375* (Kptlt. Jürgen Könenkamp).
1 Feb 43	HMS *Welshman* (M 84)	Sunk east-northeast of Tobruk, Libya, by two torpedoes from *U-617* (Kptlt. Albrecht Brandi)
10 Sep 43	HMS *Abdiel* (M 39)	Sunk in Taranto Harbour, Italy, by mines laid by the German motor torpedo boats *S-54* and *S-61*

HMS *MANXMAN* DAMAGED BY TORPEDO

After joining the Mediterranean Fleet in November 1942, *Manxman*, under the command of Capt. Robert Kirk Dickson, RN, received orders to carry stores and personnel to Malta. She took passage from Alexandria, on 10 November under escort by six destroyers. The group's arrival on the 12th marked the first resupply by surface ships from Egypt in twelve weeks. The following day, *Manxman* embarked and prepared mines for her first minelay with great difficulty because of local conditions. On 25 November, she sailed for Algiers.[1]

Manxman laid a minefield off Cani rocks in the Sicilian Channel on 29 November and then returned to Algiers. Although an E-boat had passed near her, the lay was successfully completed. *Manxman* stood out of Algiers on 1 December 1942, bound for Gibraltar to collect a supply of mines for future operations.[2]

Map 17-1

Central Mediterranean

That evening at 1701, Kptlt. Jürgen Könenkamp, commanding *U-375*, fired a spread of four torpedoes at the minelayer, which he believed to be a *London*-class cruiser. Könenkamp heard two hits after 51 seconds despite the fact that *Manxman* was zigzagging at 21 knots. Hit on the port side, the explosions disabled the minelayer and seawater entering her hull caused a 12-degree list.[3]

Manxman was initially taken in tow by the destroyer HMS *Pathfinder*, escorted by a second destroyer, HMS *Eskimo*. Passed to the rescue tug HMS *Restive*, she arrived at Mers-el-Kebir on the Algerian coast southwest of Algiers on 2 December. Later that month she was towed to Gibraltar for temporary repairs. In June 1943, *Manxman* left Gibraltar under tow by another rescue tug (HMS *Bustler*), arriving on 8 July at Newcastle-upon-Tyne, in northeast England, for permanent repairs that took almost two years. Returning to service in April 1945, she joined the British Pacific fleet.[4]

WELSHMAN SUNK BY U-BOAT TORPEDO ATTACK

Two months later, HMS *Welshman* (Capt. William H. D. Friedberger, RN) was returning to Alexandria from Malta on 1 February 1943, when she was sunk east-northeast off Tobruk, Libya, by two torpedoes from the German submarine *U-617*. Her commander, Kptlt. Albrecht Brandi had, at 1745 in early evening, fired a spread of four torpedoes. Two of the torpedoes hit, causing a boiler explosion and flooding of the mine deck which could not be corrected. The ship capsized and sank by the stern two hours later with much loss of life: eight officers, 144 ratings and 13 passengers, including two civilians and four air crewmen that had been badly burnt in a plane crash on Malta. Some of the casualties resulted from the explosions of depth charges.[5]

The escort destroyers HMS *Tetcott* and *Belvoir* recovered five officers and 112 ratings five hours later and took them to Alexandria. Another six survivors were rescued by small craft from Tobruk.[6]

HMS *Welshman*, like other fast minelayers, had supported the island of Malta during its long siege. *Welshman* brought food and essential supplies many times, and her role was later featured in the 1953 British war movie, *The Malta Story*. On the day that she was lost, the minelayer was on her way back to Alexandria after having laid a field west of Sicily. The Italian corvette *Procellaria* was sunk in the field on 31 January, as was the Italian torpedo boat *Generale Marcello Prestinari* later that day. Sent from Bizerte in mid-afternoon with orders to assist the *Procellaria*, she was mined herself at 1630, eighteen nautical miles from Ile de Cani (an island off Tunisia), and went down about an hour later.[7]

ABDIEL SUNK BY MINES AT TARANTO, ITALY

HMS *Abdiel* (Capt. David Orr-Ewing, RN) was mined and sunk in Taranto Harbour on 10 September 1943, while supporting Operation SLAPSTICK, an amphibious operation by the British 1st Airborne Division at Taranto, giving the Eighth Army a second foothold in Italy. The division was carried to Taranto on six British warships, in addition to HMS *Abdiel* and the cruiser USS *Boise* (CL-47), rather than landing from the air.[8]

As the Allied force approached Taranto, it was passed by Italian ships that had been based there, heading to surrender at Malta. Because these ships included the battleships *Andrea Doria* and *Caio Duilio*, there were some nervous moments as the two forces passed each other. The 3,600 airborne troops landed without resistance at Taranto on 9 September. They found no Germans, and were instead welcomed by the Italians as part their armistice agreement with the Allies.[9]

The British soon secured the city and its port and the nearby airfield at Grottaglie, and thereby gained control of the Adriatic coast around Bari and Brindisi. HMS *Abdiel* was sunk at anchor on the 10th by mines laid just a few hours earlier by the German motor torpedo boats *S-54* and *S-61* while they escaped from the harbour. She had taken a berth which had been declined earlier by the commanding officer of the *Boise*. Shortly after midnight, two ground mines detonated beneath her and she sank in three minutes with great loss of life among the sailors and soldiers aboard. The 6th Royal Welsh Battalion of the division suffered 58 dead and around 150 injured, while 48 crewmen were lost.[10]

Photo 17-1

Troops of the British 1st Airborne Division aboard the light cruiser USS *Boise* (CL-47) en route to make a landing at Taranto, Italy, on 8 September 1943.
National Archives photograph #80-G-85577

18

Operation MAPLE

Operation "Maple" was a long-term commitment designed to assist in the protection of the Allied forces in the Channel—especially the bombarding and assault forces—from attack by E-Boats and R-Boats based at Cherbourg and Le Havre.

The minelaying forces consisted of the Apollo and Plover based at Plymouth and Dover respectively, 4 M.L. Flotillas (22 M.L.s) and five M.T.B. Flotillas (36 M.T.B.s) distributed between the south coast commands, and heavy bombers of Bomber Command. The disposition of these forces was kept flexible in order to take advantage of any intelligence of enemy movements, or of the laying of defensive minefields by him.

—From *Battle Summary No. 39, Operation "Neptune" Landings in Normandy June, 1944*, published by HMSO in 1994.[1]

Planning for an Allied invasion of the continent of Europe had proceeded throughout 1943, and on 28 February 1944 the Naval Plan, Operation "Neptune," was issued to all authorities. The document included a detailed minelaying plan under the title Operation "Maple." From 17 April until D-day of the Normandy invasion on 6 June, all minelaying in northwestern Europe was carried out in support of this programme. The minelaying operations were divided into six phases, during which the employment of special mines was gradually introduced in a manner so that unobtrusive concentration on "Neptune" targets was not evident to the enemy. In the table, "to D-45," for example, refers to the period up to forty-five days before D-Day on 6 June.

Phases	Operations
I (to D-45)	Routine offensive laying by coastal force and aircraft using standard mines
II (D-45 to D-24)	As in Phase I with introduction of special type mines by surface minelayers
III (D-24 to D-3)	As in Phase I with laying of special type mines by surface minelayers and aircraft

IV (D-3 to D-1)	Laying of special type mines only by available coastal force minelayers, the main concentration off Le Havre, Cherbourg, Calais, and Boulogne, and by aircraft off Ijmuiden, Hook, West Scheldt, Chenal de Four, and Brest
V (D-1 to D-Day)	Laying of special type mines only by coastal force minelayers off Le Havre, Cherbourg, Entretat, and Brittany coast
VI (after D-Day)	As requisite[2]

In early 1944, surface minelaying remained the province of Coastal Forces until 11 March, when the newly built *Abdiel*-class fast minelayer HMS *Apollo* first participated in Operation MAPLE. On that day, she embarked mines at Milford Haven on the southwest coast of Wales. Six days later she began a series of minelays in the western approaches to the Normandy assault areas as part of the "Hostile" series. At the completion of each operation, she returned to Milford Haven to reload mines to continue the programme.[3]

The coastal minelayer HMS *Plover* joined in these operations on 18 April, when she carried out a minelay in the NL 126 series (MAPLE) off the Dutch coast. In total, *Apollo* sowed 1,170 mines in fields covering the Channel western approach routes, while *Plover* laid 250 mines in three fields in the Outer Ruytingen Pass, in order to help guard the centre of the eastern approaches.[4]

During the pre-MAPLE period, the Plymouth, Dover, and Nore Flotillas of Coastal Forces laid a total of 512 mines in their respective mining areas. A further 1,471 mines were added from 17 April to the end of June, as required by the mining plan. The Dover and Nore commands received motor torpedo boat reinforcements for this heavy programme, with three additional flotillas of MTBs from Portsmouth to conduct mining on the immediate flanks of the landing beaches.[5]

From the commencement of MAPLE, various varieties of special firing circuits were employed in British ground and moored mines laid by surface craft. This was done on a prearranged schedule in order to attempt to confuse the Germans by creating sweeping problems in the final days before D-Day. The special mines were sown in conjunction with the routine laying that the Germans were aware of to try to conceal the intended date and area of the assault. A concentration of fields aimed at "Neptune" targets was thus brought about unobtrusively within the normal programme. These targets were the assumed paths of enemy ships and submarines once they became aware of the D-Day assaults. Full use was also made of delayed-release sinkers to produce new fields of moored mines over a short period prior to the landings.[6]

To minimise the risk of discovery of new mine types by the enemy, their deployment by aircraft in areas near the German-held coast was delayed until 20 May. Exceptions to this rule were made in the case of Le Havre and Cherbourg. At these locations, minelaying was advanced by five days to take full advantage of suitable conditions.[7]

The special ground mines laid in the English Channel were designed to be effective against German E- and R-boats to water depths of 15 fathoms below the surface, and moored mines to be effective during all states of the tide. The life of all mines was tightly controlled, particularly on the wings of the assault area. Mines off Cherbourg were set to become inactive by 12 June, and those at Le Havre by the 20th.[8]

For Operation MAPLE, Coastal Forces and Bomber Command were assigned specific mining areas. The Portsmouth command was responsible for mining operations in the vicinity of the Normandy assault area, designated "KN." The Plymouth (Hostile), Dover (NL), and Nore (QU) commands concentrated on the eastern approaches in their respective areas, while Bomber Command sowed mines as required between the Frisian Islands and Brest, France. (The Frisian Island archipelago in northwestern Europe stretches from northwest of the Netherlands through Germany to west of Denmark, see Map 3-6).[9]

LOSS OF DUTCH MOTOR TORPEDO BOAT *MTB 203*

Photo 18-1

Earlier built Vosper 70-feet-type MTB 23 launching two torpedoes.
Australian War Memorial photograph 302504

On the evening of 17 May 1944, motor launches *ML 101*, *102*, *107*, *123*, and *125*, together with motor torpedo boats *MTB 202*, *203*, *204*, *229*, and *231*, sailed from Dover to lay mines off Etaples in northern France. At 2245, the MLs were unable to continue due to unsuitable weather, but the Dutch 9th MTB Flotilla proceeded to the initial position. As the boats drew close to it, the order "Prepare to lay mines" was given at 0115. *MTB 203* misunderstood it to mean "Lay mines," and disposed of hers in the direction 103°T from point 50°33'06" N, 01°28'34" E.¹⁰

Five minutes later, the planned lay commenced in the same direction from a position (50°32'55" N, 01°29'50" E) a little farther to the northeast. As *MTB 229* was emplacing her second mine, *MTB 203* blew up and sank after striking a mine. Mindful of the grave danger presented by the high explosives still aboard, *MTB 229* jettisoned the remainder of her mines, as did *MTB 204* before she went to the assistance of her sister boat. The other three MTBs withdrew from the area. The last survivor was picked up at 0155, and the group returned to Dover. One rating was missing and five of the crew wounded.¹¹

MTB 203 had entered service with the *204*, *235*, and *240* between March and June 1942, when the four boats were commissioned for service with the 9th MTB Flotilla. MTB *203* and *204* were White 73-feet-type boats built by J.S. White & Co. (Cowes, England), while the *MTB 235* and *240* were of the Vosper 72-feet-type class (also constructed in England at Berthon Boat, Lymington; and Morgan Giles, Teignmouth, respectively).¹²

AIR MINING OPERATIONS

Operations in northwest Europe were, from April to June, linked directly to the planned Allied invasion of Europe. Surface mining in the English Channel and North Sea came to an end in August, shortly before the retreat of the main German forces from France. However, the air mining effort continued in strength, concentrating on the German North Sea ports, the Skagerrak, Baltic, and southern Norway. The long neglected Norwegian west coast came under mining attack by carrier aircraft. The latter area was also occasionally visited by minelaying submarines, as was the Biscay coast twice on two occasions by the Free French submarine *Rubis* during the first quarter of 1944.¹³

The air mining effort in NW Europe was substantially increased in 1944 over the previous years. Bomber Command averaged 1,458 mines a month over the entire year, and 2,700 a month in April and May as part of Operation MAPLE. At the same time, the number of sorties decreased slightly owing to the modification of mine stowage in some aircraft. High altitude mining was almost universally employed from March onward. This practice resulted in a drastic reduction in aircraft casualties, but also resulted in many mines falling on land to disclose their secrets to the enemy.¹⁴

RESULTS ACHIEVED BY OPERATION MAPLE

> *These operations made an effective contribution to the general immunity from surface and U-boat attack enjoyed by the Assault Forces.*
>
> *From information so far available, it is evident that a considerable number of casualties were inflicted on the enemy and that his minesweeping organisation was stretched to the limit. In addition, the minefields in the vicinity of Ushant and off the Brittany coast had the desired effect of driving U-boats into open waters where they could be dealt with by our A/U [air and undersea] forces.*
>
> —Adm. Sir Bertram H. Ramsay, Allied Naval Commander-in-Chief, Expeditionary Force in this report on Operation NEPTUNE.[15]

Between 17 April and 30 June 1944, Bomber Command aircraft and surface forces laid a total of 7,372 mines in support of Operation NEPTUNE. Aircraft made 313 visits to the six areas identified in the following table, accounting for 61 percent of this total. HMS *Apollo*, HMS *Plover*, and Coastal Forces craft carried out 68 operations in three of these areas, accounting for the other 39 percent of the mines laid. *Apollo* sowed 1,170 mines, *Plover*, 250, and the flotillas of small, determined Coastal Forces craft collectively, 1,441, totaling 2,891 mines.

Area	Surface	Air	Total
A - Baltic, Kattegat and Kiel Canal		684	684
B - Heligoland Bight and Frisian Islands		847	847
C - Texel to Etaples	972	625	1,597
D - Cap d'Antifer to Cap de la Hague	536	481	1,017
E - Cap de la Hague to Brest	1,383	1,207	2,590
F - Biscay ports south of Brest		637	637
Total[16]	2,891	4,481	7,372

In total, 122 vessels of 82,351 tonnage were sunk or damaged during the period April through June in these areas, although some may have been due to mines previous laid. Thirty-seven of the casualties were merchant ships and miscellaneous vessels. The remainder were warships, mostly minesweepers, but the total also included seven E-boats damaged, and another two sunk, and three U-boats damaged. Surface-laid mines accounted for twenty-three of the casualties.[17]

Not surprisingly, the bulk of medals garnered during the war by officers and men of the Coastal Forces for minelaying operations were earned during this period. The dates of the awards listed in the following tables reflect when notice of such honours was reported in *The London Gazette*. A few of the medals are associated with earlier or later events, but are included for readers interested in this information. The letters "CO" and "SO" in the first column stand for commanding officer and senior officer.

COASTAL FORCES AWARDS FOR VALOUR

10th Motor Launch Flotilla (All Gazetted on 29 August 1944)	
A/LS Frederick Charles Fildew (*ML 157*)	DSM for bravery and undaunted devotion to duty in important and hazardous minelaying operations Operation HOSTILE - NW coast of Europe
AB James Frederick Shiret (*ML 159*)	MID for bravery and undaunted devotion to duty in important and hazardous minelaying operations Operation HOSTILE - NW coast of Europe
Stoker First Class Arthur Vincent Kelly (*ML 180*)	MID for bravery and undaunted devotion to duty in important and hazardous minelaying operations Operation HOSTILE - NW coast of Europe
T/Sub. Lt. Donald Louis Cranefield (*ML 181*)	MID for bravery and undaunted devotion to duty in important and hazardous minelaying operations Operation HOSTILE - NW coast of Europe
T/A/Lt. Harold Thomas Kemsley, RNVR (SO 10th ML Flotilla in *ML 181*)	Bar to DSC awarded for bravery and undaunted devotion to duty in important and hazardous minelaying operations. Operation HOSTILE - NW coast of Europe. First DSC for minelaying from light coastal craft April-November 1943, while assigned to HMS *Defiance*.
Lt. Frank Otto Stoe Man, RNVR (*ML 181* & HMS Black Bat, No 13 Wharf, Devonport)	DSC for bravery and undaunted devotion to duty in important and hazardous minelaying operations Operation HOSTILE - NW coast of Europe
Stoker First Class Frank Godfrey Sketchley (*ML 181*)	MID for bravery and undaunted devotion to duty in important and hazardous minelaying operations Operation HOSTILE - NW coast of Europe
PO Motor Mechanic John Robert Bateman (*ML 184*)	MID for bravery and undaunted devotion to duty in important and hazardous minelaying operations Operation HOSTILE - NW coast of Europe
T/Lt. Robert William Winter, RNVR (*ML 184*)	MID for bravery and undaunted devotion to duty in important and hazardous minelaying operations Operation HOSTILE - NW coast of Europe
AB William McLean (*ML 186*)	MID for bravery and undaunted devotion to duty in important and hazardous minelaying operations
Stoker First Class Edwin George Stanley Buck (*ML 259*)	MID for bravery and undaunted devotion to duty in important and hazardous minelaying operations Operation HOSTILE - NW coast of Europe

Operation MAPLE

51st Motor Launch Flotilla (All Gazetted on 1 August 1944, with the exception of Wood who was thus honoured on 2 July 1943)

AB John McInally (ML 100)	DSM for courage, skill, and devotion to duty in many hazardous minelaying operations
Leading Telegraphist Geoffrey Ernest Wood (ML 105)	DSM awarded for courage and endurance in remaining at his post during minelaying operations in spite of severe injuries (at Nore, 5 May 1943)
AB Albert Cecil Rossiter (ML 105)	DSM for courage, skill, and devotion to duty in many hazardous minelaying operations
Lt. Walter Charles Drake, RNZNVR (ML 106)	DSC for courage, skill, and devotion to duty in many hazardous minelaying operations
T/Lt. Robert George Eburah, RNVR (ML 106)	DSC for courage, skill, and devotion to duty in many hazardous minelaying operations

4th Motor Torpedo Boat Flotilla (Gazetted on 1 August 1944)

Leading Telegraphist Charles Tatham (MTB 238)	DSM awarded for courage, skill, and devotion to duty in many hazardous minelaying operations

9th Motor Torpedo Boat Flotilla (Dutch) (Gazetted on 1 Aug 1944)

AB Raymond Barnard (MTB 235)	MID for courage, skill, and devotion to duty in many hazardous minelaying operations (Nore Command)

21st Motor Torpedo Boat Flotilla (All "Gazetted" on 1 August 1944, with the exception of Wolfe who was thus honoured on 2 January 1945)

A/Leading Stoker William Greenland (MTB 224)	DSM for courage, skill, and devotion to duty in many hazardous minelaying operations
A/Leading Stoker Robert McLean Goodall (MTB 232)	MID for courage, skill, and devotion to duty in many hazardous minelaying operations
LS John Thomas Valentine (MTB 232)	DSM for courage, skill, and devotion to duty in many hazardous minelaying operations
T/Lt. John Alfred Wolfe, RNVR (CO, MTB 232)	DSC for courage, skill, and devotion to duty in many hazardous minelaying operations. Bar to DSC awarded for attacks on Nore on 14 September and 8 October 1944
Telegraphist John Clunie Dargie (MTB 233)	DSM for courage, skill, and devotion to duty in many hazardous minelaying operations

22nd Motor Torpedo Boat Flotilla (All Gazetted on 1 August 1944, with the exception of Hunt on 10 Oct 1944, and Heys on 26 Apr 1945)

AB Norman Heys (*MTB 88*)	MID for courage, skill, and devotion to duty in many hazardous minelaying operations
AB Eric Guest (*MTB 245*)	MID for courage, skill, and devotion to duty in many hazardous minelaying operations
Lt. Douglas Eric James Hunt, RNVR (*MTB 245*)	DSC for courage, skill, and devotion to duty in many hazardous minelaying operations Bar to DSC for action off the Dutch coast 8 July 1944
Ord Signalman Frederick Arthur Parker (*MTB 245*)	MID for courage, skill, and devotion to duty in many hazardous minelaying operations
AB Norman Heys (*MTB 497*)	Second MID for action against German E-boats on 6 Apr 1945

52nd Motor Torpedo Boat Flotilla (All Gazetted on 29 Aug 1944, with the exception of Cartwright on 29 Feb 44, and Pearson on 16 Jan 1945)

T/A Lt. Comdr. Thomas Nelson Cartwright, RNR	MID awarded for Operation HOSTILE - NW coast of Europe.
T/A Lt. Comdr. Thomas Nelson Cartwright, RNR (SO 52nd MTB Flotilla, and CO, *MTB 673*)	Bar to DSC for bravery and undaunted devotion to duty in important and hazardous minelaying operations. Operation HOSTILE - NW coast of Europe
Leading Telegraphist James William Pearson (*MTB 673*)	Second MID for bravery and undaunted devotion to duty in important and hazardous minelaying operations. Operation HOSTILE - NW coast of Europe DSM for coastal actions Plymouth Approaches July-August 1944

Based at HMS Hornet, Gosport (Gazetted on 1 August 1944)

PO Michael Murray (*MTB 93*)	DSM for courage, skill, and devotion to duty in many hazardous minelaying operations
PO Motor Mechanic Daniel Welsh	DSM for courage, skill, and devotion to duty in many hazardous minelaying operations[18]

19

U.S. Navy Readies Auxiliary Minelayers for Flagship Duties

Photo 19-1

Auxiliary minelayer USS *Barricade* (ACM-3) at anchor, circa 1945.
Naval History and Heritage Command #NH 79737

In early April 1944 when the U.S. Navy acquired three ex-Army 188-foot, 1,300-ton mine planter vessels, they commissioned them that same day. They then underwent conversion and fitting out to serve as auxiliary minecraft in the capacity of buoy layer, minelayer, and flagship/mother ship for a group of minesweepers. The ships' commissioning dates may be found in the first column of the table.

Minelayer	Ex-Mine Planter	Commanding Officer
USS *Chimo* (ACM-1) 7 April 1944	*Col. Charles W. Bundy* (MP-15)	Lt. Comdr. John Winston Gross, USNR
USS *Planter* (ACM-2) 4 April 1944	*Col. George Ricker* (MP-16)	Lt. Theodore Thomas Scudder Jr., USNR
USS *Barricade* (ACM-3) 7 April 1944	*Col. John Storey* (MP-8)	Lt. Charles Percy Haber, USN

Planter was placed in commission at St. Helena Annex, Navy Yard, Norfolk, Virginia. *Barricade* and *Chimo* were commissioned at the Mine Planters Pier, Fort Hancock, New Jersey, and India Wharf, Boston, Massachusetts, respectively. They then proceeded to Norfolk to join *Planter*, stopping at Section Base, Thompkinsville, Staten Island, New York, en route. The ships underwent conversion and repairs at Norfolk before operating in local waters conducting tests and drills. These included compensating compasses and calibrating degaussing systems, followed by General Quarters and emergency drills, practicing refueling at sea, and test firing 20mm and 40mm anti-aircraft guns.[1]

On 13 May, the three auxiliary minelayers joined Convoy UGS-42 in lower Chesapeake Bay to make the Atlantic crossing. *Chimo* was bound for England and duty with the Twelfth Fleet, while *Barricade* and *Planter* were headed to the Med to join the Eighth Fleet. The convoy was made up of ninety cargo vessels and tankers, three tank landing ships (LSTs), and three infantry landing craft (LCIs). The three minelayers were assigned rear positions in the convoy and detailed to pick up survivors—should this become necessary if there was an enemy attack.[2]

Escorting the convoy were the destroyer USS *MacLeish* (DD-220)—carrying commander, Task Force 63—and twelve destroyer escorts. The fleet oiler USS *Cowanesque* (AO-79) accompanied them.[3]

The transit east at 9.5 knots was uneventful for the former planter ships until the morning of 22 May, when *Planter* was forced to decrease speed to 7.5 knots in an attempt to stop excessive feed water loss to her boilers. Unsuccessful, she then stopped main engines owing to a loss of vacuum. She got under way at 0812 and tried to catch the convoy. Two hours later, the task force commander detached her by visual signal from the convoy and ordered her to proceed to Horta, Faial Island, Azores. *Planter* was accompanied by sister ship *Chimo*, the Liberty ship SS *John Fairfield*, and the destroyer escort USS *Thornhill* (DE-195).[4]

BARRICADE REMAINS WITH CONVOY UGS-42

When *Chimo* and *Planter* left the convoy, *Barricade* continued with it to the Mediterranean. She anchored off Bizerte, Tunisia, in late afternoon on 1 June and the following day proceeded up the channel to the Naval Operating Base. That afternoon, Rear Adm. Frank J. Lowrey, commander, Eighth Amphibious Force, and Comdr. R. C. Brown, commander, Mine Squadron Six, came aboard for an informal inspection of the ship.[5]

Lowrey, who would command the landings in southern France, had directed the Allied landings at Anzio in January 1944. For the landings on the southern coast of France on 15 August 1944, code named Operation DRAGOON, Lowry would command Task Force 84, also known as Alpha Force. This deployment of ships transported Maj. Gen. John W. O'Daniel's U.S. Army 3rd Infantry Division. *Barricade* and *Planter* both supported the invasion of southern France; details about their involvement follows later in this book.

CHIMO AND *PLANTER* PART WAYS AT AZORES

On 24 May, *Planter* and *Chimo* entered Horta, Azores. *Chimo* berthed starboard side to the British tanker *Empire Garden* and *Planter* outboard *Chimo*; both received fuel. (*Empire Garden* was the former German supply ship *Gedania*, which had been captured on 4 June 1941 near Iceland by the ocean boarding vessel HMS *Marsdale*.)[6]

Shortly after arrival, Lieutenant Scudder sent a message to commander, Eighth Fleet, reporting that repairs to *Planter* would require 48 hours making it impossible for her to proceed. He further indicated that she would join the next Mediterranean-bound convoy unless otherwise directed.[7]

Map 19-1

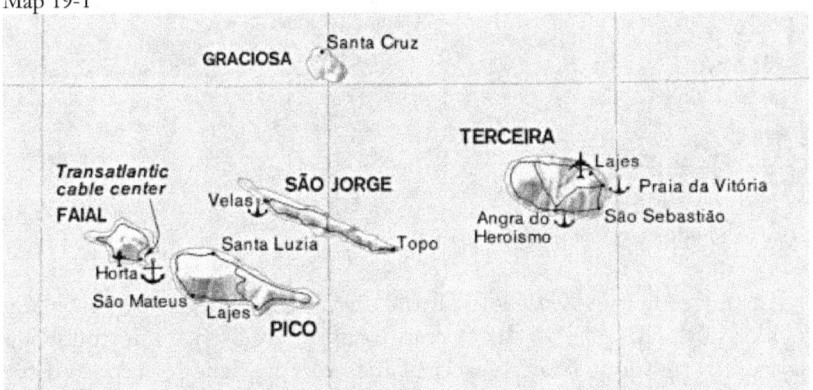

The Portuguese Azores are located in the North Atlantic about 730 nautical miles west of continental Portugal. Four of the islands are not shown; Corvo and Flores lie to the northwest, and Sao Miguel and Santa Maria to the southeast of the central islands.

Chimo left Horta the following day, bound for Belfast, Ireland. Once clear of land, her crew was exercised at general and emergency drills, which included test firing the guns at practice targets. On 31 May, Rathlin Island was sighted. Lying six miles off the coast of Northern Ireland, it marked the northernmost point of Northern Ireland. Passing through the Northern Channel and Irish Sea, *Chimo* arrived at Wilford

Haven and anchored alongside the harbour oiler RFA *War Hindoo*. After refueling, she stood out for Plymouth, England.[7]

Sighting Trevose Head Light on the north Cornish coast on 2 June, *Chimo* entered Plymouth Harbour and moored to Buoy #11. Her crew loaded ammunition and food stores and Lt. Comdr. Gross reported to commander, Twelfth Fleet for duty. Four days later, *Chimo* would sail from Plymouth for the Normandy assault beaches.[8]

Map 19-2

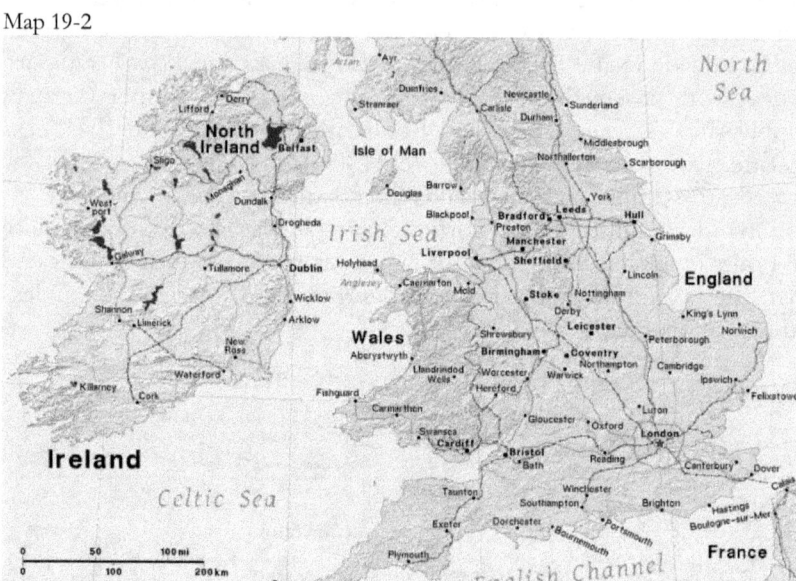

Ireland and other areas of the UK

After repairs, *Planter* left Horta on 2 June, under escort by the trawler HMS *Kingston Amber* to join Convoy UGS-43. The following day, upon sighting escort vessels of the convoy, Scudder reported for duty by visual signal to commander, Task Force 64, in the destroyer USS *McCormick*. *Planter* was ordered to position No. 89 near the rear of the convoy, and detailed to pick up survivors if necessary.

On the evening of 11 June, the convoy began forming in four columns preparatory to entering the swept channel between the Galite Islands and northern Tunisia. *Planter* left the convoy shortly before noon the next day, and proceeded independently into Bizerte Harbour, and berthed at the Advance Amphibious Training Base.

The following chapter introduces the Canadian Mine Force, before sixteen of its *Bangor*-class minesweepers sail to England to take part in the Normandy invasion.

20

Royal Canadian Navy Mine Force

In total, [in World War II] Canadian warships destroyed 42 enemy surface ships and, either alone or with other ships and aircraft, sank 33 submarines. The RCN lost 33 ships and suffered 1,990 fatalities. At the end of the war, the RCN was the fourth largest fleet in the world—behind only those of the US, Britain and the Soviet Union—with more than 400 warships. Although the RCN had no battleships or submarines, Canadian sailors served with distinction on both types of vessels in the Royal Navy.

—Article titled "Royal Canadian Navy" by Richard Gimblett[1]

WWII II Canadian Naval Reserve Recruitment poster
Library and Archives Canada No. R1300-72, 1939-45

HMCS *Fundy*

HMCS *Bras d'Or*

HMCS *Llewellyn*

Up to this point in the book, discussion of the Royal Canadian Navy's contributions to operations in Enemy Waters has been limited to a few references to Canadians commanding or otherwise serving aboard RN minesweepers. This chapter introduces the Royal Canadian Navy's Mine Force as a primer to discussion, in the following chapter, of the participation of sixteen RCN *Bangor*-class fleet minesweepers in the invasion of Normandy.

At the outbreak of war in September 1939, the Royal Canadian Navy had only six *River*-class destroyers, four minesweepers, and two small training vessels; bases at Halifax and Victoria; and a mere 145 officers and 1,674 men. The RCN expanded substantially with its larger ships transferred or purchased from the American and British navies (many of the American-sourced were through the Destroyers for Bases Agreement). Smaller ships such as corvettes, frigates, and minesweepers were built in Canada for both the RCN and RN. By the end of the Battle of the Atlantic, the RCN was the primary navy in the northwest sector of the Atlantic. Under the command of Rear Adm. Leonard W. Murray, it was responsible for the safe escort of innumerable convoys and the destruction of many U-boats.[2]

Convoy duty in the frigid western North Atlantic demanded lots of ships, resulting in the RCN employing both *Bangor-* and *Algerine*-class minesweepers in this role in addition to corvettes, frigates, and destroyers. Because *Algerine*s built for Canada were used solely for convoy escort and were never fitted with minesweep gear, it would be easy to presume that Canada contributed little to Allied mine warfare in Enemy Waters. Such a belief would be mistaken; Canadians in the Naval Reserve (RCNR) and Naval Volunteer Reserve (RCNVR) served aboard Royal Navy minesweepers (and, perhaps, minelayers as well). The remainder of the chapter is devoted to describing the development of Canada's mine forces in World War II.

In 1939, with war imminent and the possibility that it would soon face enemy-laid minefields blocking Canadian harbours and threatening sea routes, the Royal Canadian Navy had only four minesweepers. These were HMCS *Comox*, *Fundy*, *Gaspé* and *Nootka* (four relatively new ships comprising the *Fundy*-class). These vessels, modified versions of the British *Basset*-class trawler minesweeper, were strengthened for ice conditions and named for bays in Canada. Two had been built in inland shipyards, *Gaspé* along the St. Lawrence and *Fundy* on the Great Lakes.[3]

The four ships, all commissioned in 1938, served out the war and were decommissioned in 1945. Spanning 163 feet in length, the coal-burning vessels displaced 460 tons, drew 14.4 feet of water, and had a modest speed of 12 knots. Their austere armament consisted of only one 12-pounder gun—which presumably provided only a measure of comfort to their crews of 3 officers and 35 ratings. Two of the minesweepers (*Fundy* and *Nootka*) were assigned to the west coast, and the remaining two to the east coast (*Comox* and *Gaspé*).[4]

Photo 20-1

HMCS *Fundy* (J88) under way, location and date unknown.
www.readyayeready.com/ships/shipview.php?id=1144&ship=FUNDY%20(1st)

Fundy-class Minesweepers

Ship	Builder	Comm.	Paid Off
Comox (J64)	Burrard Dry Dock Co. Ltd., Vancouver	23 Nov 1938	27 Jul 1945
Fundy (J88)	Collingwood Shipyards Ltd., Collingwood	1 Sep 1938	27 Jul 1945
Gaspé (J94)	Morton Engineering & Dry Dock Co., Quebec City	21 Oct 1938	23 Jul 1945
Nootka (J35)	Yarrows Ltd., Esquimalt	6 Dec 1938	29 Jul 1945[5]

In recognition of its severe shortage of mine clearance vessels, the Canadian government authorised the construction of a large number of steel-hulled *Bangor*-class minesweepers. Like corvettes, these light draught ships did not require the same degree of shipbuilding expertise as cruisers and destroyers. In the interim, the *Bras d'Or*, a trawler laid down at Sorel, Quebec, in 1919, was pressed into service as an auxiliary minesweeper. The long serving and marginally seaworthy, former lightship was commissioned on 15 September, five days after Canada went to war on 10 September 1939.[6]

SHORT SERVICE OF HMCS *BRAS D'OR*

> *While sailing some six miles off Southwest Point, Anticosti Island, at 0350 on 19 October 1940, the second officer of the Romanian merchantman,* Inginer N. Vlassopol, *turned to his helmsman and remarked: "the little boat has just put her lights out." But his assumption that the auxiliary minesweeper, HMCS* Bras d'Or, *had simply darkened ship was incorrect. Instead, something had gone terribly wrong as thousands of tons of seawater brought the small ship and her crew of thirty to a mysterious end.*
>
> —Richard Oliver Mayne, "The little boat has just put her lights out:" The Life, Fate and Legacy of HMCS *Bras d'Or*, The Northern Mariner, XIV No 4, (October 2004)

Before delving into the *Bangor*-class minesweepers, the centre piece of the Royal Canadian Navy's Mine Force, it's worthwhile to devote a page or so to HMCS *Bras d'Or*, which was tragically lost in a storm at sea on 19 October 1940 with all hands. The challenges faced by her crew during arduous service were shared by many other officers and men aboard marginal vessels pressed into naval service. This was also true of the crews of similar small vessels of the Royal Navy and U.S. Navy which "day in and day out" swept harbours and approaches in all types of weather and performed sundry other duties.[7]

Photo 20-2

HMCS *Bras d'Or* at anchor, location and date unknown.
readyayeready.com/ships/shipview.php?id=1047&ship=BRAS%20D'OR%20(1st)

The 265-ton *Bras d'Or* had been laid down at Sorel, Quebec in 1919 as a fishing trawler. The New York ship owner for whom this vessel was ordered went bankrupt soon after her launching in 1919, and she and five sisters were sold incomplete. She was completed in 1926 for service as Lightship *No. 25* with the Department of Marine and Fisheries. Later requisitioned on 15 September 1939 as an auxiliary minesweeper and renamed *Bras d'Or*, she patrolled the Halifax approaches from 1939 to 1940. She joined the St. Lawrence patrol in June 1940, based at Rimouski, Quebec, at the mouth of the Rimouski River.[8]

Designed for civilian purposes, she and other auxiliaries were far from ideal warships. In fact, even those who sailed in them would on occasion complain about their seaworthiness. A crewman aboard another vessel observed about *Bras d'Or*, "That ship should never have been at sea. She had no [watertight] bulkheads. Anytime the 4-inch gun was fired everything of glass in the ship broke." But, with precious few ships available early in the war, *Bras d'Or* and her sisters were at the vanguard of Canada's coastal defence.[9]

Operational records reveal that between January and May 1940, the average auxiliary was sent on patrol every other day; in extreme cases, some went to sea thirty times in a given month. On top of her contribution to this hectic patrol schedule, *Bras d'Or* was also required to periodically sweep the approaches to both Halifax and Sydney harbours for mines. Other auxiliaries pulled double duty as well, escorting Sydney-to-Halifax convoys because "no destroyers [were] available." All Canadian destroyers were constantly at sea screening larger British warships or shepherding convoys along the 400-mile approach to Halifax. Auxiliaries were doing the same in littoral waters, and *Bras d'Or* gave much-needed assistance.[10]

HMCS *Bras d'Or*'s last assignment with the St. Lawrence patrol involved the surveillance of a Romanian freighter of concern to the British Admiralty. With Romania falling into the Axis orbit, the Admiralty had asked the RCN to detain the *Inginer N. Vlassopol*, while she was loading pulpwood in Clarke City, Quebec. Lacking the resources to seize her in that port, *Bras d'Or* was told to shadow the freighter to Sydney at which point the navy would act on the Admiralty's request. In late afternoon 18 October, *Bras d'Or* reported the *Vlassopol* was in sight. It was her last signal. An early winter storm, which some described as "the worst in twenty years," hit the area the following morning. Although both vessels proceeded down the St Lawrence, it was only the *Vlassopol* that arrived in Sydney. It was assumed that the

Bras d'Or had been lost on the 19th with her entire ship's complement of thirty men.[11]

Four armed yachts, HMCS *Vison*, *Reindeer*, *Husky*, and *Elk* were all damaged in the same storm. This led Capt. Howard E. Reid, commanding officer Atlantic Coast, to express the conclusion that "vessels... built for pleasure purposes and for use in calm waters, are unsuitable for convoy during winter months." Nonetheless, due to a shortage of warships, auxiliaries and armed yachts were in high demand. During the winter of 1940 the British Commander-in-Chief, American West Indies, virtually begged the Canadian Naval Service Headquarters to supply him with as many of these types as possible because he was "seriously concerned about the situation regarding A/S [anti-submarine] vessels at my disposal."[12]

BANGOR-CLASS MINESWEEPERS

Photo 20-3

Canadian minesweeper HMCS *Kenora*, during War II.
Naval History and Heritage Command photograph #NH 60538

The *Bangor*-class minesweepers were a class of 111 warships operated by the Royal Navy (42), Royal Canadian Navy (54), Royal Indian Navy (13), and Imperial Japanese Navy (2) during the war. *Bangor*s in the Canadian and Indian navies included ships transferred by the Royal Navy which had previously been part of the RN. The two units of the IJN had been laid down at Hong Kong for the Royal Navy. They were still under construction when Hong Kong fell to the Japanese and were completed as *W-101* and *W-102* for the Imperial Japanese Navy.[13]

The lead ship, HMS *Bangor*, from which the class derived its name, was commissioned on 7 November 1940. Royal Navy ships were named after coastal towns of the United Kingdom; most of the RCN *Bangor*s honoured Canadian towns, cities, and bays. The requirement for rapid, war-construction resulted in three variations based on the availability of propulsion machinery. The ships all had twin-screws; but some were driven by diesel-engines, others by steam turbines or reciprocating engines. The steam turbine-powered *Bangor*s were also known as the *Ardrossan*-class and the reciprocating engine-powered versions as the *Blyth*-class. The diesel-powered ships were shorter than the others owing to the absence of boiler rooms. None of the fifty-four *Bangor*s of the Royal Canadian Navy were steam-turbine powered.[14]

Diesel-engine *Bangor*-class Minesweepers (10)
(162 feet, 592 tons, 8.25-foot draught, speed 16 knots, 60 ship's complement)

HMCS *Brockville* (J 270)	HMCS *Lachine* (J 266)	HMCS *Trois Rivières* (J 269)
HMCS *Digby* (J 267)	HMCS *Melville* (J 263)	HMCS *Transcona* (J 271)
HMCS *Esquimalt* (J 272)	HMCS *Noranda* (J 265)	HMCS *Truro* (J 268)
HMCS *Granby* (J 264)		

Reciprocating-engine *Bangor*-class Minesweepers (44)
(171.5 feet, 673 tons, 8.25-foot draught, speed 16 knots, 70 ship's complement)

HMCS *Bayfield* (J08)	HMCS *Goderich* (J 60)	HMCS *Nipigon* (J154)
HMCS *Bellechasse* (J170)	HMCS *Grandmere* (J258)	HMCS *Outarde* (J161)
HMCS *Blairmore* (J314)	HMCS *Guysborough* (J52)	HMCS *Port Hope* (J280)
HMCS *Burlington* (J250)	HMCS *Ingonish* (J69)	HMCS *Quatsino* (J152)
HMCS *Canso* (J21)	HMCS *Kelowna* (J261)	HMCS *Quinte* (J166)
HMCS *Caraquet* (J38)	HMCS *Kenora* (J281)	HMCS *Red Deer* (J255)
HMCS *Chedabucto* (J168)	HMCS *Kentville* (J312)	HMCS *Sarnia* (J309)
HMCS *Chignecto* (J160)	HMCS *Lockeport* (J100)	HMCS *Stratford* (J310)
HMCS *Clayoquot* (J174)	HMCS *Mahone* (J159)	HMCS *Swift Current* (J254)
HMCS *Courtenay* (J262)	HMCS *Malpeque* (J148)	HMCS *Thunder* (J156)
HMCS *Cowichan* (J146)	HMCS *Medicine Hat* (J256)	HMCS *Ungava* (J149)
HMCS *Drummondville* (J253)	HMCS *Milltown* (J317)	HMCS *Vegreville* (J 57)
HMCS *Fort William* (J311)	HMCS *Minas* (J165)	HMCS *Wasaga* (J162)
HMCS *Gananoque* (J259)	HMCS *Miramichi* (J169)	HMCS *Westmount* (J318)
HMCS *Georgian* (J144)	HMCS *Mulgrave* (J313)[15]	

The modest-sized *Bangor*s were poor sea keepers—reportedly worse even than the 205-foot *Flower*-class corvettes. The poorest were the diesel ships, because of their shorter length. With their shallow draughts and resultant poor stability, they had a propensity to bury their bows in head seas.¹⁶

ALGERINE-CLASS USED FOR CONVOY ESCORT

An improved version of the *Bangor*-class vessels was also built in Canada. Forty-one *Algerine*-class ships were constructed, mostly at Port Arthur shipyards. Designed with more sheer at the bows, they performed better in heavy seas. Intended from the beginning as convoy escorts, the twelve *Algerines* that served in the RCN were used solely in this role and did not carry minesweeping gear. The Royal Navy, however, employed its *Algerines* extensively as minesweepers.¹⁷

ISLES-CLASS MINESWEEPERS AND MINELAYERS

Photo 20-4

Minesweeping trawler HMS *Anticosti* under way, location and date unknown.
www.readyayeready.com/ships/shipview.php?id=1013&ship=ANTICOSTI

The Royal Canadian Navy also counted amongst its mine force, ten *Isles*-class vessels. Eight of these were minesweeping trawlers and two, minelayers. The 164-foot ships could make 12 knots, propelled by a reciprocating engine, driving a single shaft. They were fitted with one 12-pound anti-aircraft gun, and three smaller 20mm AA guns in a single triple-mount. Ship's complement was 4 officers and 36 men.¹⁸

The ships used by the RCN for minesweeping were all products of Ontario shipyards; a total of sixteen were built. All eight minesweeping trawlers were built for the Royal Navy and immediately loaned to the Royal Canadian Navy, though never commissioned in the RCN. They were mostly manned by British personnel and were returned to the Royal Navy following the war. In contrast the two minelayers, HMCS

Whitethroat and HMCS *Stonechat*, were built for the RN but transferred to the RCN upon completion, commissioned, and manned by Canadian personnel. It appears that they were received too late to perform minelaying, so were used for other duties including removal of defences. The dates in the "Paid Off" column in the tables, indicates when ships were decommissioned and returned to the Royal Navy or retained by the Royal Canadian Navy.[19]

Isles-class Minesweeping Trawlers

Ship	Builder	Comm.	Paid Off
HMS *Anticosti* (T274) ex-HMS *Anticosti*	Collingwood Shipyards Ltd. (Collingwood, Ontario, Canada)	8 Oct 1942	Returned to Royal Navy 17 Jun 1945
HMS *Baffin* (T275) ex-HMS *Baffin*	Collingwood Shipyards Ltd. (Collingwood, Ontario, Canada)	26 Aug 1942	Returned to Royal Navy 20 Aug 1945
HMS *Cailiff* (T276) ex-HMS *Cailiff*	Collingwood Shipyards Ltd. (Collingwood, Ontario, Canada)	17 Sep 1942	Returned to Royal Navy 17 Jun 1945
HMS *Ironbound* (T284) Ex-HMS *Ironbound*	Kingston Shipbuilding Co. (Kingston, Ontario, Canada)	16 Oct 1942	Returned to Royal Navy 17 Jun 1945
HMS *Liscomb* (T285) ex-HMS *Liscomb*	Kingston Shipbuilding Co. (Kingston, Ontario, Canada)	8 Sep 1942	Returned to Royal Navy 17 Jun 1945
HMS *Magdalen* (T279) ex-HMS *Magdalen*	Midland Shipyards Ltd. (Midland, Ontario, Canada)	24 Aug 1942	Returned to Royal Navy 17 Jun 1945
HMS *Manitoulin* (T280)	Midland Shipyards Ltd. (Midland, Ontario, Canada)	28 Sep 1942	Returned to Royal Navy 17 Jun 1945
HMS *Miscou* (T277)	Collingwood Shipyards Ltd. (Collingwood, Ontario, Canada)	17 Oct 1942	Returned to Royal Navy 17 Jun 1945

Isles-class Minelayers

Ship	Builder	Comm.	Paid Off
Stonechat (M25)	Cook, Welton & Gemmill (Beverley, UK)	12 Nov 1944	Returned to RN in 1946
Whitethroat (M03)	Cook, Welton & Gemmill (Beverley, UK)[20]	7 Dec 1944	6 May 1946

MINELAYER HMCS *SANKATY*

The Royal Canadian Navy also boasted one large minelayer, HMCS *Sankaty*, a former Stamford-Oyster Bay, Massachusetts, ferry. The elderly vessel, launched in 1911, was commissioned on 24 September 1940 at Halifax as a minelaying, loop-laying, and maintenance vessel.

Paid off on 18 August 1945, she became the Prince Edward Island ferry *Charles A. Dunning*. Following this service, she was sold for scrap, and sank on 27 October 1964 while en route to Sydney, Canada.[21]

Photo 20-5

Ferry *Sankaty* at Union Wharf, Vineyard Haven, Massachusetts; and sailors aboard her in March 1941, while she was the minelayer HMCS *Sankaty* (Library and Archives of Canada/Dept. of National Defence/PA 105292)
http://www.forposterityssake.ca/GALLERIES/SANKATY.htm

LLEWELLYN-CLASS COASTAL MINESWEEPERS

Photo 20-6

Coastal minesweeper HMCS *Llewellyn* (J278) under way, location and date unknown.
www.readyayeready.com/ships/shipview.php?id=1224

Ten much smaller *Llewellyn*-class coastal minesweepers were constructed in Canadian shipyards late in the war. Two of the ships were built in Quebec, and the others in yards on the western coast of Canada. Based on the British Admiralty Type I Motor Mine Sweeper (MMS), their hulls were of wood for self-defence against magnetic-influence mines.[22]

The 119-foot *Llewellyn* had a 22-foot beam, 8.7-foot draught, and 228-ton displacement. Intended to tow ponderous sweep gear, her propulsion was designed for torque and, thus, she had a modest top speed of 12 knots. She was fitted with two twin 0.5-inch (12.7mm) machine guns; and equipped with "Double L" magnetic minesweeping gear. Her ship's complement was 3 officers and 20 ratings.[23]

Royal Canadian Navy *Llewellyn*-class Coastal Minesweepers

Ship	Builder	Comm.	Paid Off
Llewellyn (J278)	Chantier Maritime de St. Laurent, Île d'Orléans, Quebec	24 Aug 1942	31 Oct 1951
Lloyd George (J 279)	Chantier Maritime de St. Laurent, Île d'Orléans, Quebec	24 Aug 1942	16 Jul 1948
Revelstoke (J373)	Star Shipyards, New Westminster, British Columbia	4 Jul 1944	23 Oct 1953
Cranbrook (J372)	Star Shipyards, New Westminster, British Columbia	12 May 1944	3 Nov 1945
Coquitlam (J364)	Newcastle Shipbuilding, Nanaimo, British Columbia	25 Jul 1944	30 Nov 1945
St. Joseph (J359)	Newcastle Shipbuilding, Nanaimo, British Columbia	24 May 1944	8 Nov 1945
Rossland (J358)	Vancouver Shipyards, Vancouver, British Columbia	15 Jul 1944	1 Nov 1945
Daerwood (J357)	Vancouver Shipyards, Vancouver, British Columbia	22 Apr 1944	28 Nov 1945
Lavallee (J371)	A.C. Benson Shipyard, Vancouver, British Columbia	21 Jun 1944	27 Dec 1945
Kalamalka (J395)	A.C. Benson Shipyard, Vancouver, British Columbia[24]	2 Oct 1944	16 Nov 1945

LAKE-CLASS MINESWEEPERS

The RCN ordered sixteen *Lake*-class minesweepers from Canadian yards but none were commissioned. Only ten were completed, VJ-Day having intervened, and these were transferred to the USSR.[25]

MINESWEEPING AUXILIARIES

As they needed auxiliary minesweepers at St. John's, Sydney, and Halifax to perform various tasks—including guardship, loop-laying, and coil skid towing (related to defensive mining)—in addition to minesweeping, the RCN took whatever was at hand. Amongst the eleven vessels pressed into service as "maids of all duties" were a former fishing trawler, a rum runner, two wooden-hulled coasters, and a World War I vintage ex-Royal Navy minesweeper. The remaining six were

Norwegian whale catchers, which were unable to return to their homeports because of the German occupation.

The Norwegian factory ship *Suderøy*, with catchers *Suderøy IV*, *V*, *VI*, and *Star XVI* had been homeward bound from the South Atlantic in Spring 1940, when directed by Naval authorities at Kingston, Jamaica, to follow the U.S. and Canadian coast to Newfoundland before turning east across the Atlantic toward Finnmark (the northernmost and easternmost county of Norway, just north of Finland). It's unclear when *Suderøy I* and *Suderøy II* joined this group, but they accompanied the others northward. Upon learning on 9 April of the German invasion of Norway, the ships entered Hampton Roads, Virginia. to confer with British authorities; which sent them to Halifax, for further orders.[26]

Map 20-1

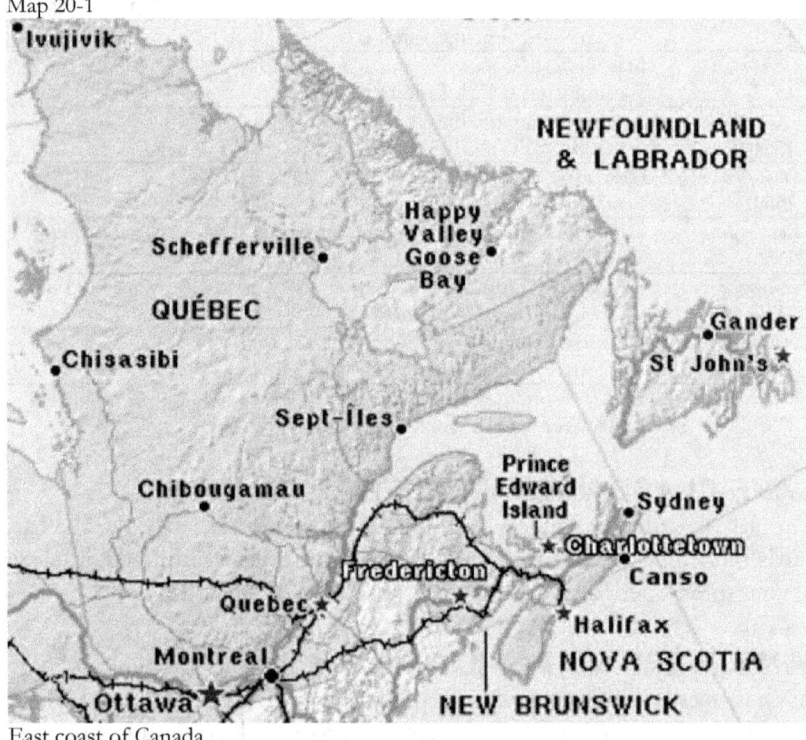

East coast of Canada

It is believed that *Suderøy I* and *II* also accompanied the whale factory and catchers to Halifax. Stranded overseas, the six whale chasers were chartered by the Royal Canadian Navy. *Suderøy I* was requisitioned by the South African Navy on 23 November 1942 for use as a minesweeper in South African waters; and renamed *Parktown* (T55). *Suderøy II* had been taken earlier, on 20 August, as *Johannesburg* (T56).²⁷

Auxiliary Minesweepers

Ship	Duty	Comm.	Paid Off
Rayon d'Or (ex-fishing trawler)	Based at Halifax and Sydney; duties included minesweeping, loop-laying, and maintenance.	1939	1945
Reo II (ex-rum runner *Reo II*)	Served as a minesweeper, an examination vessel, and a coil skid towing vessel.	23 Jan 1941	19 Oct 1945
Ross Norman (ex-wooden-hulled coaster)	Auxiliary minesweeper, coil skid towing craft, and mobile deperming craft with Halifax Local Defence Force.	19 Jun 1940	8 Apr 1946
Standard Coaster J10 (ex-wooden-hulled coaster)	Based at Halifax; functioned primarily as a coil skid towing vessel.	11 Feb 1942	25 Mar 1946
Star XVI (ex-Norwegian whale-catcher)	Assigned to St. John's Local Defence Force; transferred in June 1942 to Sydney.	Aug 1941	31 Aug 1945
Suderøy I (ex-Norwegian whale-catcher)	Later South African Navy minesweeper *Parktown* (T55).		Nov 1942
Suderøy II (ex-Norwegian whale-catcher)	Later South African Navy minesweeper *Johannesburg* (T56).		Aug 1942
Suderøy IV (ex-Norwegian whale-catcher)	Served with the Halifax Local Defence Force until returned to her owners.	Jun 1941	31 Aug 1945
Suderøy V (ex-Norwegian whale-catcher)	Assigned to St. John's Local Defence Force; transferred in June 1942 to Sydney.	2 Jun 1941	7 Aug 1945
Suderøy VI (ex-Norwegian whale-catcher)	Served with Halifax Local Defence Force throughout the war as an auxiliary minesweeper.	19 Mar 1941	31 Aug 1945
Venosta J11 (ex-RN minesweeper in WWI)	Gate vessel and auxiliary minesweeper at Halifax, and later employed at Sydney.²⁸	17 Nov 1939	22 Jan 1942

CANADIAN MINESWEEPERS LOST IN THE WAR

Five Royal Canadian Navy minesweepers were lost in the war: four fleet minesweepers and auxiliary minesweeper HMCS *Bras d'Or*. The latter vessel was lost in a storm, and HMCS *Chedabucto* to a collision at sea. German U-boats torpedoed the remaining three—fleet minesweepers

HMCS *Clayoquot*, *Guysborough*, and *Esquimalt*—sending them down into the abyss with great loss of life.

RCN Minesweepers Lost in World War II

Date	Ship	Fate
19 Oct 1940	*Bras d'Or* Auxiliary minesweeper	Foundered in the Gulf of St. Lawrence on 19 Oct 1940; 5 officers and 25 ratings perished.
31 Oct 1943	*Chedabucto* (J 168) Fleet minesweeper	Lost in a collision with the cable vessel *Lord Kelvin* on the St. Lawrence River; 1 officer perished.
24 Dec 1944	*Clayoquot* (J 174) Fleet minesweeper	Sunk by U-*806* outside the approaches to Halifax Harbour while escorting convoy XB-139; 4 officers and 4 ratings perished.
17 Mar 1945	*Guysborough* (J 52) Fleet minesweeper	Sunk by U-*868* about 210 miles north of Cape Finisterre in the Bay of Biscay; 53 officers and ratings perished.
16 Apr 1945	*Esquimalt* (J 272) Fleet minesweeper	Sunk by *U-190* in the harbour approaches to Halifax; 39 officers and men perished.[29]

BATTLE HONOURS

It is not possible to provide within this 15-page overview of the Canadian Navy's Mine Force details about the operations of scores of minesweepers. However, a review of a list of those that received Royal Navy Battle Honours ATLANTIC and NORMANDY during the war, and later, the special Canadian Battle Honour GULF OF ST. LAWRENCE provides some insights. While a majority of the mine force was assigned to Canada's east coast, a few ships were allocated to the west coast for the entire war, thus earning no battle honours.

The officers and men aboard ships and craft that fought in the Battle of the Gulf of St. Lawrence sacrificed much in their efforts to help maintain peace and freedom in Canada. (Many of those whose lives were claimed in the waters of the Gulf of St. Lawrence—from Québec City to Cabot Strait and in the Strait of Belle Isle—have no known grave.) Unrecognised by the British Crown during the war which bestowed battle honours, Canada remedied this oversight in 1992, when it approved the unique Battle Honour Gulf of St. Lawrence. Since then, this honour has been awarded to 129 warships—corvettes, *River*-class frigates, *Bangor* minesweepers, wooden minesweepers, armed yachts, auxiliaries, and *Fairmile* motor launches.[30]

The periods of service eligibility for the battle honour were May to October 1942 and/or September and November 1944, when German submarines were active in the Gulf. For some of the "Gulf of St.

Lawrence" entries in the tables, the associated year(s) are missing because the authors were unable to ascertain them during research.[31]

Two *Isles*-class minesweeping trawlers and one *Isles* minelayer, two *Fundy*-class minesweepers, and thirty-nine *Bangor*-class minesweepers earned Battle Honour ATLANTIC. There are many books and articles devoted to the Battle of the Atlantic, which is not within the scope of this book. Desperate for escorts to screen convoys from U-boat attack, the Allies pressed minesweepers into convoy escort duties, to augment or substitute for scarce, overtaxed destroyers and frigates.

As previously mentioned, sixteen *Bangors* which participated in the Allied landings at Normandy, received Battle Honour NORMANDY.

Isles-class Minesweeping Trawlers and Minelayers

Minesweeper	Battle Honour	Minelayer	Battle Honour
Baffin	Atlantic 1944	*Whitethroat*	Atlantic 1945
Miscou	Atlantic 1944		

Fundy-class Minesweepers

Ship	Battle Honour	Ship	Battle Honour
Comox	Atlantic 1940-45	*Gaspé*	Atlantic 1940-45

Llewellyn-class Minesweepers

Llewellyn	Gulf of St. Lawrence	*Lloyd George*	Gulf of St. Lawrence

Bangor-class Minesweepers

Bayfield	Atlantic 1943-44 Normandy 1944	*Lockeport*	Gulf of St. Lawrence 1944 Atlantic 1943
Blairmore	Atlantic 1943-45 Normandy 1944	*Mahone*	Atlantic 1942-1945
Brockville	Gulf of St. Lawrence	*Malpeque*	Atlantic 1941-1942 Normandy 1944
Burlington	Gulf of St. Lawrence 1942 Atlantic 1942-44	*Medicine Hat*	Gulf of St. Lawrence 1942 Atlantic 1943
Canso	Atlantic 1944 Normandy 1944	*Melville*	Gulf of St. Lawrence
Caraquet	Atlantic 1943-44 Normandy 1944	*Milltown*	Gulf of St. Lawrence Atlantic 1942-1944 Normandy 1944
Chedabucto	Gulf of St. Lawrence Atlantic 1942-1943	*Minas*	Atlantic 1941-1944 Normandy 1944
Cowichan	Atlantic 1941-45 Normandy 1944 English Channel 1944-45	*Mulgrave*	Gulf of St. Lawrence Atlantic 1943-1944 Normandy 1944

Clayoquot	Gulf of St. Lawrence Atlantic 1942/1944	*Nipigon*	Gulf of St. Lawrence Atlantic 1941-1945
Digby	Gulf of St. Lawrence 1942, 1944 Atlantic 1942-44	*Noranda*	Gulf of St. Lawrence
Drummondville	Gulf of St. Lawrence 1942 Atlantic 1942-45	*Port Hope*	Gulf of St. Lawrence Atlantic 1943-1945
Esquimalt	Gulf of St. Lawrence Atlantic 1943-44	*Quinte*	Atlantic 1942
Fort William	Gulf of St. Lawrence 1942 Atlantic 1943 Normandy 1944	*Red Deer*	Gulf of St. Lawrence Atlantic 1942-1945
Gananoque	Gulf of St. Lawrence	*Sarnia*	Gulf of St. Lawrence Atlantic 1942-1943
Georgian	Gulf of St. Lawrence Atlantic 1941-42/44 Normandy 1944	*Stratford*	Gulf of St. Lawrence Atlantic 1942-1944
Goderich	Atlantic 1942-45	*Swift Current*	Gulf of St. Lawrence Atlantic 1943-1944
Granby	Gulf of St. Lawrence Atlantic 1942-44	*Thunder*	Atlantic 1941-42/44 Normandy 1944
Grandmere	Gulf of St. Lawrence 1942 Atlantic 1943, 1945	*Trois Rivières*	Gulf of St. Lawrence
Guysborough	Atlantic 1943-44 Normandy 1944	*Truro*	Gulf of St. Lawrence
Ingonish	Gulf of St. Lawrence Atlantic 1944	*Ungava*	Gulf of St. Lawrence Atlantic 1942/1944
Kenora	Gulf of St. Lawrence 1942 Atlantic 1942-1945 Normandy 1944	*Vegreville*	Gulf of St. Lawrence Atlantic 1944 Normandy 1944
Kentville	Gulf of St. Lawrence Atlantic 1944-1945	*Wasaga*	Atlantic 1944 Normandy 1944
Lachine	Gulf of St. Lawrence	*Westmount*	Gulf of St. Lawrence Atlantic 1944

21

D-Day at Normandy

It can be said without fear of contradiction that minesweeping was the keystone of the arch in this operation. All of the waters were suitable for mining, and minesweeping plans of unprecedented complexity were required. The performance of the minesweepers can only be described as magnificent. The passage of the Western Task Force to the assault area, and of the assault waves and supporting ships up to the beaches, without loss from mines, is the best testimonial to the effectiveness of their work. An equally high standard was maintained in the unremitting daily labour of sweeping the assault area during the build-up phase.

—Rear Adm. Alan G. Kirk, USN, commander, Western Naval Task Force, during the Allied invasion at Normandy.[1]

Photo 21-1

Under the Enemy's Nose by Dwight C. Shepler. Canadian Minesweeping Flotilla 31 clearing a lane for assault forces across the English Channel to the Normandy coast the night before the D-Day landings on 6 June 1944.
Naval History and Heritage Command Accession #88-199-ES

On 5 June 1944, the eve of the invasion of the continent of Europe, out of every port indenting the southern coast of England, from Bristol to the Thames, had emerged, or were emerging nearly 5,000 ships and craft. The sea-borne movement of forces across the Channel to the Normandy beaches, code named Operation NEPTUNE, was the kick-off, the first step of Operation OVERLORD, aimed at the liberation of Western Europe and the defeat of Nazi Germany. OVERLORD plans stretched out to D+90 Day. By that time, early September 1944, the drive on the Seine and Paris were to be completed, and swelling Allied armies regrouping for their march to the Rhine commenced. But, if NEPTUNE failed, there would be no march on Germany, only the possibility of another Dunkirk disaster.[2]

American general Dwight D. Eisenhower, the supreme commander of Allied Expeditionary Forces, was overall in charge of OVERLORD. His principal subordinates, the commanders of Allied Armies, Navies, and Air Forces, were all British. They were General Sir Bernard L. Montgomery, Admiral Sir Bertram Ramsay, and Air Marshall Sir Trafford Leigh-Mallory. Under Ramsay, were Rear Adm. Alan G. Kirk, USN, commander, Western Naval Task Force, and Rear Adm. Sir Philip Vian, RN, commander, Eastern Naval Task Force.[3]

Many published works exist, detailing the heroics of the flotillas of British minesweepers that led British and Canadian amphibious and supporting forces to the Gold, Juno, and Sword assault beaches at Normandy. Much less material is available concerning the efforts of American and Canadian sweepers in support of the American landings at Utah and Omaha Beaches. Serving under Kirk, rear admirals Don P. Moon, USN, and John L. Hall, USN, were directly responsible for the assault forces landing on those beaches.[4]

This chapter describes the role that Canadian and American minesweepers played in the landings at Utah and Omaha, the two westernmost of the five allied beaches in the Bay of the Seine. The bay stretches eastward from the Cotentin (Cherbourg) peninsula forming its western boundary, to the broad estuary of the Seine River near Le Havre. The allied western flank lay on the east coast of the peninsula at Utah Beach. To the immediate east, separated by the Carentan Estuary, was Omaha Beach. Extending eastward from Omaha were the three British-Canadian beaches, Gold, Juno, and Sword. The eastern boundary of Sword Beach ended at Ouistreham, a small port city at the mouth of the River Orne which guarded the allied left flank.[5]

As part of the defences of their "Atlantic wall," the Germans had sown a dog-leg shaped belt of mines in waters through which the assault forces had to transit, stretching northeast toward the Straits of Dover.

(This term referred to an extensive system of coastal defence and fortifications built by Germany between 1942 and 1944 along the coast of continental Europe and Scandinavia as a defence against an anticipated Allied invasion of Nazi-occupied Europe from the United Kingdom.) This belt of mines and smaller mined areas are identified on the map, below, by dot-filled geometrical shapes.

Map 21-1

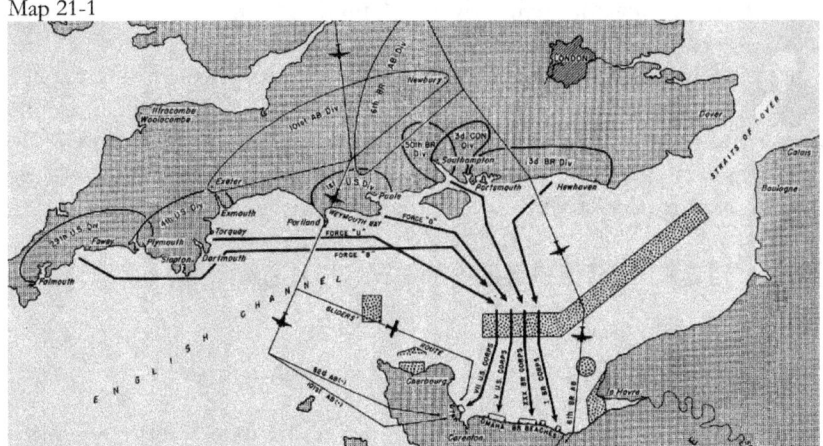

Main Allied sea routes and air routes to the Normandy beaches in the Bay of Seine. The 1st British Corps, which used the far-right approach route, included the Canadian forces that landed on Juno Beach.
https://history.army.mil/books/wwii/utah/maps/MAP2.JPG

MINESWEEPERS SORTIE FROM BRITISH PORTS

As hundreds, and eventually thousands of Allied assault and support force vessels moved out of English ports on the eve of the planned invasion, minesweeping flotillas led the way, clearing wide paths along the coast to Area Z. Located thirteen miles south of the Isle of Wight, this rendezvous point marked where the Channel crossing started. Soon crowded beyond description with milling craft, Area Z would become known as "Piccadilly Circus."[6]

Under the cloak of darkness and by sunrise the following day, 6 June, Allied minesweepers had to clear an invasion path from Area Z to Normandy. Twenty-two flotillas were to sweep ten approach channels, each 800 yards wide by dawn's first light. Included among the scores of British minesweepers were 32 American (11 AMs and 21 YMSs) and 16 Canadian *Bangor*s. Five swept channels would lead from Area Z to the Bay of Seine; midway, each would split into a slow lane and a fast lane, then merge again at the Transport area off Normandy.[7]

When the sun went down about 2200 on 5 June, the wind increased and the sea turned rough in the Channel, as a pale full moon broke through the overcast, and gunfire flickered along the horizon. By that time, Sweep Units 1 and 2—British and Canadian ships with Oropesa sweep gear streamed—were some forty miles north of Normandy. By 0300 on D-Day, Units 1 and 2 had completed their assault sweeps and moved seaward to the Transport Area.[8]

Shortly after midnight, Sweep Unit 3, composed of American AMs—steel hulled minesweepers—had begun sweeping the Fire Support Channel that would be used by bombardment ships. By 0330, they had finished their initial tasking and joined the British in the Transport area. Dawn was about to break over Normandy, and with it, bitter fighting ashore by combat troops delivered to the beachhead by the largest invasion fleet the world had ever known.[9]

RCN 31ST MINESWEEPING FLOTILLA

There is no doubt that the mine is the greatest obstacle to success.

—Diary entry made by Adm. Sir Bertram Ramsay, RN, naval commander for Operation NEPTUNE, in March 1944, in reference to the planned invasion of Normandy.[10]

In the largest such operation ever undertaken, 247 minesweepers were employed to sweep ten approach lanes across the English Channel, clear the disembarkation and fire support sectors of the assault area and sweep the final paths to the beaches. Among this flotilla of vessels were sixteen *Bangor*-class minesweepers of the Royal Canadian Navy. Minesweeping had been a serious concern to the RCN since World War I when German U-boats had laid mines off the Halifax approaches. Yet in World War II, apart from one brief scare in 1943, there had been no need for this service.[11]

Accordingly, their fleet minesweepers had spent all their time on anti-submarine patrols and escort duties in the North Atlantic off the East Coast of Canada. Their minesweeping gear had been removed for their work as escorts. When located, it was found to be, in many cases, rusty, as was the ships crews' ability to use it. In preparation for their invasion assignment, the *Bangor*s were refitted in Canada with Mk I Oropesa wire sweeping gear and with the SA Type C sweep for use against acoustic mines. The existing "pom pom" anti-aircraft guns were removed and replaced with twin Oerlikons. The ships retained their anti-submarine capability with ASDIC (sonar) but with a reduced complement of operators and depth charges. Out of necessity, training required to gain proficiency in sweeping would take place in England.[12]

The sixteen *Bangor*s left Canada in four divisions in the latter half of February 1944. Fifteen arrived at Plymouth between 7 and 13 March. The sixteenth, HMCS *Mulgrave*, had to undergo repairs at Ardrossan in southwestern Scotland. Crossing the Atlantic, she grounded at Horta, Azores, and had to be towed to Greenock, Scotland. *Mulgrave* rejoined the others at Plymouth in April. Operating from Plymouth, the ships carried out exercises in Tor Bay; each day streaming and recovering sweeps, laying dan buoys, and practising precise station-keeping which would be so vital in Operation NEPTUNE.[13]

Although he expected the ships and their crews to require some familiarisation, Comdr. John B. G. Temple, RN (Commander, Minesweepers West, and the British officer responsible for their instruction) was aghast at the job that confronted him. He observed that the Canadians "were not minesweeping minded as they had been employed solely on escort work," and that "some were also under the impression that minesweeping was child's play." For his part, the Canadians' senior officer, Comdr. Antony H. G. Storrs, RCNR, later recalled the British "sucking their teeth wondering if these Canadians are really up to it."[14]

An intensive training programme, which the British termed "Pious Hope," stretched from mid-March until late May. The training team led by Temple included lieutenant commanders Dudley Hoare, RNR and Ronald Gresham, RNVR; and lieutenants John Nicholas, RNVR, Burns, and Harry Preston. (The latter two individuals are possibly Lt. Wallace Bruce Burns, RN, of HMS Vernon; and T/Lt. Henry Francis Morrison Preston, RNR, who had previously served with Temple in HMS *Cadmus* and, like Temple, was attached to the French battleship FS *Paris* moored at Devonport as a depot ship.) During training, sweeping was practiced vigourously and the *Bangors*' equipment was brought up to scratch. Once they were considered competent, the Canadians proceeded to Portsmouth to take part in a number of rehearsal exercises, including the six-day Operation GANTRY, during which the *Bangor*s and their crews were tested. Despite poor visibility and some equipment failures, they successfully located and swept a large field of 150 dummy mines. At completion, Temple declared them "efficient, keen and competent."[15]

The Canadians had been looking forward to leading the 3rd Canadian Division into Juno Beach, but, in what Storrs described as a "great disappointment," they were assigned to support the American landings, perhaps because planners "thought that probably we'd manage better with the Americans than the Brits." Moreover, they had hoped to form their ships into two complete RCN flotillas but, instead, were split up. Ten *Bangor*s were assigned to the 31st Canadian Minesweeping Flotilla under Storrs, and the remaining six ships were dispersed among three British flotillas. On a positive note, all sixteen ships were attached to the Western Task Force, and so would be in close proximity while leading the way into the American assault beaches Utah and Omaha.[16]

The identities of the flotillas to which the *Bangor*s were assigned are provided in the following table, as well as the assignment of the flotillas' duties during the assault and post-assault phases, and the identities of the ships' commanding officers. A danlayer is a ship which follows minesweepers as they work an area, and lays dan buoys to define the clear channels swept. Force O and Force U refer to Omaha Beach, and Utah Beach, respectively.

Assignments of RCN Minesweepers for Normandy Invasion

Canadian 31st Minesweeping Flotilla
(Assault: Force O, sweep Channel No. 3; post-assault: Fire Support Ships' Channel and as required)

Ship	Commanding Officer
HMCS *Blairmore* (J 314)	Lt. Joseph Charles Marston, RCNR
HMCS *Caraquet* (J 38)	A/Comdr. Antony Hubert Gleadow Storrs, RCNR (Senior Officer, 31st M/S Flotilla)
HMCS *Cowichan* (J 146)	Lt. Kenneth William Newman Hall, RCNR
HMCS *Fort William* (J 311)	A/Lt. Hugh Campbell, RCNR
HMCS *Malpeque* (J 148)	Lt. Donald Davis, RCNVR
HMCS *Milltown* (J 317)	Lt. Edward Henry Maguire, RCNVR
HMCS *Minas* (J 165)	Lt. James Barrett Lamb, RCNVR
HMCS *Wasaga* (J 162)	Lt. James Henry Greene, RCNR
HMCS *Bayfield* (J 08) danlayer	T/Lt. Stanley Pierce, RCNR
HMCS *Mulgrave* (J 313) danlayer	T/Lt. Ralph Morton Meredith, RCNR

Danlayer for the British 4th Minesweeping Flotilla
(Assault: Force O, sweep Channel No. 4; post-assault: Area between Channels 3 and 4 and as required)

HMCS *Thunder* (J 156)	A/Lt. Comdr. Herman Dwight MacKay, RCNR

British 14th Minesweeping Flotilla
(Assault: Force U, sweep Channel No. 2; post-assault: Areas between Channels 3 and 4 and as required)

HMCS *Guysborough* (J 52)	Lt. Benjamin Thomas Robert Russell, RCNR
HMCS *Kenora* (J 281)	Lt. Douglas Wilson Lowe, RCNVR
HMCS *Vegreville* (J 257)	Lt. Thomas Bottrell Edwards, RCNR
HMCS *Georgian* (J 144) danlayer	T/Lt. Alexander Grant, RCNVR; relieved on 20 Jun 1944 by T/Lt. Douglas Wharrie Main, RCNR

Danlayer for the British 16th Minesweeping Flotilla
(Assault: Force U, Sweep Channel No. 1)

HMCS *Canso* (J 21)	T/Lt. John Kincaid, RCNR[17]

HMCS *Mulgrave* was mined in the English Channel off Le Havre, France on 8 October 1944. Heavily damaged, she was decommissioned on 6 July 1945.[18]

LEADING THE ASSAULT ACROSS THE CHANNEL

> *We were to hold our course, no matter what was ahead - there must be no holes in our sweeping as ships loaded with troops would be following us, and would be depending on us.*
>
> —An officer aboard the Canadian minesweeper HMCS *Georgian*, recalling instructions to the 31st Minesweeping Flotilla before it commenced sweeping a path across the English Channel on the night before D-Day of the Normandy Invasion.[19]

The 31st Flotilla had responsibility for sweeping Approach Channel No. 3 into Omaha Beach. Its ten ships weighed anchor at Portland in the early hours of 5 June and proceeded east in the English Channel toward Z-Buoy—the junction area for the assault convoys south of Portsmouth which acquired the moniker "Piccadilly Circus." Getting to Z-Buoy took most of the day, and it was not until evening that the flotilla and its escort vessels formed up to commence sweeping the 70 nautical mile path to Omaha Beach.[20]

Two Fairmile "B" motor launches preceded Commander Storr's HMCS *Caraquet*, the lead ship in a formation which, followed by *Fort William*, *Wasaga*, and *Cowichan*, was formed in an echelon to port. In this type formation, as shown in the following diagram (which depicts three ships in a starboard echelon formation), the minesweepers proceed in echelon, each overlapping the sweep wire of the one ahead, with the lead ship supposedly in clear water (no danger from mines). In this case, *Caraquet* did not enjoy that luxury. It was known that the Germans had sown mines in the channel, and the waters along her projected track had not been previously swept. (The ships each had a single sweep streamed to port since the tide was in flood. This arrangement sweeping along the western side of the swept channel ensured the sweeps maximum lateral deployment and that ships' propellers would not be fouled.) [21]

HMCS *Bayfield* and *Mulgrave* laid dimly lit dan buoys to mark the swept channel. Astern of them were *Minas*, *Malpeque*, *Blairmore*, and *Milltown* ready to replace any of the sweepers that fell out due to breakdowns or enemy action.[22]

Diagram 21-2

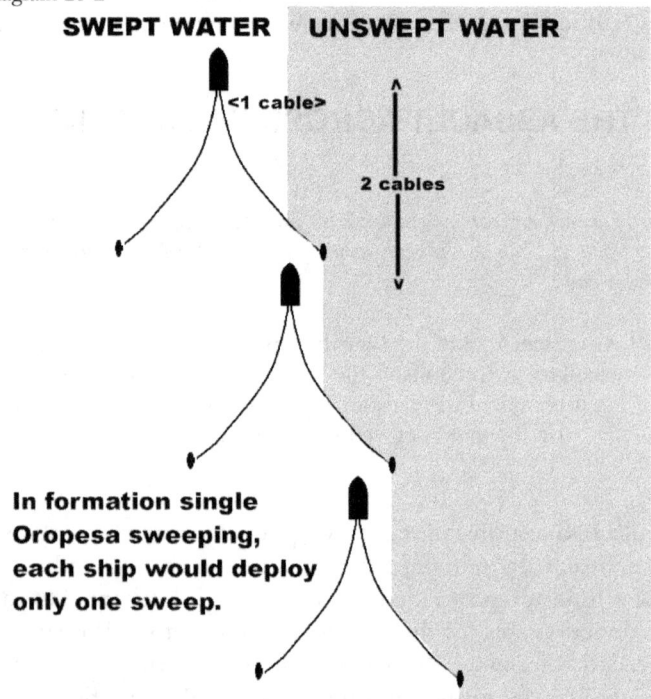

When echelon minesweeping was performed by the Royal Navy, the approximate area covered by each ship's sweep was one cable (220 yards) and the approximate distance of one ship behind another was two cables.
Collection of Rob Hoole

British-invented Oropesa sweeping was also used by the Americans and Canadians at Normandy. In it, sweep wires were streamed off a ship's stern to either the port or starboard side, or both. A float (termed a "pig") marked the position of and supported the end of each sweep wire. "Otters" and "kites" attached along the sweep wires caused them to veer away from the ship, and kept them at the required depth, respectively. The mooring cable of a mine, upon encountering a sweep wire, would slide along it until mechanical or explosive "cutters" positioned near the end of the sweep severed it. The mine body would then float to the surface where it could be destroyed by gunfire from danlayers or other ships in trail. In this case, mines that rose to the surface were not to be exploded unless they imperiled other ships, for fear of alerting the enemy.[23]

The minesweepers advanced at 7.5 knots clearing a lane about a thousand yards wide. Complicating matters, planners had laid out a channel that was not perfectly straight, but followed a series of gentle doglegs, each requiring minor course changes. Further, when the tidal current reversed itself in the middle of the passage, the formation had to recover sweep gear, reverse course and head in the opposite direction. This was done to avail themselves of safe, previously swept waters, while redeploying their sweeps to starboard and resuming where they had left off now following the eastern side of the swept channel. Except for one dan buoy being inadvertently cut, this difficult, and time-consuming manoeuvring was completed without incident.[24]

Leading the assault flotillas, the minesweepers were the most vulnerable vessels in the invasion armada. Planners had thought they would be detected by the enemy, and suffer heavily for it. Yet the night remained peaceful, and the only outward sign of enemy activity visible was a barrage of anti-aircraft fire put up over Normandy in response to Allied bombing. To the west of the 31st, the British 14th Flotilla, which included the Canadian ships *Guysborough*, *Kenora*, *Vegreville* and *Georgian*, swept close enough to the French coast to distinguish individual houses in the dim light but were still not detected.[25]

DUTY OFF THE ASSAULT BEACHES

> . . . the dark of the night began to give way to twilight [and] one could begin to appreciate the immensity of the operation. As light increased more and more dark shapes began to make distinctive forms, and what had appeared to be just a few ships in close proximity now became a whole panorama of sea power stretching as far as the eye [could see] to seaward.
>
> —Lt. Joseph Charles Marston, RCNR, commanding officer of HMCS *Blairmore* recalling the awesome sight off Normandy as darkness began to give way to dawn on 6 June 1944.[26]

Twenty minutes past midnight on 6 June, the 31st Flotilla reached the transport area about eight nautical miles off the French coast, formed into line abreast, and turned south. Sweeping shoreward to the ten-fathom line (60-foot water depth off Omaha Beach), they cleared the way for the assault craft preparing to make the shore. They would later witness the awe-inspiring sight described above.[27]

Commander Storrs had told his ships to hold their formation even under bombardment. The sweeping, off the beach in darkness and poor weather, with a strong cross-tide, required superb seamanship as he led the 31st Flotilla close to the German guns. Only when rounds from shore battery fire began to fall in the sea nearby did he order "in sweeps." During recovery of sweep gear, *Caraquet* and *Fort William* had to cut theirs free after that of each ship became entangled in wrecks.[28]

Blairmore, initially a reserve ship, was still recovering her gear forty minutes before H-Hour (the scheduled time at which assault forces were to begin storming ashore), when she found herself close under the guns of the Allied shore bombardment force. Lt. Joseph Charles Marston later described the effect that 12-inch gun rounds passing overhead the minesweeper had on the efficiency of his crew:

> ... the big guns of [the battleship USS] *Arkansas* spoke for the first time, and right on schedule. We were in a position a few hundred yards from her and in her line of fire. The resultant blast, concussion, and general commotion shivered one's timbers more than a little, and the sweeping party on the quarter deck were prodded into such a frenzy of work and efficiency, that we were soon all clear to get the hell out of it, which we did by drawing off seaward, there to mosey around keeping clear of the vast numbers of ships all seemingly being drawn into the maw of the invasion beaches.[29]

Photo 21-2

Opening the Attack by Dwight C. Shepler, 1944. D-Day morning broke over the Normandy coast to find the battleship USS *Arkansas*, matriarch of the battle fleet, banging away at the beachhead with her main battery guns. To seaward, the French cruisers *George Leygues* and *Montcalm* sent shells hurtling into their captive homeland. Assault landing craft streamed toward the beaches while transports filled the horizon. Naval History and Heritage Command #88-199-EW

AWARDS FOR VALOUR

Comdr. Antony Hubert Gleadow Storrs, RCNR, commanding officer HMCS *Caraquet* and senior officer, 31st MS Flotilla, remained in the naval service after the war. He retired as a rear admiral in 1962. During his career, he was awarded the Distinguished Service Cross & Bar, Legion of Merit, Legion d'Honneur, Croix de Guerre, and made an Honourary Commodore, Canadian Coast Guard. His citation for exemplary leadership of the 31st Flotilla at Normandy reads in part, "Storrs's zeal, patience and cheerfulness, together with the skill and judgement he displayed in the execution of this complex task, contributed significantly to the success of Operation Neptune, the naval contribution to the invasion." The Americans also recognised his accomplishments with the Legion of Merit.[30]

Summary information about Storrs' DSCs, and French medals awarded to other commanding officers of *Bangor*-class minesweepers (for duty at Normandy or elsewhere in Europe) follows. The term "Gazetted" in the table refers to the date when notice of the award first appeared in *The London Gazette* or *Canada Gazette*; "CO" and "SO" denote commanding officer and senior officer, respectively.

Distinguished Service Cross (DSC)

Individual	Gazetted	Narrative
A/Lt. Comdr. Hugh Campbell, RCNR CO HMCS *Fort William*	26 Dec 44 L 20 Jan 45 C	For gallantry, skills, determination, and undaunted devotion to duty during the landing of Allied Forces on the coast of Normandy.
T/Lt. Joseph Charles Marston, RCNR CO HMCS *Blairmore*	3 Jul 45 L	For consistent zeal, courage, and good seamanship whilst serving in HM 104th and HM 31st Mine-Sweeping Flotilla and the 159th Trawler Group and in HMCS Blairmore in arduous operations along the coast of Southern England and Northern France.
A/Lt. Comdr. Ralph Morton Meredith, RCNR CO HMCS *Mulgrave*	11 Dec 45 L 30 Mar 46 C	For distinguished service during the war in Europe.
A/Comdr. Antony H. G. Storrs, RCNR CO HMCS *Caraquet*; SO, 31st M/S Flotilla	1 Jan 45 L 13 Feb 45 L	DSC: Minesweeping - New Year Honours 1945. Bar to DSC for courage and skill in minesweeping operations during the landing of Allied Forces in Normandy.[31]

Croix de Guerre 1939-1945 (French award for bravery)

Croix de Guerre avec Etoile en Vermeil:
Silver-gilt star, for those who had been mentioned at the corps level

Croix de Guerre avec Palme en Bronze:
for those who had been mentioned at the army level

A/Comdr. Herman Dwight MacKay, RCNR CO HMCS *Thunder* until 25 June 44; medically unfit on 29 January 46	31 Aug 45 L 24 Nov 45 C	For the precision and complete success of dredging operations in connection with the action of Naval Forces against the enemy in the Battle of the Atlantic.
Lt. Donald Davis, RCNVR CO HMCS *Malpeque*	16 Jul 49 C	For services in the invasion of Normandy.[32]

The French awarded Comdr. Herman Dwight MacKay, RCNR, the Croix de Guerre (avec Etoile en Vermeil). Based on the medal citation, it appears that it was for mine clearance. The reference to "dredging operations" is likely due to a misunderstanding arising from the French-

English translation of minesweeping. In French, a minesweeper is a 'dragueur de mines' (dredger of mines) and minesweeping is 'dragage de mines' (dredging of mines).

WESTERN TASK FORCE MINESWEEPERS

It was considered essential that within each task force area, the minesweeping forces should be under the command of an experienced minesweeping officer from the completion of sweeping of the assault approach channels onward. Rear Admiral Kirk had requested that a British officer be assigned to carry out this duty in his area. Commander Temple was appointed as commander, Minesweeping West, with the USS *Chimo* as his headquarters ship. Acting Captain Richard Borthwick Jennings, RN, was appointed as captain, Minesweeping East.[33]

Western Task Force

Force U (55 minesweepers, plus motor launches and danlayers)

Flotilla	Country	Type Minesweepers
14th MSF	British/Canadian	8 Fleet minesweepers, with two minesweeping motor launches and three or four danlayers also attached
16th MSF	British/Canadian	8 Fleet minesweepers, with two minesweeping motor launches and three or four danlayers also attached
132nd MSF	British	10 Motor minesweepers (MMS)
"A" MSF	American	11 Fleet minesweepers (AMs)
"Y1" MSF	American	11 Yard minesweepers (YMS)
"Y2" MSF	American	7 Yard minesweepers (YMS)

Force O (37 minesweepers, plus motor launches and danlayers)

Flotilla	Country	Type Minesweepers
4th MSF	British/Canadian	9 Fleet minesweepers, with two minesweeping motor launches and three or four danlayers also attached
31st MSF	Canadian	8 Fleet minesweepers, with two minesweeping motor launches and two danlayers also attached
104th MSF	British	10 Motor minesweepers (MMS)
167th MSF	British	10 British Yard minesweepers (BYMS)[34]

U.S. NAVY MINESWEEPER CHANNEL CROSSINGS

By late afternoon on 5 June, Comdr. Henry Plander's "A" Squadron was well out in the English Channel, heading south. The eleven 220-foot *Raven*-class and 221-foot *Auk*-class minesweepers comprising it were following a path cleared a day earlier by the 14th British Minesweeping Flotilla, or, at least, thought they were. Because of strong three- to six-knot currents setting ships off their intended track, at any particular moment courses made good might be well off those ordered steered. Ships following swept channels might have veered out of them, and the channels might not be exactly where planned.[35]

"A" Squadron (Comdr. Henry Plander, USN)

Auk (AM-57)	Lt. Daisy Louis Brantley, USN	*Raven* (AM-55)	Lt. Comdr. John Francis Madden, USN
Broadbill (AM-58)	Lt. Comdr. Oscar Berndherd Lundgren, USN	*Staff* (AM-114)	Lt. Joseph Henry Napier, USN
Chickadee (AM-59)	Lt. William Donald Allen, USN	*Swift* (AM-122)	Lt. Comdr. Richard Kendall Cockey, USNR
Nuthatch (AM-60)	Lt. Comdr. Robert A. L. Ellis, USN	*Threat* (AM-124)	Lt. Comdr. Homer Ewald Ferrill, USNR
Osprey (AM-56)	Lt. Charles H. Swimm, USNR	*Tide* (AM-125)	Lt. Comdr. Allard Barnswell Heyward, USNR (KIA)
Pheasant (AM-61)	Lt. Comdr. Harold Irving Pratt, USNR[36]		

LOSS OF MINESWEEPER USS *OSPREY*

Photo 21-3

USS *Osprey* (AM-56) under way, circa April 1941.
Naval History and Heritage Command photograph #NH 84026

Toward dusk, *Osprey* hit a mine and sank, becoming the first ship lost in the Normandy invasion. (A French gunboat had warned the American squadron at 1610 that they were heading out of the swept channel, but the minesweepers held their course, intending to make best speed and arrive as soon as possible at the point at which they were to commence minesweeping.) As *Osprey* sank, she wallowed out of column, a blazing wreck, 6 men missing and 29 wounded. A squadron mate, *Chickadee*, helped *Osprey* put out fires and a passing PT boat picked up some of the crewmen who had been blasted overboard. However, all actions to save her proved futile. As seawater surged through her broken hull, she developed a heavy port list, and at 1815, Lt. Charles H. Swimm, USNR, ordered his crew to abandon ship. *Osprey* went down soon after, the first of many Allied ships sunk by German mines during the Normandy invasion.[37]

The remaining minesweepers continued across the channel in a column formation, leading a huge convoy stretching out behind them as far as the eye could see and, in the evening as dusk deepened, streamed gear to clear a channel toward the beach. The first sweep was uneventful; no mines were found. As the ships approached waters nearer shore in which transport and fire support ships would operate, they again streamed "O" type gear to perform a clearance sweep of the area. The former executive officer of USS *Tide* recalled subsequent events of that day in an interview on 31 August 1944:

> Starting at six hours before H-hour, we went in close to the beach and carried out the final phase of our assault sweeping preparation for the actual landings. We completed this assault sweep shortly before the loading of the landing craft was started and when we had completed this phase, we left the immediate area and formed a line of defense against E-boat and U-boat operations, protecting and screening the transports and other ships taking part in the invasion.
>
> When the first landings started … it was obvious that the Germans had concentrated their mining efforts … employing small obstacle mines to destroy landing craft and personnel as they came in. These were, of course, too far inshore for ships of our draft to reach and they could not have been destroyed by any sweeping, by minesweepers. As the day wore on, enemy planes came over and during the afternoon and night of D-Day the waters in the area of Utah beach and adjacent to Utah beach were heavily mined from the air. The Germans dropped ground mines and nothing could be done about that during the night of D-Day.[38]

YARD MINESWEEPERS (YMS) OFF UTAH BEACH

The few books and articles which include information about the 21 YMS minesweepers at Normandy typically cite their parent organisations as "Sweep Unit 4" and "Sweep Unit 5," or "Y1" and "Y2" Squadrons. In actuality, the organisation on D-Day, 6 June, was more fluid. Six YMSs were assigned to Y1 Squadron and seven to Y2, with the remaining ships designated as spare sweepers, danlayers, or part of a special sweeping group. This allowed for redundancy in the event of ship breakdowns or other unforeseen difficulties; for example, *YMS-377* replaced *YMS-375* in Y1 Squadron on 6 June.

Y1: Lt. Henry J. White, USNR	Y2: Lt. C. I. Rich, USNR
YMS-305	*YMS-346*
YMS-377 Lt. Lewis C. Bryan, USNR	*YMS-347*
YMS-378	*YMS-348*
YMS-379	*YMS-349*
YMS-380	*YMS-350*
YMS-406	*YMS-351*
	YMS-352

Spare Sweepers: *YMS-356, 358, 375*   **Danlayers:** *YMS-381, 382*
Special Sweeping Group: *YMS-231, 247, 304*[39]

The 136-foot yard minesweepers (YMSs) had arrived off Utah Beach about an hour behind the AMs and taken up their duties. At 0100 on 6 June, at a point some fifteen miles off the coast, the six wooden-hulled ships of Y1 Squadron commenced sweeping shoreward. At 0330, when about three miles off Omaha Beach, the group turned starboard and began sweeping westward along the coast toward the fishing village of St. Vaast on the Cotentin peninsula. While on this leg, the ships passed within a mile of enemy batteries on the St. Marcouf Islets, unchallenged.[40]

Sweeping Fire Support Channel No. 3, the ships of Y1 Squadron (*YMS-406, 378, 305, 379, 380,* and *377*) were in a starboard echelon formation, with "O" gear streamed to cut moored mines. At 0512, upon completion, they began recovering their starboard sweep gear in preparation for streaming port gear and widening Fire Support Channel No. 1, seaward. The YMSs were a couple of miles off St. Vaast when dawn broke and they came under fire of German shore batteries while still bringing the gear aboard. HMS *Black Prince* rushed in to open counter-battery fire and as the enemy guns shifted to her, the minesweepers were caught in the cross fire of the British cruiser and shore batteries, dueling with heavy calibre guns.[41]

Photo 21-4

USS *YMS-406* conducting builder's trials on 19 November 1943; she was built by Henry C. Grebe and Co., Chicago, Illinois. She was one of eighteen such 136-foot, wooden-hulled minesweepers at Normandy, for the Allied invasion of the continent. Naval History and Heritage Command photograph #NH 43511

Lt. (jg) Lewis Calvin Bryan, USNR, commanding officer of USS *YMS-377*, later described D-Day morning at Normandy:

> We swept a fire support channel towards the beach, turned right and proceeded along the beach for a time, while making our turn with gear streamed astern, and then right again to head seaward. As we made the final turn, about a mile off the beach, our battleships and cruisers began shelling the shore, which drew return fire from German coastal guns. This resulted in, at one point, rounds from both the ships and the shore batteries passing overhead, which whistled or whined by depending on their caliber. As we headed outbound, the shore bombardment stopped as tow planes with gliders came in low right over our mast, which on departure, after releasing the gliders, passed over us again only higher. (The Waco CG-4A gliders built by the Ford Motor Company were essentially wooden boxes with wings. The duty of the pilots who flew the motorless aircraft was to deliberately crash land them in the drop zone and fight as combat infantrymen.) Then the landing craft came in, and after that, all hell began to break loose.

Photo 21-5

Painting *The Sea Wall at the Eastern American Beach* by Mitchell Jamieson, 1944. Scene at Utah Beach at about 1500 on D-Day. The fighting had moved inland, but an enemy artillery battery, some distance inland but still in range, sent shells steadily over the Americans whose work was on the beach itself.
Naval History and Heritage Command #88-193-IC

Y1 SQUADRON SKIPPERS HONOURED

Photo 21-6

Depicted from left to right, Lieutenants Ralph P. Fiebach Jr., William E. Beckham Jr., Henry J. White, and Franklin Duerr Buckley; Lt. (jg) Lewis C. Bryan, and Lt. George D. Harrelson, receiving the Bronze Star Medal.
Courtesy of Lewis C. Bryan

For their bravery under fire while sweeping mines off Utah Beach, the commanding officers of *YMS-305, 377, 378, 379, 380*, and *406* later received Bronze Star Medals.

ALLIED CASUALTIES AT NORMANDY ON D-DAY

The Allied assault at Normandy began shortly after midnight on 6 June with more than 2,200 bombers attacking targets along the coast and inland. Cloud cover hindered the air strikes, and the coastal bombing at Omaha Beach was particularly ineffective. More than 24,000 American, British, and Canadian airborne assault troops and 1,200 planes followed the air bombardment. At 0530, six Allied divisions and numerous small units began landing on the beaches. The Allies landed more than 160,000 troops at Normandy, of which 73,000 were Americans assaulting Omaha and Utah beaches. A combined 83,115 British and Canadian forces landed on Gold, Juno, and Sword beaches.[42]

Allied casualties on 6 June have been estimated at 10,000 killed, wounded, and missing in action: 6,603 Americans, 2,700 British, and 946 Canadians. Over the following days the Allies gradually expanded their tenuous foothold as the fighting moved inland. The Allies finally broke out of Normandy on 15 August, and once out, advanced quickly liberating Paris on 25 August. The German forces' retreat across the Seine River, five days later, marked the end of the Normandy campaign with its great loss of lives and injured bodies and minds on both sides.[43]

FLAGSHIP USS *CHIMO* ARRIVES OFF UTAH BEACH

> *We have got to lay more mines and still more mines in the Seine Bay with the tenacity of a bulldog .... It is incomparably more effective to sink a whole cargo at sea than to have to fight the unloaded personnel and material on land.*
>
> —Adolf Hitler expressing the importance of continuing to mine the waters off the assault beaches at a conference on the Normandy situation a couple of weeks after D-Day, 6 June 1944.[44]

On 6 June, fueled and provisioned—with commander, Minesweepers West, Comdr. John Bruce Goodenough Temple, RN, and his staff embarked—auxiliary minelayer USS *Chimo* (AMC-1) departed Plymouth at 0714. Out in the Channel, she joined Convoy EBM-2 bound for the western assault area. She arrived there at 0930 the following morning. Anchoring off Utah Beach in the vicinity of the attack transport USS

Bayfield (APA-33) and other assault force ships, *Chimo* commenced operations as tender to British and U.S. minesweeper forces, while also serving as flagship and operational headquarters for Commander Temple.[45]

The Allies lost some ships and landing craft to mines during the Channel crossing and assault on the Normandy beaches on D-Day. Following these, losses would continue to mount as sea forces helped to consolidate and support the beachhead. Early that morning, 7 June, the transport *Susan B. Anthony*, with 2,317 combat troops aboard, hit a mine at 0750 off Omaha Beach. Fortunately, HMS *Mendip*, HMS *Norbo*, USS *Pinto*, *LCI-496*, and *LCI-489* were nearby and took off everyone aboard without any loss of life, before she went under at 1010.[46]

The minesweeper USS *Tide* was not so lucky. That same morning while over Caronnet Banks, with *Threat*, *Swift*, and *Pheasant* searching for contact mines believed sown by German E-boats during the night, a magnetic mine exploded under her forward engine room. Details about the loss of *Tide*, and other minesweepers and the minelayer USS *Miantonomah* (CM-10), follow in the next chapter.[47]

22

Consolidation of the Beachhead

There was a big explosion, and gray smoke and white water all mingled. We saw first those who had been blow farthest by the explosion. They were all dead.... And then from all over the sea around us, sounding small and childlike in the wild world of waters, came cries of 'Help! Help! and one startling plea of 'Please help me!' We fished six men out of the water, two uninjured. Other craft had come to the rescue and we searched among them, taking only the living and leaving the dead bobbing and ebbing and awash like derelicts in an unwitting sea. One of the men we got was naked. Every stitch of clothing had been blown off him.

—Description of the sinking of the minesweeper USS *Tide* (AM-125) by news correspondent Ira Wolfert, who watched her go down from the deck of the attack transport USS *Bayfield* (APA-33).[1]

Diagram 22-1

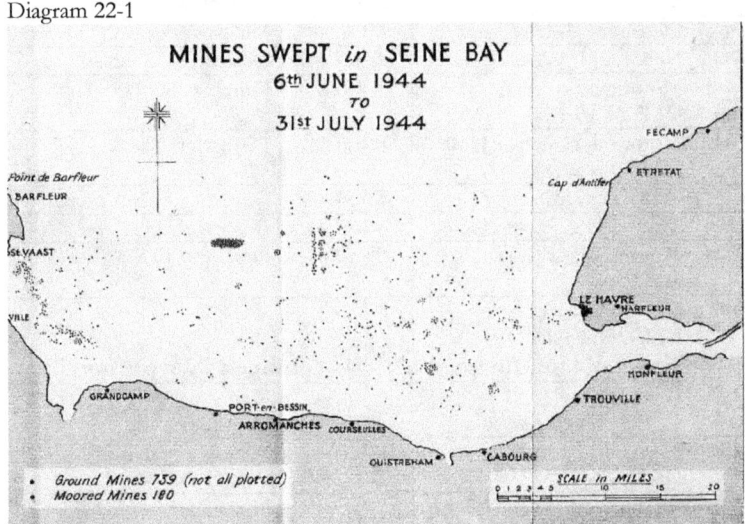

Plot of German mines swept by Allied forces in Seine Bay, 6 June-31 July 1944. As indicated, not all of the 739 ground, and 180 moored mines, are plotted. Courtesy of Michael Friend, from a collection of pages and photographs belonging to his father, Lt. Peter Donald Friend, GM, RNVR. It also appears in *The Minesweepers' Victory* by Hilbert Hardy.

During the period from the assault landings on D-Day to the withdrawal of the task and assault group commanders in early July (which marked the conclusion of Operation NEPTUNE), the naval effort was primarily devoted to building up the army to the required size and providing it supplies and reinforcements. Bombarding ships supported the army as and when required but, naturally, as the troops advanced farther inland, calls for fire became less frequent.[2]

Enemy attempts to disrupt naval traffic were intensified, but for the most part ineffective. Frequent attempts on the "Spout" (the series of lanes swept clear of mines crossing the English Channel) and anchorage by surface craft were usually intercepted and driven off, and in no case was success achieved by them in the assault area. Enemy submarine efforts were also handily mastered. The only measure of success obtained by the Germans was through minelaying by low flying aircraft. This form of attack proved very difficult to counter and took its toll of Allied shipping.[3]

A listing of Royal Navy, U.S. Navy, and Free French naval vessels lost to enemy mines in June and July 1944 follows. These ships were sunk, or sufficiently damaged to be considered total losses.

Allied Warships Lost to Mines in June 1944
(Operation NEPTUNE)

Date	British	Date	American/French
6 Jun 44	Destroyer HMS *Wrestler*	6 Jun 44	Minesweeper USS *Osprey*
8 Jun 44	Netlayer HMS *Minister*	6 Jun 44	Destroyer USS *Corry*
11 Jun 44	Motor gunboat HMS *MGB 17*	7 Jun 44	Minesweeper USS *Tide*
13 Jun 44	Motor minesweeper HMS *MMS 229*	8 Jun 44	Destroyer USS *Rich*
24 Jun 44	Destroyer HMS *Swift*	8 Jun 44	Destroyer USS *Meredith*
24 Jun 44	Trawler HMS *Lord Austin*	8 Jun 44	Destroyer USS *Glennon*
24 Jun 44	Motor minesweeper HMS *MMS 8*	9 Jun 44	Destroyer FFS *Mistral*
27 Jun 44	Motor gunboat HMS *MGB 326*		

Allied Warships Lost to Mines in July 1944 (grouped by type, not date)

Date	British	Date	American/French
2 Jul 44	Landing Ship HMS *Empire Broadsword*	2 Jul 44	Minesweeper USS *YMS-350*
2 Jul 44	Motor Torpedo Boat HMCS *MTB 460*	30 Jul 44	Minesweeper USS *YMS-304*
8 Jul 44	Motor Torpedo Boat HMCS *MTB 463*	30 Jul 44	Minesweeper USS *YMS-378*[4]

These totals do not include merchant ships and auxiliaries lost to mines, nor the scores of landing craft that fell victim to them. Three hundred four landing craft were lost or disabled in the assault at Normandy. It is estimated that fifty percent of the casualties were due to beach obstacles in combination with Teller mines (anti-tank devices which derived their name from the German word for "plate," after the mine's flat, circular shape). These type mines were widely employed, attached to obstacles, such as poles and tetrahedra. Otherwise, they functioned as land mines, detonated by pressure from landing craft.[5]

In the following table, the capital letters S, J, G, O, and U denote the Sword, Juno, Gold, Omaha, and Utah assault beaches.

Allied Landing Craft Casualties at Normandy (Destroyed or Disabled)

Craft	Eastern Task Force			Western Task Force	Total
LCT	S-18	J-45	G-34	O&U combined-34	131
LCI (S)	S-15	J-7	-----		22
LCI (L)	S-9	-----	-----	O&H combined-12	21
LCS (M)	-----	J-2	-----		2
LCA	S-29	J-36	G-52		117
LCM	S-7	-----	-----		7
LCP (L)	S-1	-----	G-3		4
Total	79	90	89	46	304[6]

Photo 22-1

Painting *The Tough Beach* by Dwight C. Shepler, 1944. Studded with beach and underwater obstacles, mines, and German fortified positions and pillboxes, Omaha Beach proved deadly to many American soldiers and sailors on 6 June 1944. Naval History and Heritage Command #88-199-EU

USS *TIDE* SUNK BY MINE OFF UTAH BEACH

While mine clearance by Allied minesweepers undoubtedly saved the lives of untold numbers of soldiers and sailors, the ships could not do anything about mines lying in waters too shallow for them to operate. Compounding the difficulties of the landings, German aircraft continued to sow mines throughout the Normandy invasion. On the night of 6 June, a group of steel-hulled AMs took up station off the Carentan Estuary to prevent the possible egress of German E-boats based up the Vire River. They did not encounter any enemy forces but did receive a report that German patrol craft had attempted to come down the river under the cover of darkness and had been driven back.[7]

Photo 22-2

USS *Tide* (AM-125) sinking off Utah Beach after striking a mine during the Normandy invasion, 7 June 1944. Motor torpedo boat *PT-509* and USS *Pheasant* are standing by. U.S. Navy Photograph, now in the collections of the National Archives.

On the morning of 7 June, USS *Tide* (AM-125), *Threat* (AM-124), and *Swift* (AM-122) received orders to sweep an area near shore. The ships streamed "O" type gear but did not cut any moored mines. At 0937, having recovered her gear, *Tide* prepared to leave the area. Her former executive officer described on 31 August 1944 the subsequent events of that day, D+1:

The *Tide* had just taken her gear aboard and had come to six knots speed when a tremendous explosion below decks lifted the entire ship completely out of the water. Officers and men watching from other ships stated that the ship was lifted a full five feet into the air.⁸

The force of the explosion broke *Tide*'s back, tore a huge hole in her hull, and ruptured bulkheads below the waterline. When the executive officer went below to assess the damage, he found mattresses from the forward crew's compartment in the after-engine room, carried there through the crew's compartment, forward engine room, and refrigeration spaces by the deluge of seawater rushing into the ship. He then went to the bridge and found there were practically no men on their feet. All had been killed or wounded, most seriously. The commanding officer, Lt. Allen B. Heyward, USNR, ordered the executive officer to take over immediately and died.⁹

George Crane, now acting commanding officer, remembered efforts made to save the minesweeper:

> The *Tide* was obviously sinking rapidly, and since all of our radios had been knocked out by the force of the explosion, it was necessary to call for assistance by the use of megaphones.... [T]he *Pheasant*, another AM of our squadron, and the *Swift* were nearby and came in to help evacuate the casualties, as did the PT boat commanded by Commander John Bulkeley. In addition, a Higgins landing boat from the USS *Bayfield* came alongside and evacuation of the wounded started immediately.
>
> At about this time, a serious fire broke out aft on the ship, and I ordered the magazines flooded in order to prevent explosion. Shortly thereafter, we were able to bring the fire under control, the magazines were flooded and the ship, which at first listed badly to starboard, began to settle on an even keel. I ... gave the order to prepare to abandon ship .... The ships alongside, having taken as many survivors as they could take, then pulled away and the USS *Threat* ... came alongside to evacuate the remaining men.¹⁰

The commanding officer of *Threat* placed his ship alongside *Tide*'s starboard side so that, as the sinking vessel began to capsize, it leaned into *Threat* and remained nearly level long enough to allow the evacuation of those still aboard. *Swift* then attempted to tow the minesweeper to the nearby beach. However, after she took a strain on the line made fast to *Tide*'s bow, the crippled ship broke in two and sank in approximately 40 feet of water.¹¹

ASSAULT FORCE U SHIP CASUALTIES

More ships were lost following the mining of *Tide*. From 5-17 June, Assault Force U suffered casualties to thirty vessels, all occurring, with the exception of *Osprey*, *LCI(L)-219*, and *Susan B. Anthony*, in the Utah Beach assault area. Despite horrific ship losses, conditions for the Allied assault at Normandy would have been much grimmer had German military planners better guessed the date of the landings. During early 1944, in anticipation of an Allied assault on continental Europe, German forces had planted huge defensive minefields along the coasts of northern France, Belgium, and Holland. However, believing the invasion would occur before late May, they programmed the mines to sterilise themselves at that time in order to leave the waters safe for their own operations. Therefore, six days before the Allies crossed the English Channel, thousands of German mines obligatorily neutralised themselves, only a few remained effective, one of which got *Osprey*.[12]

Hitler also erred in not directing the use of deadly pressure mines, the deployment of which he personally controlled. He finally ordered their use on 7 June, and the Luftwaffe began four days later to plant other mine types, as well, throughout the invasion—reflecting the Fuhrer's respect for their capabilities.[13]

SUPPORT BY *CHIMO* AT NORMANDY

On 7 June, when USS *Chimo*, with commander, Minesweeping West and staff embarked, arrived off Utah Beach, the ship went to General Quarters day and night, and fired at enemy planes. That same day, her crew issued minesweeping supplies to the *YMS-378* and effected repairs to USS *Pheasant*. The following day, an aircraft bomb fell about 300 yards off *Chimo*'s port bow, shaking the ship severely but no damage or casualties resulted. Similar to the previous day, she issued minesweeping gear and performed repairs to *ML 117*, HMS *Poole*, and *YMS-356*.[14]

On 9 June, *Chimo*'s guns fired at enemy planes coming in, and she experienced another near bomb miss, 250 yards off her starboard bow. As before, the detonation shook the ship severely but caused no damage. Her work load on that day increased with the provision of spare parts or repairs to ships of the British 14th Minesweeping Flotilla, HMS *Romney*, *Seaham*, *Whitehaven*, *Guysborough*, and *Poole*.[15]

Chimo remained at anchor off Utah Beach until 19 June, ready at all times to get under way, with two boilers on the line and engine room personnel standing by. During this period, she was subject to numerous attacks by enemy planes, and on 16 June a mine was dropped 200 yards off her starboard bow. Each day, up to ten minesweepers received minesweeping supplies and/or repairs from her. On 15 June, seven recipients were sister ships. The American *YMS-305, 347, 349, 350*, and *375*, and the British *BYMS 2070* and *2173* of the 159th M/S Flotilla.[16]

The largest production run of any World War II warship was not—as one might imagine—a particular class of destroyer, frigate, or submarine, but rather 561 scrappy little 136-foot wooden-hulled vessels characterised by Arnold Lott in *Most Dangerous Sea* as belligerent-looking yachts wearing grey paint. Of these 561 products of American yards, 150 were transferred by the builders directly to the Royal Navy under the provisions of "Lend-Lease." Legislation enacted on 11 March 1941 authorised the U.S. government to abandon its existing policy of neutrality and embark on an all-out military and naval defence programme by issuing orders to private firms and by constructing new government plants.[17]

In late morning on 19 June 1944, *Chimo* left Utah Beach to proceed to Plymouth, to refuel, obtain supplies, and make some repairs to her engineering plant. Arriving at Victoria Wharf the next day, she issued minesweeping gear to the ships of "A" Squadron. The minelayer returned to the Bay of Seine on 3 July and, from anchorage off Utah Beach, tended minesweepers until the 21st. Her first day back, the crew was at General Quarters night and day. On 8 July, enemy planes were reported in the near vicinity. Heavy anti-aircraft fire was sighted five miles distant, but her guns did not open fire. On 30 July, survivors from *YMS 304* came aboard for treatment.[18]

YMS-304 SUNK, AND *YMS-378* DAMAGED BY MINES

On the morning of 30 July, the units of Y1B Division—*YMS 378, 304, 381, 382, 380* and *231*—were carrying out a magnetic and acoustic sweep in a starboard echelon formation. The ships were proceeding at 7½ knots, with a 1,000-yard interval between them and sweeping distance of 300 yards (the path width covered by each ship's sweep). Magnetic sweep gear streamed off the fantail of each ships was pulsed 5½ seconds on, 9½ seconds off at 2,600 amps, 180 volts. To counter acoustic mines, the ships had an acoustic hammerbox (sound maker with 19-inch diaphragm), streamed but were not pulsing. This was likely because of the risk of unintentionally actuating acoustic mines underneath or near ships in a nearby anchorage area. The ships did have parallel pipe-noise makers—which produced less intense, shorter range sound—also towed astern, to detonate acoustic mines.[19]

The ships were off St. Vaast, proceeding toward the harbour. Their tasking was to sweep L channel to the L3 buoy and then M channel to St. Vaast Harbour. At 0927, an acoustic mine exploded directly under the keel of *YMS-304*, amidships, and she went down in 65 seconds.

About that time, *YMS-378* detonated a mine 50 yards distant on her port bow, a second mine midway between her and *YMS-304*, and a third 100 yards astern. All were believed to be acoustic mines with very course settings. As soon as this happened, *YMS-381*, the third ship in formation behind *378* and *304*, hauled out and launched her wherry (a type boat traditionally propelled by oars) to recover survivors. The boat picked up three men.[20]

YMS-378, badly damaged, manoeuvred to pick up survivors until her engine room flooded causing a loss of propulsion. By this time, she had settled about three feet deeper in the water. A RAF crash boat (No. *2869*) which had previously appeared on the scene and taken the three survivors from the *381*, came alongside *YMS-378* and collected all seriously injured personnel from *YMS-304*—two officers and seventeen men. The RAF boat landed the survivors ashore at Arromanches in northwestern France—one died en route.[21]

Meanwhile, *YMS-381* retrieved another two officers and eight men from the sea, which she delivered to the *Chimo*. These survivors were later landed at Omaha Beach for evacuation to the UK. Two seriously injured men were picked up by a French fishing boat and landed at St. Vaast, and later taken to the hospital in Cherbourg. Of *YMS-304*'s complement of 4 officers and 32 men, one died while being transported in the crash boat and five went missing—killed outright by the blast, or who perished when the ship went down, or in the sea.[22]

Practically at the same time that *YMS-304* and *378* were mined, the *YMS-380* and *231* detonated two magnetic mines, but were apparently unscathed. An LCM assault craft towed *YMS-378* into shallow water in St. Vaast Harbour where she could rest on the bottom. Her bow was then down about twelve feet, and she was listed to port. Gasoline pumps from the other ships, as well as ones from a Norwegian coaster and the harbour master at St. Vaast were used to pump water out of the ship until holes in her hull could be plugged. *YMS-378* was then towed to the *Chimo* and placed alongside her where further repairs to her hull could be made in preparation to being towed to the UK.[23]

YMS-382 towed *378* to England on 3 August. A survey of the ship identified broken beam and ribs, bulkheads out of line, and significant machinery derangement. Following a determination that permanent repairs would be cost prohibitive, she was struck from the Naval Register on 16 September 1944.[24]

23

Opening the Port of Cherbourg

Hennecke performed a feat unique in the history of coastal defense; he carried out an exemplary destruction of the harbor of Cherbourg.

—Remarks made by Adolf Hitler when he presented the Knights Cross of the Iron Cross to Rear Adm. Walter Hennecke, the naval commander at Cherbourg prior to the capture of the French port city by Allied forces

All the sweepers were acutely aware of the pressing need to open the port so that heavy equipment and supplies could reach our troops, and we all crowded the edge to sweep the largest possible area in each successive pass. If it had been a clearance sweep around an operating harbor, we would have confined sweeping to periods of slack water. At Cherbourg the USS YMS-350 kept sweeping when mines were exploding close astern, and when under less urgent conditions, we would have turned out to sea or even cut our sweep lines.

—Mead B. Kibbey, former executive officer of the yard minesweeper *YMS-350*, explaining working in mined waters off the Cherbourg breakwater, knowing that cross currents could set the ship down on a mine in unswept adjacent waters, in order to open the port as soon as possible to shipping delivering vital combat cargo.[1]

After first gaining a foothold at Normandy, more than 100,000 allied soldiers had begun the march across Europe to defeat Hitler and they required logistic support landed from sea. Thus, the Allies' capture of Cherbourg—the nearest major port to the Normandy beaches—became of great urgency. This need became more acute with the failure of the artificial harbour Mulberry A in a storm on 19 June. The port at Cherbourg was guarded along the sea approaches by a series of outlying forts along breakwaters and jetties. The port finally fell to VII Corps of the U.S. First Army on 29 June. Despite naval bombardment from a combined force of American and British ships, German defence forces inside the forts had held out to the end. Two minesweeping units swept in ahead of the bombardment group. Twenty-one British sweepers comprised one unit under Comdr. Roger Thomson, RN. The other unit under Commander Plander consisted of USS *Auk*, *Broadbill*, *Chickadee*, *Nuthatch*, *Pheasant*, *Swift*, and *Threat*; the Canadian minesweeper HMCS *Thunder*; and the British motor launches *ML 139, 142, 257*, and *275*.[2]

Three days after the U.S. 79th Infantry Division captured Fort du Roule—the fort which had dominated the city and its defences—Commodore William A. Sullivan, USN, and Commodore Thomas McKenzie, CB, CBE, RNR, the heads of American and British salvage sections, flew to the French city to survey the port area and develop a plan for its rehabilitation. Because of the immense destruction wrought by the Germans and the substantial numbers of different type mines they had sown in harbour waters, their inspection led to the belief that restoring the port to use, in a reasonable amount of time, would be a truly daunting task. This work was begun as soon as resources could be mobilised. It included minesweeping; employment of Royal Navy volunteer divers trained to render bombs and mines safe; removal of sunken vessels and other obstructions; and hydrographic work, necessary to enable safe maritime navigation within the port.[3]

Photo 23-1

Aerial view of port of Cherbourg showing the outer and inner harbours.
Source: http://www.ibiblio.org/hyperwar/USA/USA-E-Logistics1/img/USA-E-Logistics1-p291.jpg

The first challenge was to sweep approach channels through the two main entrances, and clear areas for anchorages and landing points in the vast outer harbour. The British 9th and Canadian 31st Minesweeping Flotillas—made up of steel-hulled 162-foot *Bangor*-class fleet minesweepers—began this work on 30 June. The first sweep by these ships was followed by multiple passes through the same waters by BYMSs (British Yard Minesweepers) of the British 159th Flotilla. As part of this operation, YMSs of the American Y Flotilla began sweeping off the eastern entrance of the outer harbour, in order to permit safe passage of salvage ships into it.[4]

YMS MINESWEEPING OPERATIONS COMMENCE

Photo 23-2
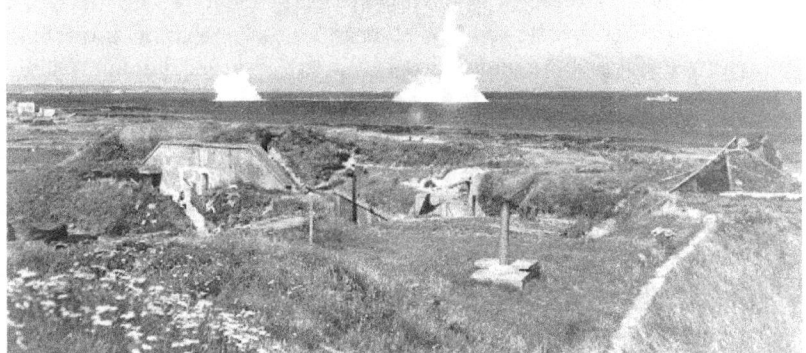
U.S. Navy YMS minesweepers exploding German magnetic mines near Cherbourg, France, 30 July 1944.
Navy photograph #80-G-256041, now in the collections of the National Archives

The last of the American YMSs departed the Bay of Seine on 29 August for the Cherbourg area. During the first month of the Normandy invasion, American, British, and Canadian minesweepers had disposed of 261 mines off Utah Beach and 291 off Omaha, but there were more at Cherbourg. The harbour had been mined with magnetic, acoustic, pressure, and contact mines—with and without snaglines. These were sown throughout the inner harbour and in extensive fields outside the great breakwater that protected the entrance. To rid the waters of them, Allied forces swept for eighty-five days before the approaches to the port were completely cleared. Time was of the essence; the task was dangerous and one which had to be completed as soon as possible.[5]

A story of two ships illustrates these conditions. *YMS-350* of the United States Navy was mined and many of her survivors were saved by *ML 137* of the Royal Navy. *ML 137* was a Fairmile "B" type motor launch, of which more than 500 had been built in Canada and the UK during the war. The 112-foot, twin gasoline engine powered wooden vessels were capable of 20 knots. Carrying a ship's company of two or three officers and fourteen to sixteen men, they had served in many theatres of war. Their diverse duties included convoy escort, air-sea rescue, minelaying and minesweeping. The relatively shallow draught of a ML (4' 9") made it especially suited for inshore mine clearance, when fitted out for sweeping moored mines.[6]

In the late morning on 2 July, MLs *137*, *143*, and *257* arrived at Cherbourg to await orders to begin sweeping. *ML 143* anchored, and *ML 137* and *ML 257* made up alongside her. Almost immediately, the boom of an exploding mine nearby signaled the destruction of British motor minesweeper *MMS 1019*—sunk by a snagline contact mine. The motor launches got under way and sped to the scene to rescue survivors. They were greeted by the sight of litter and wreckage rising and falling on the oily surface of the undulating sea, with survivors clinging to large debris. A minesweeper's boat was picking them up. In the middle of the spreading patch of oil was the single remaining survivor, his face and hair blackened with oil, shouting desperately, "Help, help, for God's sake." *ML 137* plucked him from the water, and delivered him to the minesweeper to which the others had been taken.[7]

ML 137 returned to her mooring, and passed the afternoon against a backdrop of intermittent explosions of mines being swept by U.S. Navy minesweepers a short distance away. That evening after supper, her commanding officer T/A/Sub. Lt. Brendan A. Maher, RNVR, was watching mines explode astern of a YMS, when suddenly he noticed that the sweeper had begun to heel over. The vessel was *YMS-350*. A searchlight blinked an SOS from her bridge. As MLs *137* and *257* rushed to get under way to help, she slowly listed to port, and then settled by the stern.[8]

Within two minutes after a snagline mine detonated under her stern, the wooden minesweeper turned on her side and started to sink by the stern. The executive officer, Lt. (jg) Mead Kibbey, USNR, and Electrician's Mate Third Class John Hamilton, who both suffered slight injuries from the blast, had crawled over the starboard gunwale opposite the wardroom and sat on the side of the ship waiting for it to sink out from under them. Adrift nearby was a section of half-inch line buoyed by two-inch floats about a foot apart. It was the detonating line from the snagline mine which had caused the ship's destruction.[9]

Kibbey and Hamilton had heard stories of the suction of a sinking ship pulling survivors under, but luckily for them such things must have been associated with larger vessels. As the YMS sank slowly by the stern, the two men were deposited gently in the sea. As they swam away, their eyes burning from the diesel fuel on the surface, they could see that they were about equal distance from an American PT boat and a small British motor launch providing two options for rescue. The latter vessel had rigged a climbing net over the side, and was helping survivors aboard, but the deciding factor for the two Americans was when Kibbey pointed out to Hamilton that the British undoubtedly had rum aboard. Aboard the *ML 137*, they were given the anticipated "medicinal" and blankets. In this instance, rum has won out over Coca Cola, but in a larger sense the story represents the cooperation between the two navies in executing this huge endeavour.[10]

Before saving himself, Kibbey had saved the lives of two of his shipmates, as attested by the citation for the Navy and Marine Corps Medal he was later awarded by the Secretary of the Navy:

> For heroic conduct in effecting the rescue of two men serving with him on board the U.S.S. *YMS 350*, when that vessel was sunk off the Coast of France on July 2, 1944. After swimming clear of the sinking ship, Lieutenant Kibbey unhesitatingly returned to the dangerous area to save a man who had sustained a severe back injury and was unable to move. Risking his life in a second daring attempt, he again returned to effect the rescue of a drowning fellow officer. By his cool courage, outstanding fortitude and grave concern for others, Lieutenant Kibbey saved the lives of two men who otherwise might have perished, and his selfless efforts throughout were in keeping with the highest traditions of the United States Naval Service.[11]

WORK CONTINUES OUTSIDE AND INSIDE HARBOUR

While the larger fleet minesweepers continued operations outside the harbour, the YMS, BYMS, and MMS flotillas began sweeping the Grande Rade (outer harbour), which contained wrecks of every size and shape, as well as moored-contact, acoustic, and magnetic mines. By 13 July, one hundred thirty-three mines had been swept, but many more remained. In spite of these adversities, concurrent salvage operations began due to the pressing requirement to open the port. In clearing the Grande Rade, the MLs swept ahead of YMSs and BYMSs, and behind them followed the MMSs, although only the MLs could sweep in all the corners. Within the smaller and more confined Petite Rade (inner harbour), landing craft functioning as "snagline sweepers" led the way, followed by MLs and MMSs. In the Petite Rade, the MLs found mines believed to be the new Katy type fitted with snaglines. Although marked as to location if found, these potentially very lethal mines were not disturbed as sweeping methods to counter them had not yet been developed.[12]

Katy mines were seven-feet tall, and sat quietly on the sea floor waiting for unsuspecting ships. They could be triggered in two ways: The first was by contact with the single horn on top of the Katy during low tide or while anchoring. The other was by a propeller ensnaring the snagline affixed to the firing mechanism when passing above one of the mines. The propeller would draw the snagline tight, thereby exploding the ordnance. In July and August, seventeen ships and craft were sunk or damaged by mines during the Cherbourg harbour clearance, mostly of the Katy variety. They are listed below:[13]

Date	Vessel	Location/Cause of Loss or Damage
2 July	YMS-347	Damaged by a mine detonation outside breakwater
2 July	YMS-350	Lost outside breakwater; 11 of 34 crewmen killed and 9 injured
2 July	MMS 1019	Lost outside breakwater
8 July	LCP(L)-267	Sunk inside Petite Rade while clearing snag mines with Senior Royal Marine officer of the flotilla lost
20 July	Army Small Tug ST-344	Blew up and sank in Grande Rade after striking a mine
20 July	British Hopper barge	Blew up and sank in Grande Rade after striking a mine
20 July	Army barge	Keel broken by a mine in Grande Rade, but towed to shallow water
21 July	British Hopper barge Dredge 36	Sunk in Querqueville Bay by a Katy mine
21 July	Army Small Tug ST-253	Holed and sank, presumably due to a Katy mine
30 July	HMT Sir Geraint	Damaged by a mine
2 August	LCG-1062	Mined and sank off outer approaches to Cherbourg Harbour; three crewmembers missing
4 August	British Dredger	Mined in vicinity of Fort de l'Quest, probably by a Katy mine; raised on 21 September by a Salvage Force lifting craft
5 August	MMS 279	Damaged by a near miss from an acoustic mine detonation
12 August	British LBO 68	Sunk near Fort de l'Quest, probably by a Katy mine
13 August	BYMS 2034	Damaged by the explosion of a ground mine, probably a Katy, in Grande Rade
16 August	LST-391	Struck a mine, believed to be a Katy, in east end of Petite Rade
26 August	British Coaster	Struck a mine in channel outside harbour, sank in three minutes with fifteen crewmen missing[14]

Because of the abundance of mines remaining to be cleared, the Royal Navy began despatching additional sweepers to Cherbourg—entire flotillas in some instances or, one or more ships drawn from separate ones. An exact accounting of the ships that plied the mined waters is difficult to ascertain. The flotillas identified by a single asterisk in the following table were confirmed to be present at Cherbourg. Those not so marked were likely present and there were likely other flotillas and/or ships of other flotillas as well. Eric David Minett, in *The Coast is Clear the Story of the BYMS*, indicates that a number of BYMSs drawn from south coast flotillas for reserve duty, were standing by in Portland or Portsmouth for minesweeping operations at Cherbourg, if needed. The ships he cited were a part of the flotillas identified with two asterisks in the table. The remaining flotillas listed (without asterisks) are those which Rear Adm. John E. Wilkes, USN, commander, U.S. Ports and Bases, France, (then headquartered at Cherbourg) identified in war diary entries as either being under his command, or ready to be sent to him when needed. Because Wilkes was responsible for mine clearance in other captured French ports as well as Cherbourg, these flotillas may or may not have swept at Cherbourg.

Country	Flotilla	Minesweepers
United States	Y*	Y1 Squadron: Yard Minesweepers *YMS-305, 356, 358, 375, 377, 378, 379, 380, 381, 382, 406*
United States	Y*	Y2 Squadron: *YMS-231, 247, 304, 346, 347* (damaged at Cherbourg), *348, 349, 350* (sunk at Cherbourg), *351, 352*
Canada	31st*	Bangor-class Fleet Minesweepers: *Blairmore* (J314), *Caraquet* (J38), *Cowichan* (J146), *Fort William* (J311), *Malpeque* (J148), *Milltown* (J317), *Minas* (J165), *Wasaga* (J162) Danlayers: *Bayfield* (J08), British MS Trawler *Green Howard* (FY632), British MS Trawler *Gunner* (FY568), *Mulgrave* (J313), British Motor Launches: *ML 345, 454, 465, 473*
Britain	9th*	Bangor-class Fleet Minesweepers: *Bangor* (J00), *Blackpool* (J27), *Boston* (J14), *Bridlington* (J65), *Bridport* (J50), *Eastbourne* (J127), *Sidmouth* (J47), *Tenby* (J34) Danlayers: *Bryher* (T350), *Dalmatia* (FY844), *Ijuin* (FY612), *Sigma* (FY1709) Four Motor Launches
Britain	15th	Bangor-class Fleet Minesweepers: *Ardrossan* (J131), *Bootle* (J143), *Dunbar* (J53), *Fort York* (J119), *Fraserburgh* (J124), *Llandudno* (J67), *Lyme Regis* (J193), *Worthing* (J72) Danlayers: *Calvay* (T383), *Dorothy Lambert* (FY558), *James Lay* (FY667), *Niblick* (FD77)
Britain	102nd	Based at Sheerness: Motor Minesweepers: *MMS 8, 19, 40, 44, 45, 71, 110, 113, 115, 181*
Britain	131st	*MMS 15, 56, 59, 78*

Britain	157th		Based at Great Yarmouth: *BYMS 2034* (damaged at Cherbourg), *2038, 2039, 2076, 2078, 2141, 2213, 2214, 2221, 2230*
Britain	159th*		Based at Grimsby: *BYMS 2032, 2052, 2055, 2070, 2071, 2157, 2173, 2211*
Britain	163rd*		Based at Lowestoft: *BYMS 2079, 2167*
Britain	165th**		Based at Harwich: *BYMS 2035, 2041, 2058, 2202, 2205, 2206, 2233, 2252*
Britain	167th*		Based at Ardrossan and Liverpool: *BYMS 2047, 2051, 2061, 2069, 2155, 2156, 2182, 2210*
Britain	168th**		Based at Portsmouth: *BYMS 2234*
Britain	169th**		Based at Dover: *BYMS 2154, 2255*
Britain	170th**		Based at Swansea: *BYMS 2050, 2188, 2256*
Britain	206th*		*MMS 1002, 1004, 1019* (sunk at Cherbourg), *1020, 1047, 1048, 1077*
Britain	139th		Three flotillas of minesweeping "LL" trawlers: HMT *Conway Castle* (FY509), *Courtier* (FY592), *Georgette* (FY804), *Northcoates* (FY548), *Perdrant* (FY1714), *Probe* (T186), *Proctor* (T185), *Prowless* (T196), *Sir Agravaine* (T230), *Sir Gareth* (T227), *Sir Geraint* (T240; damaged at Cherbourg), *Sir Kay* (T241), *Sir Lamorak* (T242), *Sir Tristram* (T229)[15]

MINE CLEARANCE BY PORT ("P") PARTIES

As minesweepers cleared the profusion of mines from the approaches, channels and anchorages, "P" Parties (Port Parties including clearance divers) searched Bassin Des Flots, the Avant Port de Commerce and the shallows of the Nouvelle Plage, working outward to meet the sweepers. The clearance divers (please see Postscript for details about their dangerous activities) and minesweepers sometimes worked in the same waters, but not at the same time. To avoid danger to themselves from exploding mines, the divers dove only for three hours either side of low water, the time span when the minesweepers were most at risk (and thus not engaged in sweeping) due to the minimum water column separating them from the mines. Conversely, the minesweepers swept three hours either side of high water when they were least vulnerable. Once safe entry to a landing point ashore was assured, clearance was extended by the divers to the docks and basins, which had been thoroughly blocked by shallow-water mines, booby-traps, wreckage, and other items the Germans employed for that purpose.[16]

U.S NAVAL ADVANCED BASE, CHERBOURG

Despite all the challenges and setbacks, a naval base was soon established at Cherbourg. U.S. Naval Advanced Base, Cherbourg, was placed in commission on 15 July. The following day, four Liberty ships entered the port and berthed. The first men and supplies were landed by lighters (flat-bottomed barges) on 18 July. By month's end, facilities for the handling of twelve to sixteen Liberty ships were available. The first ships using the harbour were limited to an anchorage in its western end, but by 29 July, the eastern end had been opened to traffic as well. In the next few months, Cherbourg would become of critical importance as a port of logistic supply to the U.S. Army in Europe, second only to Marseilles. Mines still remained, though, and minesweeping continued. That autumn, a BYMS or YMS flotilla was usually stationed at Cherbourg, plus a number of MMSs.[17]

MINESWEEPERS LAUDED

British Prime Minister Winston Churchill visited Cherbourg on 20 July while the minesweeping and salvage efforts were in progress. He later asserted that the mine clearance at the French port was "the greatest minesweeping operation in history." Following Germany's surrender, he sent a message to the officers and the men of the British minesweeping flotillas, praising their vitally important and generally unsung efforts throughout the war:

> Now that Nazi Germany has been defeated I wish to send you all on behalf of His Majesty's Government a message of thanks and gratitude.
>
> The work you do is hard and dangerous. You rarely get and never seek publicity; your only concern is to do your job, and you have done it nobly. You have sailed in many seas and all weathers.... This work could not be done without loss, and we mourn all who have died and over 250 ships lost on duty.
>
> No work has been more vital than yours; no work has been better done. The Ports were kept open and Britain breathed. The Nation is once again proud of you.[18]

THE ALLIES MOVE FORWARD

On 11 October 1944, Rear Adm. John E. Wilkes left Cherbourg and the following day reestablished his U.S. Ports and Bases, France, headquarters at Le Havre, which lay to the east across the Baie de Seine at the mouth of the River Seine. His challenge remained the same: to run the captured ports and unload materials from vessels so that the extended supply lines might be kept operating with sufficient volume to support the advancing armies.[19]

LOSS OF MINELAYER USS *MIANTONOMAH*

Two-and-a-half weeks before Wilkes set up his new base, the minelayer USS *Miantonomah* had fallen victim to a pressure mine off the harbour at Le Havre. She sank with great loss of life.

Her story, including that preceding her loss, illustrates the danger posed by mines throughout mine clearance operations in French waters. Like *Chimo*, *Barricade*, and *Planter*, she had departed the U.S. in May 1944. Spanning 292 feet in length with a top speed of 17.5 knots, she was more seaworthy than the smaller 188-foot auxiliary minelayers and five knots faster. Sailing with a different convoy to support the Allied invasion of Europe, *Miantonomah* left Gravesend Bay, New Jersey, on 5 May, as a unit of Task Group 27.1.[20]

Photo 23-3

Minelayer USS *Miantonomah* (CM-10) near the Norfolk Navy Yard on 29 April 1944 after a major reconfiguration of her mine laying facilities. Norfolk Navy Yard photo 7509(44)

She arrived at Bristol, England, on 16 May and began duty with the Twelfth Fleet. *Miantonomah* operated out of Bristol until D-Day, 6 June, when she proceeded southwest to Plymouth (sailing around Land's End, the southwest tip of Great Britain) to carry out despatch and escort duties in British waters. She arrived off Grandcamp, France, near Utah Beach, on 25 June, and the following day embarked Rear Adm. John E. Wilkes, USN. Wilkes was then serving as Flag Officer, West, responsible for basic naval organisations on the "Far Shore" and their liaison with Army authorities.[21]

Miantonomah arrived at Cherbourg on 9 July. On the 18th, Wilkes, newly appointed commander, U.S. Naval Bases France, hauled down his flag and moved ashore prior to the minelayer's departure for England. After arriving at Plymouth later that day, *Miantonomah* returned to Cherbourg on the 20th carrying supplies and materials for port clearance operations. For the next two months, she made runs between English and French ports, providing valuable support for salvage and clearing operations.[22]

On 22 September 1944, *Miantonomah* arrived at Le Havre, which had been liberated less than two weeks earlier, with supplies of food and gasoline for a "Seabee" (Navy Construction Battalion) party. Entry into the harbour was impeded by two "block ships" positioned at the entrance, which left only an 80-foot wide channel. Unloading was completed the next day, following which preparations were made to get under way early afternoon on 24 September for Plymouth. Departure was postponed until the 25th due to strong northwest winds and heavy seas, which precluded negotiation of the narrow channel between the two block ships.[23]

On the morning of 25 September, the commanding officers of two Royal Navy minesweepers anchored nearby, came aboard *Miantonomah* to brief Comdr. Austin E. Rowe, USNR, on the mine threat. Prior to getting under way, Rowe ordered all hands not on watch to don lifejackets and lay (station themselves) topside, and the highest degree of watertight integrity be set throughout the ship. The harbour master and French pilot boarded at 1300, and as soon as a net tender working at the entrance was clear, the minelayer proceeded out of the harbour. While passing between the block ships, speed was increased in order to manoeuvre properly. Once clear, speed was reduced to that prescribed for safety in the mined area.[24]

At 1415, while proceeding outbound in the port side of the channel in accordance with existing instructions to avoid oyster mines, and approximately 3,000 yards from the entrance, *Miantonomah* was rocked by a tremendous underwater explosion under her engine room. This blast dazed or injured nearly all of the crew. A U.S. Coast Guard cutter with aid from the British 66th Motor Torpedo Boat Flotilla began evacuating the injured from the ship, but a large, and sudden list to starboard and settling by the stern, halted these actions. The two craft as well as other craft and a French fishing boat aided in rescuing survivors found in the water. Later, the wounded were treated ashore at the 28th Construction Battalion sickbay, and aboard the British minesweeper HMS *Franklin*.[25]

Miantonomah sank about twenty minutes after the explosion with a loss of fifty-eight officers and men. The only defence against oyster pressure ground mines was to proceed at dead slow speed while maximizing the depth of water between the mine and ship passing near or directly above it. Ideally, suspect areas would be transiting at high tide. After the sinking, Commander Rowe expressed his belief that *Miantonomah* had made safe entry into the harbour because the sea was smooth, which allowed stopping the screws as much as possible, just giving slow kicks ahead to maintain steerage way. Upon departure, a strong westward wind made it necessary to maintain a speed of about 5½ knots. This higher speed he thought caused the mine to explode.[26]

UNIT AND PERSONAL AWARDS

USS *Chimo* and USS *Miantonomah* earned battle stars for the Normandy invasion, and Chief Commissary Steward Kermit Luther Bergum, USN, the Navy and Marine Corps Medal for heroism in assisting the removal of an officer from the *Miantonomah* at the time of her sinking. His medal citation reads in part:

> Due to an underwater explosion the ship listed heavily to starboard and sank in sixteen minutes. Chief Bergum was taken with the Captain from the sinking ship by a small fishing boat. While leaving the ship the Captain noticed the Navigator unconscious and entangled in the signal halyards. Despite a broken leg sustained in the disaster, Chief Bergum rendered assistance to the Captain in carrying the unconscious officer to safety. The courage, devotion to duty, and outstanding loyalty displayed by Chief Bergum on this occasion was in keeping with the highest traditions of the United States Naval Service.

Southern France – Operation DRAGOON

Three months after Normandy...in August 1944, we joined in an amphibious assault force to land in the South of France, on the French Riviera. We left from North Africa, and strangely enough we took French troops that had left France in 1940 when the Germans pushed them out of France, so they were going back to their own homeland. We also took some Americans and we operated from North Africa to the South of France and it was to a place called St. Raphael on the French Riviera. It was called the Champagne landing this one, I think, there was very little opposition because the Germans had pulled back and when we were on the beach some of the lads, some of the sailors went up and picked grapes out of the vineyards, and it was one of the easier or better operations that we had.

—Alwyn Thomas, crewmember of the
tank landing ship HMS *Bruiser* (F127)[1]

In August 1944, two months after the Normandy landings, an amphibious assault was made on the southern coast of France between Hyères and Cannes. The objectives of the operation were to secure the beachheads, capture neighboring ports, and strike north, up the Rhône Valley, to link up with Patton's Third Army. This captured area would form the right wing of the Allied force invading Germany, and cut off German forces in the west of France. During its early planning, its code name was Operation Anvil to complement Operation Sledgehammer, that for the invasion of Normandy. Subsequently, Anvil became Dragoon while Sledgehammer became Overlord. It was reported Winston Churchill, who opposed the plan and claimed to having been "dragooned" into accepting it by Roosevelt and Eisenhower, chose the latter name. The British prime minister believed that the operation would divert military resources that could be employed supporting ongoing operations in Italy. He preferred, instead, an invasion of the oil-producing regions of the Balkans. Churchill reasoned that by attacking the Balkans, the Allies could deny Germany oil, as well as hinder the advance of the Red Army of the Soviet Union; thus achieving a superior negotiating position in post-war Europe with that nation.[2]

When the invasion of southern France was initially considered, the Allied landing at Anzio had gone badly, and planning was "put on hold." The operation—now named DRAGOON—was revived, following the successful Normandy invasion. After the capture of Rome on 5 June 1944, it was possible to allocate troops to the invasion of southern France but it could not be carried out concurrently with the Normandy invasion because of the lack of necessary shipping and naval support. This became available for DRAGOON after the landings at Normandy. As a matter for the entire invasion force, there was increasing urgency to bring naval ships into the Mediterranean to join the Eighth Fleet and capture additional ports and open up shipping, as the Allies were struggling to resupply their armies in France. Since the Germans had destroyed the port facilities at Cherbourg, and a violent storm had damaged the artificial harbour "Mulberry A," it was important to seize the ports at Marseilles and Toulon. Also, Allied Free French leaders were pressing for an invasion from the south.[3]

The weather and tides in the western Mediterranean were favourable for an amphibious landing in southern France in late summer 1944. By this time, the German U-boat menace was rapidly waning because of improved countermeasures by Allied ships and aircraft. As a final inducement, after attrition during the Normandy invasion requiring the redeployment of its squadrons, the German air force was no longer able to operate in strength in the Mediterranean.[4]

Although the assault beaches of southern France were well protected by German coast defences, enemy fortifications lacked depth and there were few reserves for counterattack. Allied planners chose the Cavalaire Bay–Rade d'Agay area for the assault because it was within range of supporting Allied aircraft operating from fields in Corsica which they then controlled. Also, it offered a favourable sea approach with only a narrow coastal area suitable for enemy mining, and it had the fewest German coastal gun batteries capable of reaching the approach, transport areas, and landing beaches. A bridgehead in this area would provide a suitable base for a landward attack on Toulon and Marseilles, and rapid movement up the Rhône Valley.[5]

ALLIED ASSAULT FORCE

On D-Day, 15 August 1944, the Allied naval force arrived off the southern France assault beaches. They found low clouds over the entire coast, with local bands of thick fog in the eastern part of the Golfe du Lion. The winds were light and seas calm, and, in the remainder of the western Mediterranean, the skies were clear with visibility of ten miles. The lower clouds over the beaches thinned by mid-morning and dissipated by noon.[6]

Map 24-1

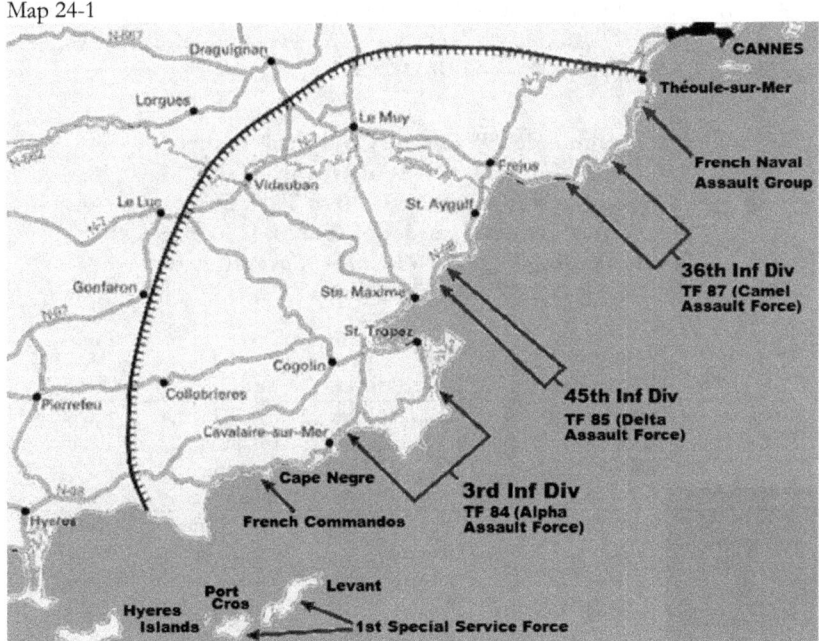

Three divisions of the U.S. Seventh Army and two divisions of French Army B landed on the French Riviera of southeast France as part of Operation DRAGOON.

The naval force under the command of Vice Adm. Henry K. Hewitt, Commander, Eighth Fleet and Commander, Western Naval Task Force, included 503 U.S. ships and craft, 252 British, 19 French, 6 Greek, and 63 merchant ships of various nations—a total of 843 vessels. Ship-borne on this fleet were 1,267 landing craft.[7]

The Allied Assault Force comprised three divisions of the U.S. Seventh Army: assault force Alpha (U.S. 3rd Infantry), Camel (U.S. 36th Infantry), and Delta (U.S. 45th Infantry) under their respective commanders—and two divisions of Free French Army B under the command of Gen. Jean de Lattre la Tassigny. In support of the main landing, Rear Adm. Lyal A. Davidson's Support Task Force (TF 86) carried out assaults with commando forces (Operations SITKA and ROMEO), and Capt. Henry C. Johnson's Task Group 80.4 (Special Operations) handled related diversionary operations.[8]

Actor-turned-naval-officer Lt. Comdr. Douglas E. Fairbanks Jr., USNR—an Eighth Fleet Staff Special Operations officer—commanded a task unit of this task group. Details about the sinking of two German corvettes in battle due to actions taken by Fairbanks and Lt. Comdr. John Bulkeley, USN—who won the Medal of Honor for taking Gen. Douglas MacArthur, his family and staff off the Philippine Islands in 1942—may be found in the author's book, *We Are Sinking, Send Help!: The U.S. Navy's Tugs and Salvage Ships in the African, European, and Mediterranean Theaters in World War II*.

Commander, Western Naval Task Force:
Vice Adm. Henry Kent Hewitt, USN
Commanding General Western Task Force:
Maj. Gen. Alexander M. Patch, USA
Commanding General VI Corps, Seventh Army:
Maj. Gen. Lucian K. Truscott Jr., USA

Assault Force	Naval Commander	Army Commander	Troops	Assault Areas
Alpha TF 84	Rear Adm. Frank J. Lowry, USN flagship: *Duane*	Maj. Gen. John W. O'Daniel, USA	U.S. Army 3rd Infantry Division (reinforced)	Cavalaire and Pampelonne
Delta TF 85	Rear Adm. Bertram J. Rodgers, USN flagship: *Biscayne*	Maj. Gen. William W. Eagles, USA	U.S. Army 45th Infantry Division (reinforced)	Baie de Bougnon and Golfe St.-Tropez
Camel TF 87	Rear Adm. Spencer S. Lewis, USN flagship: *Bayfield*	Maj. Gen. John E. Dahlquist, USA	U.S. Army 36th Infantry Division (reinforced), French Army units	Frejus, St.-Raphaël, and Calanque d' Actheor
Sitka TG 86.3	Rear Adm. Theodore E. Chandler, USN flagship: HMCS *Prince Henry*	Col. Edwin A. Walker, USA	Joint Canadian-American 1st Special Service Force	Îles d'Hyères
Romeo TG 86.3	Rear Adm. Theodore E. Chandler, USN	Lt. Col. Georges-Régis Bouvet	First Groupe Commandos d'Afrique	Cap Nègre and le Rayol
Special Operations TU 80.4.2	Lt. Comdr. Douglas E. Fairbanks Jr., flagship: HMS *Aphis*	Capitaine de Frégate R. Seriot	French Groupe Navale d' Assaut	Théoule-sur-Mer and Baie de La Ciotat[9]

MINESWEEPING IN ALPHA FORCE ASSAULT AREA

The Allied landings of Operation DRAGOON in the South of France hastened the driving of German forces from France. With only a small and dwindling force of U-boats to challenge sea supremacy in the main theatre, general offensive mining operations were limited for the remainder of 1944 to two surface-lays off the northwest coast of Italy. In April, the Italian torpedo boat *Sirio* (now a part of the Allied forces) laid a field of twenty-eight Italian mines off the island of Capraia, which sank an enemy minelayer. In May, motor launches of the British 8th Motor Launch Flotilla sowed thirty mines on an enemy coastal route south of Vada Rocks, without success. HMS *Teviot Bank* also laid a defensive field, containing 119 moored mines, off Capri in June 1944.[10]

Following Operation MAPLE, conducted in preparation for the Normandy invasion, offensive mining requirements decreased dramatically throughout the remainder of the war. Minesweeping support of amphibious assault operations, and mine clearance associated with the opening of French ports, remained and became the primary focus of offensive mine warfare. Thus, previous emphasis in this book on minelaying shifted in Chapters 21-23 to minesweeping. This trend continues in the next few pages, which describes minesweeping conducted in Cavalaire Bay, off the Task Force 84 beaches, which was representative of that in other assault areas as part of DRAGOON.

Comdr. William L. Messer, USN (Commander, Task Group 84.8) was responsible for the Escort Sweeping Group charged with carrying out assault sweeping into the Alpha Force (Task Force 84) "Red" and "Yellow" assault beaches.

Task Group 84.8 (Escort Sweeping Group)	
TU 84.8	Comdr. Messimer embarked in USS *Barricade* (ACM-3)
TU 84.8.1 Red Beach Sweeping Unit	Comdr. A. V. Wallis, USNR (CoMinDiv 16), in USS *Prevail* (AM-107)
Section 1 (Mine Division 16):	USS *Pioneer* (AM-105), *Prevail* (AM-107), *Seer* (AM-112), *Dextrous* (AM-341), *SC-535*, *SC-979*
Section 2 (YMS Section Five)	Lt. Howard Johnson, USNR: *YMS-18*, *YMS-21*, *YMS-34*, *YMS-82*, *YMS-355*
Section 3 (Shallow Water Section)	Lt. (jg) D. V. Edmundson, USNR: *BMS-2*, *BMS-4*, *BMS-14*, BMS-15, and one LCC (control)
TU 84.8.2 Yellow Beach Sweeping Unit	Comdr. Alister Angus Martin, RNR, in HMS *Rothesay* (J19)
Section 1 (13th MS Flotilla)	HMS *Rothesay* (J19), HMS *Stornoway* (J31), *Polruan* (J97), *Brixham* (J105), *Bude* (J116), *Aries* (J284)
Section 2 (SC Section Two)	Lt. Robert Belknap, USNR, *SC-498*, *SC-655*, *SC-770*, *SC-978*

Section 3 (YMS Section Three)	Lt. Frank Morley, USNR: *YMS-13, YMS-20, YMS-27, YMS-199, YMS-251*
Section 4 (Shallow Waters Section)	Ens. G. L. Greene, USNR: *BMS-10, BMS-11, BMS-12, BMS-16*, and one LCC (control)[11]

Legend:
ACM: Auxiliary Minelayer
AM: Minesweeper
BMS: Boat Minesweeper (LCVP landing craft)
BYMS: British Yard Minesweeper
J-series: British Fleet Minesweepers
LCC: Control Landing Craft
SC: Submarine Chaser
YMS: Yard Minesweeper

Task Group 84.8 was comprised of USN and RN minesweepers, operating under minelayer/flagship USS *Barricade*. (USS *Planter*, served as flagship for the Task Force 85 Sweep Unit, whose activities are not discussed herein).

BRITISH-DESIGNED, CANADIAN-BUILT, *ALGERINE*-CLASS FLEET MINESWEEPER TRANSFERRED

Photo 24-1

Algerine-class minesweeper HMS *Persian*, sister ship to HMS *Aries*, 31 March 1944. Imperial War Museum Admiralty Official Collection photograph IWM A 22676

The excellent cooperation between the Allies, in the supply and manning of ships and conduct of minesweeping in Europe is demonstrated by HMS *Aries* (J284) which participated in the Normandy assault and landings. It is an interesting aside to the main story. She was a British designed *Algerine*-class 225-foot minesweeper, laid down on 23 March 1942 as AM-327 by Toronto Shipbuilding, Ltd., one of fifteen built in a Canadian shipyard for the U.S. Navy. Under lend-lease she was turned over to the Royal Navy on 17 July 1943, and renamed.[12]

Her first two commanding officers were Royal Navy, as likely were all or most of her crew. In August 1944, the top two spots aboard *Aries* were filled by officers of the British Dominions. New Zealander Comdr. Roger Stannard Cameron, RNZNVR, was in command, with a Canadian, T/Lt. Comdr. Herbert Bruce Carnall, RCNVR, as executive officer/minesweeping officer. There was also a Canadian medical officer aboard, T/Surg. Lt. Charles William Eaton Howitt, RCNVR. Cameron was both an experienced and heroic officer, as witnessed by his having previously been awarded a DSC & Bar earlier in the war.[13]

Carnall previously had a variety of ship and staff assignments associated with mine warfare, including duty in 1942 as executive officer aboard HMS *Sidmouth* and 9th Flotilla Minesweeping Officer. He was present at Operation Jubilee (the Dieppe Raid) in 1942 and served with the RAF (on loan for minelaying duties). After Operation Dragoon, Carnall remained with the ship, which during post-war mine clearance took part in Operation Antagonize in the southern Adriatic. On 2 July 1945, while sweeping off Trieste, HMS *Aries* exploded a mine caught in her gear. The denotation caused extensive damage to her stern and the death of four men, identified below:

Missing, Presumed Killed	Died of Wounds
Sub. Lt. Maurice Barnes, RNVR	Able Seaman Frederick T. Spinage
Able Seaman Thomas Johnston	Leading Wireman James Eatherall[14]

But for the efforts of T/Lt. (E) Anthony Close, RNR, the ship might have been lost. He was later appointed an MBE (Member of the Most Excellent Order of the British Empire) "For skill, efficiency and leadership shown in Damage Control in H.M.S. *Aries* when she struck a mine off Trieste on 2nd July, 1945."[15]

Lt. Comdr. Herbert Bruce Carnall, RCNR, was awarded the DSC "For distinguished service during the war in Europe," as announced in *The London Gazette* of 11 December 1945.

Following three years of war and post-war service with the Royal Navy, HMS *Aries* was returned to U.S. custody in August 1946, sold to Greece the following year and renamed *Armatolos* (M 12). Transferred back to U.S. custody in February 1948, she remained on the Naval Vessel Register until eventually sunk as a target in 1977.[16]

U.S NAVY LCVP AND SUB-CHASER MINESWEEPERS

In addition to minesweeper types with which readers are already familiar, Task Group 84.8 included two new ones for shallow water clearance. These were U.S. Navy submarine chasers and LCVP landing craft fitted with sweeping gear. The first LCVP thus equipped had been introduced in the Mediterranean during the Allied landings at Anzio, Italy, in January 1944.[17]

The development of the LCVP and SC minesweepers by Comdr. William L. Messimer, permitted far more efficient sweep coverage in an amphibious assault channel than had been previously possible. The LCVPs were able to sweep shoreward to the one-fathom curve (water depth of six feet) and courageously did so off the French beaches whenever boat obstacles did not prevent it. German cement-block mines and Teller mines found emplaced among obstacles could only be cleared by countermining or hand-removal by divers. Pre-assault aircraft bombing, naval gunfire, and rocket attacks on the beaches were expected to accomplish some small, random amount of countermining in shallow water, and presumably they did.[18]

The LCVPs (36-foot wooden "Higgins boats") were essentially "jury-rigged" minesweepers, manned by personnel who had no training in sweeping other than the special training for this operation. Despite these limitations, they carried out their missions successfully. Other unmanned LCVP explosive drones, remote-controlled from an LCC (Landing Craft Control), were also employed. The LCC was a 56-foot U.S. Navy craft designed to carry scouts and raiders. As a result of the efforts of the manned and unmanned craft, and shore bombardment, only thirteen assault craft were damaged by mines (of the some 320 LCVPs, DUKWs, LCI(L)s and LCTs which beached on Alpha assault beaches). This damage occurred before hand-placed demolition charges could be placed to complete the clearance. A DUKW was a six-wheel drive amphibious vehicle.[19]

The efforts of the sub-chasers fitted with light German designed sweep gear were also satisfactory. Following the successful employment of *SC-770* at Anzio as a shallow water sweeper, five additional sub-chasers were equipped with, and their crews given intensive training with, this gear. Initially, U.S. Navy Size 5 "O" gear was tried, but proved unsuccessful owing to the delicate adjustment involved and instability of the gear at more than 6 knots. A last-minute conversion to the German sweep gear was made and proved satisfactory in every respect. The time spent experimenting with the "O" gear was not wasted, however, as it proved extremely beneficial as a training exercise to personnel not previously experienced in minesweeping.[20]

SWEEPING INTO ALPHA BEACHES AT CAVALAIRE

Photo 24-2

Allied shipping unloading on beaches in Cavalaire Bay, France, on 18 August 1944. National Archives photograph #80-G-K-1940

The "Alpha" Assault began at 0300 on 15 August, when patrol craft assigned to act as "mark boats" took station ten miles off Red and Yellow beaches. This was followed by similar stationing of "five-mile boats." Minesweeping commenced at 0440. By 0630, the fleet minesweepers of the British 13th Minesweeping Flotilla, together with the AMs of Mine Division 16 and the YMS and the SC shallow sweepers, had swept the gunfire support and transport areas and the boat lanes up to 1,000 yards from the beach. No mines were found, but in Pampelonne Bay, the British sweepers were attacked by five small enemy patrol vessels. They were all destroyed by the sweepers without themselves suffering any damage. The sweepers also engaged in counterbattery fire after taking fire from shore batteries. By 0500, the principal amphibious forces (gunfire support ships, tank landing ships and tank landing craft, and troop transport convoys) had arrived in the transport area.[21]

From 0710 to 0745, the shallow minesweepers extended the boat lanes to within 100 yards of the beaches. Concurrently, gunfire support ships took allocated targets under fire. Drone LCVPs were sent in from 0710 to 0730, nine to each beach. Those on Red beach detonated mines as planned; on Yellow beach, three failed to function properly and one exploded prematurely, damaging *SC-1029*, one of the mark boats.[22]

At H-Hour, 0800, the first wave of assault craft (support craft and the first troop wave) hit the beach. Preceding it were LCC-led boat minesweepers (manned LCVPs), to conduct a sweep of the boat lanes. Drone boats, also directed by the LCC, were in the van of the support wave. The first assault waves at 0800 were followed by eight other waves, which landed on schedule, though the ninth wave on Red beach was delayed by shallow water mines on the right flank of the beach.[23]

MINELAYER/FLAGSHIP SUPPORT OF SWEEPERS

> *The ACMs, BARRICADE and PLANTER, have very successfully fulfilled their assigned missions as minesweeper headquarters ship and tender. The ample deck and stowage space with handling gear make them ideal for carrying stores and extra sweep gear during extended operations far from bases. In addition, staff and communications facilities were provided by them.*
>
> —Commander, Task Force 84, Rear Adm. Frank J. Lowry, USN, in his report on minesweeping actions during DRAGOON.[24]

Embarked with Comdr. William L. Messimer aboard USS *Barricade* was a small staff to assist him in carrying out his commander, Escort Sweeping Group duties. Lt. Comdr. Edward R. D. Sworder, RNVR, liaison officer for the British minesweepers assigned under Messimer's control was one of these individuals. Sworder had previously been awarded the Distinguished Service Order "For outstanding bravery and enterprise in the action in the harbour at Oran in [the sloops] H.M. Ships *Hartland* and *Walney*." Being on the staff of Captain, Minesweepers, Mediterranean, he contributed much to the planning and execution of operations. During sweeping, the RN officer was continually being shuttled by *ML 564* between *Barricade* (anchored in the Bay of Cavalaire off the beaches) and British ships to assist with command and control.[25]

Another member of the staff was USN Lieutenant B. B. Stern, mine disposal officer. Stern's activities are lesser known. One that was known fortunately proved to be of little consequence. He was called upon on 17 August to deal with a mine reported to have been washed ashore, which subsequent investigation revealed to be merely a buoy.[26]

ALLIED TROOPS ADVANCE INLAND

The amphibious landing by the main assault forces was carried out with great success. Scheduling H-Hour (the specific hour on D-Day at which time the first assault elements are scheduled to land) at 0800 provided the daylight necessary to supporting forces to exert the full power of naval artillery on weakening German coast and beach defences prior to the landing, as well as aerial bombardment immediately prior to the assault. As anticipated, strong residual coast defence was encountered during movement ashore, but the Germans lacked air power and reserve ground forces for an effective counterattack. The Allied troops gained the beach without significant losses, and began to advance against varying degrees of German resistance. Admiral Hewitt summarised these post-landing operations:

> After having cracked the crust of the beach defense all our troops were able to move forward very rapidly towards the Rhône Valley and there was very little resistance. The French troops, however, pulling to the westward towards Toulon encountered very heavy resistance from the Germans and from the German Navy-manned shore defenses in that area. They, however, with the assistance of naval gunfire, made very rapid advances and were able to completely surround Marseilles and Toulon and cut off its defenders from the other German troops to the north.[27]

DRAGOON A GREAT SUCCESS

The success of the Allied assault forces in their rapid advance ashore and movement inland during the Invasion of southern France, is evidenced in messages sent by dignitaries present offshore, as paraphrased by Admiral Hewitt in August 1944 war diary entries:

- 1157 on D-Day (15 August): From Winston Churchill aboard HMS *Kimberley* (F50): Best wishes to you all for the great success of your joint enterprise and good luck to all concerned. The Prime Minister sends his compliments to Secretary Forrestal.
- 1616 on 16 August: From Admiral Hewitt to Vice Adm. Alan G. Kirk, Commander, U.S. Naval Forces, France: Landings carried out 15th at 0800 against light resistance. At the end of the first day all troops landed except on ST RAPHAEL-FREJUS Beaches. These towns occupied the 16th morning. During night and morning 16th, American troops progressed deeply inland and arrived on a line GONFARON-LE MUY where contact was made with paratroops. Ships have operated counterbattery and support fire. Light expense ammunition.

No casualties. General Lattre and French divisions arriving today under French escort will land immediately.
- 2243 on 16 August: From Secretary of the Navy James Forrestal, aboard USS *Augusta* (CA-31), to Maj. Gen. Lucian K. Truscott Jr., commanding General VI Corps: To you and your division Commander[s], I send the Navy's 'Well Done' and the Navy's appreciation of good partners and gallant fighters.[28]

UNIT AND PERSONAL AWARDS FOR VALOUR

Among the many ships and individuals awarded unit and personal honours for the invasion of Southern France were auxiliary minelayers *Barricade* and *Planter*, and Lt. Comdr. Edward R. D. Sworder, RNVR.

	Invasion of Southern France	
Barricade (ACM-3)	15 Aug-25 Sep 44	Lt. Charles Percy Haber, USN
Planter (ACM-2)	15 Aug-25 Sep 44	Lt. Theodore Thomas Scudder Jr., USNR; Lt. Richard Albert Knapp, USN

The London Gazette of 10 July 1945 announced the awarding of the Legion of Merit to Lieutenant Commander Sworder for his actions in Operation DRAGOON:

> The KING has been graciously pleased to give unrestricted permission for the wearing of the following decorations, bestowed by the President of the United States of America.
> For gallant and distinguished service during the invasion operations of the South of France:
> Legion of Merit, Degree of Officer.
>
> Acting Temporary Lieutenant-Commander
> Edward Robert Denys SWORDER, D.S.C., R.N.V.R.

25

Victory in Europe

The lights went out and the bombs came down. But every man, woman and child in the country had no thought of quitting the struggle.

—These emotive words from Prime Minister Winston Churchill on VE Day 1945 praised the struggle of all those living and working on the home front during World War II.[1]

Photo 25-1

V-E Day celebrations on Bay Street in Toronto, Canada, 7 May 1945.
Photo by John H. Boyd, City of Toronto Archives Fonds 1266, Item 96241

After the assault phase of Operation DRAGOON cracked the German beach defences, American troops began fighting their way from the southern France beaches toward the Rhône Valley and met very little resistance. However, French troops moving to the westward toward Toulon encountered very heavy resistance from German ground forces and from German Navy-manned shore defences in that area. Despite this enemy opposition, they made rapid advances with the assistance of Allied naval gunfire, and were able to surround the port cities of Marseilles and Toulon; and to cut off their defenders from the other German troops to the north.[2]

The port of Marseilles was especially desired by the U.S. Seventh Army because it was near the mouth of the Rhône River and was the terminus of the railroad system up the Rhône Valley. From Marseilles, American troops could advance north in France by rail. The port's harbour was artificial, formed by a breakwater and divided into a number of different basins with rather small deep-water entrances. The Allies also wanted to open the port of Toulon, which offered a large sheltered harbour and had been a principal naval base for the French fleet, for use by shipping.[3]

Following the surrender of Marseilles and Toulon on 28 August 1944, and the opening of these major ports, after significant minesweeping and salvage work was completed, the requirement for facilities for landing war supplies, equipment and personnel was met. Capt. William L. Messimer, (Commander, Escort Sweeper Group, Mediterranean) attributed the sweeping success to Comdr. Alister Angus Martin, RNR:

> [The] clearance of Golfe de Fos and Port de Bouc approach channel, anchorage and harbor...was accomplished, and very thoroughly, in an amazing short time, chiefly because of the careful planning of the Senior Officer, 13th Minesweeping Flotilla, and the minesweeping skill of himself and of his forces.[4]

Martin was awarded the Distinguished Service Order (DSO) and Distinguished Service Cross (DSC) with two bars, and mentioned in despatches three times during the war. He would retire in 1958 with the rank of captain.[5]

In his own report, Martin praised British and American naval officers who had operated under his direction:

> I was particularly fortunate in having as my second-in-command - until the arrival of Commander [Norman Eyre] Morley in H.M.S. "RHYL" - so able and experienced an officer as Lieutenant-

Commander C. R. [Charles Robertson] Fraser, R.N.R. of H.M.S. "ARIES". This officer knows the flotilla well and was able to take over his temporary attachment to us in his stride. Lieutenants [Frank] Morley and [Robert] Belknap of the Y.M.Ss and S.Cs respectively very soon showed me that they had a sound grasp of all the requirements and at all times had their ships on the top line for any assignment. Lieutenant Jenkins with three of his ships was attached for only a comparatively short period but during that time he gave much and willing help. Lieutenants Duffy and [D.V.] Edmundson and Ensign [G.L.] Greene of the Boat sweeper sections, gave grand service and deserve high praise for their handling of these little craft and their crews. We did what we could for them to make their lot less uncomfortable and although this was not very much I think they knew where they belonged.

The feeling of team work between the ships was complete and I could not wish to have had the honour of commanding a better unit, which I trust has played its part in this vast and brilliantly planned operation.[6]

The final battles between Allied and German forces took place in late April and the first week in May 1945. Adolf Hitler committed suicide on 30 April—as the Battle of Berlin raged above his command bunker below the city—after designating Grossadmiral Karl Dönitz as his successor. Dönitz initially vowed to fight on, but quickly changed his mind. On 7 May General Alfred Jodl, Chief of Staff of the German Armed Forces High Command, signed surrender documents for all the German military forces at Reims, France, only eight days after Hitler's death. The following day, Field Marshal Wilhelm Keitel (Chief of the General Staff of the German Armed Forces) signed a similar document in Berlin, surrendering to Soviet forces in the presence of Gen. Georgi Zhukov and other Allied and Axis representatives.[7]

ODE TO MINELAYERS AND MINELAYING AIRCRAFT

> *As regards minesweeping, it needs but little imagination to appreciate the courage and endurance of those engaged in this vitally important service. It was perhaps a poor consolation for them to see the results of their efforts from time to time, but it was at least some consolation, whereas the minelayers and the minelaying aircraft had no such reward.*
>
> —Capt. John Stewart Cowie, CBE, RN, in his seminal book, *Mines, Minelayers and Minelaying*, published in 1949.[8]

In closing this book, it is important to note and emphasise that mining was the most successful form of Allied attack on enemy shipping in northwest Europe during the war. The data in the following table do not include Russia's contribution and a few vessel casualties due to Norwegian mines. What the statistics do not reflect is the number of enemy ships or submarines (denied by mines the relative safety of coastal routes) forced further out to sea, which enabled their attack by Allied warships, aircraft or submarines.

Enemy Controlled Vessels Sunk in NW Europe by Four Main Methods of Allied Attack

#Vessels Sunk	% of Vessels	Tonnage Sunk	% of Tonnage
Surface Vessels and Submarines			
376	17%	777,554	27.3%
Direct Air Attack (RAF and RN FAA)			
485	23%	696,028	24.5%
Air Raids (RAF and USAAF)			
429	20%	581,550	20.5%
Sea Mines			
850	40%	782,915	27.7%
Total Enemy Vessel Losses and Associated Tonnage			
2,140	100%	2,838,047	100%[9]

Aircraft accounted for 72 percent of all the mines laid offensively in enemy waters, and claimed 87 percent of the total shipping casualties. Naval ships, submarines, and Coastal Forces craft collectively contributed the remaining 28 percent of the effort for 13 percent of the casualties. Of the latter 28 percent, Coastal Forces craft fared best, 53 percent of these casualties for 30 percent of the mines laid.[10]

However, the mining successes came with a cost. The British lost twelve surface ship and submarine minelayers, as well as some Coastal Forces craft, in European and Mediterranean waters, and 500 British and Dominion aircraft in all theatres of war:

- 2 auxiliary minelayers: *Port Napier* and *Princess Victoria*
- 3 of 6 cruiser minelayers: *Abdiel*, *Latona*, and *Welshman*
- 3 of 6 destroyer minelayers, *Esk*, *Ivanhoe*, and *Intrepid*
- 4 of 6 *Porpoise*-class minelaying submarines, *Narwhal*, *Cachalot*, *Seal*, and *Grampus*

The U.S. Navy lost the minelayer *Miantonomah* off the harbour at Le Havre, France.

Postscript

Photo Postscript-1

This illustration from *Deep Diving and Submarine Operations* by Sir Robert H. Davis depicts 'P' Party members performing their various duties.

Enemy Waters introduces readers to mine warfare by Britain and other Allied nations against German and Italian naval and maritime forces in World War II—with emphasis on offensive and defensive minelaying. Relatively little text is devoted to paying tribute to the heroic efforts of the Royal Naval Patrol Service, which toiled day-after-day, throughout the war, to keep coastal channels, harbour approaches and ports swept clear of mines. Repeated mining by German planes, submarines and surface craft, performed mainly at night, was undertaken in an effort to cripple the Royal Navy, and to prevent delivery by ships of food, fuel, and other vital supplies and materiel on which Britain depended—and thereby force the island nation to capitulate.

Less known than the exploits of the Royal Navy's minesweeping force, to which several books have been devoted, are those of clearance divers. Minelaying, mine disposal, and diving and salvage are often intertwined, and it is fitting to close out this book with a brief tour of the rich history of the Royal Navy's mine clearance forces during World War II. This section is most suited for readers with current or past mine warfare experience, and in particular, Royal Navy Mine Warfare specialists and Clearance Divers.

ROYAL NAVY CLEARANCE DIVING

Clearance Diving takes its name from operations carried out during and immediately after the Second World War to clear the ports and harbours of the Mediterranean and Northern Europe of unexploded ordnance and booby traps laid by the Germans. Given the opportunity, retreating enemy forces exerted every effort to deny Allied forces the use of ports and harbours for the support/resupply of troops ashore. In addition to mining and booby-trapping, the Germans sank ships and other vessels at port entries to block access, and dumped trucks, harbour cranes, other debris, and munitions into port waters, and blew up remaining harbour facilities.

Harbour clearance work was originally undertaken by Royal Navy Render Mines Safe (RMS) and Bomb Disposal units and later by Port Clearance Parties or 'P' Parties. Two of the latter, (Naval Parties 1571 and 1572) went into action soon after D-Day to clear the vast quantities of unexploded ordnance and general debris left after the Allied invasion. They were joined later by other 'P' Parties including, 1573, 1574, 1575, and 2444 (many of which had Commonwealth naval personnel) and 'P' Party 3006 manned by members of the Royal Netherlands Navy.

From their inception in late 1943, 'P' Parties were formed by the Admiralty's Director of Minesweeping, under the auspices of HMS Vernon in Portsmouth. They were, in effect, a form of minesweeping. On 1 October 1944, all Royal Navy diving, including the 'P' Parties, was transferred to the Director of Torpedoes and Mining (DTM), also under the auspices of HMS Vernon.

ABUNDANT AWARDS FOR HEROISM AND VALOUR

George Cross
For acts of the greatest heroism or for most conspicuous courage in circumstance of extreme danger

George Medal
For gallantry not in the face of the enemy, where the services were not so outstanding as to merit the George Cross

Members of RN RMS and Bomb Disposal Units, and the later 'P' Parties, were among the most highly decorated of the war and were awarded as many George Crosses and George medals as the British Army and RAF combined. The most highly decorated were RNVR, RANVR and RCNVR officers initially commissioned as Temporary Acting Sub Lieutenants (Sp) (i.e. Special Branch) straight from civilian life. They were then trained at HMS Vernon and sent out to deal with unexploded bombs and mines around the UK and farther afield. Of particular note, Lt. John Bridge, RNVR, and Lt. Hugh Syme, RANVR, remain the only two service personnel to have ever been awarded the George Cross, George Medal and a bar to the George Medal, all for their work in bomb & mine disposal.

Brief information about seven decorated, very colourful members of the Render Mines Safe and Bomb Disposal communities follows. One of these individuals later disappeared, following the war, under mysterious circumstances while reportedly working for MI6 (British Intelligence).

ROYAL NAVY-TRAINED U.S. NAVAL OFFICERS

Photo Postscript-2

Navy Cross
For extraordinary heroism in combat not justifying the Medal of Honor

Draper L. Kauffman as a Lt., RNVR, England, September 1940-November 1941; and receiving from his father, Rear Adm. James L. Kauffman, USN, a gold star in lieu of a second Navy Cross for heroism under fire during the invasion of the Marianas Islands. Naval History and Heritage Command photograph #NH 95549 (left), and to the right National Archives photograph #80-G-244700

Many more British officers and ratings passed through Royal Navy Render Mines Safe and Bomb Disposal training, than did those of Allies. To give representation to the latter's small numbers, the next few pages honour two Americans, two British, two Canadians, and one Australian. One of the Royal Navy's BSOs (Bomb Safety Officers) during the early days of the war was Draper Kauffman. The son of a U.S. Navy admiral, he had graduated from the U.S. Naval Academy in 1933 but later returned to civilian life because of weak eyesight. In 1939, he joined the French Ambulance Service and was captured by the Germans when they overran Holland and Belgium.

Upon his release as a citizen of a then neutral country (USA), Kauffman joined the Royal Navy as a Sub Lieutenant RNVR and was trained as a BSO. He performed bomb disposal duties until October 1941 when, following an accident he returned to Washington, DC. While there, he was approached by the U.S. Bureau of Ordnance and asked to establish a Bomb Disposal School. Thus, Kauffman, Lieutenant, RNVR, became Lieutenant Kauffman, USNR, and officer-in-charge of the U.S. Bomb Disposal School.

Kauffman would twice be awarded the Navy Cross for heroism (the first time for bomb disposal at Pearl Harbor) and ultimately retire from the USN as a rear admiral. He is most known for having established the U.S. Navy's first Underwater Demolition Teams (UDTs), the forerunners of the Sea, Air & Land Teams more commonly known as the SEALs. His eventful life and career are described in the book, *America's First Frogman - The Draper Kauffman Story*, written by his sister, the late Elizabeth Kauffman Bush, who was also the sister-in-law of former American President George H. W. Bush.

A YANK AND A BRIT SERVING TOGETHER

Photo Postscript-3

Lt. (jg) Gene Haderlie, USN, and Lt. Stephen Wilkinson, RNVR, attached to the Royal Navy for bomb & mine disposal duties.
Collection of Rob Hoole

Gene Haderlie, another American, had a remarkable upbringing even before his action-packed service with the Royal Navy. His rough-and-tumble childhood in Wyoming during the Depression was followed in early adulthood by an undergraduate expedition to Baja Mexico. During this trip, he crossed paths with John Steinbeck, and suffered an inflamed appendix which was taken out by a veterinarian. Presumably these experiences helped steel him for two years as a hard-hat diver in World War II, defuzing mines in the English Channel and enduring the trauma of D-Day. These experiences and his later work as a biological oceanographer are detailed in *Conversations with Marco Polo*, published in 2006.

Photo Postscript-4

Gene Haderlie, USN, dressed in MRS (Mine Recovery Suit) in 1944. Collection of Rob Hoole

Lt. Stephen Wilkinson, RNVR, had been a concert pianist before the war, attesting to the diversity of volunteers for this dangerous vocation. Wilkinson received a Mention in Despatches "for courage and undaunted devotion to duty," and was awarded the associated oak leaf on 15 August 1944. Only a few months later, he lost the use of his right hand after a flooder cutter (see Appendix B) exploded while he was examining a GP moored-influence mine. It was one of several such German mines brought to the Frater Armament Depot near Portsmouth, following their capture on a barge in Antwerp, Belgium.

CANADIAN ORGANISES AND LEADS PORT PARTIES

Comdr. James Harries, RCNVR, was instrumental in organising and leading the 'P' Parties. He was awarded the George Medal for great bravery and undaunted devotion to duty in the face of danger while engaged in bomb & mine disposal in Kent between November 1942 and March 1943. He was then awarded a bar to his George Medal for exceptional gallantry, skill and great devotion to duty, often in close proximity to the enemy, during mine-searching and clearance operations in the ports of Normandy and the Low Countries. Finally, he was appointed an OBE (Most Excellent Order of the British Empire) for courage, determination and outstanding leadership in mine clearance and mine disposal operations in North West Europe immediately before the close of the war with Germany and in the months that followed.

While the RMS and Bomb Disposal Units suffered many grievous casualties, not a single member of a 'P' Party was lost in action. (Able Bodied Seaman William Brunskill died of wounds after a German V-2 rocket hit the Cinema Rex in Antwerp where he was watching a film on 16 December 1944.) This remarkable record was despite the 'P' Parties clearing hundreds of tons of ordnance from the ports of Cherbourg, Caen, Dieppe, Le Havre, Boulogne, Rouen, Calais, Antwerp, Ostend, Terneuzen, Zeebrugge, Flushing, Amsterdam, Rotterdam, Hamburg, and Bremen. Sixty mines were cleared at Bremen alone.

AUSTRALIAN DISARMS FIRST 'OYSTER' MINE

Australian Lt. Comdr. George Gosse, RANVR commanded 'P' Party 1571. He was awarded the George Cross for his work when divers searching the Ubersee Hafen (port) in Bremen reported the presence of a mine which from their description appeared to be an entirely new type. Gosse immediately dived and verified that it was a GD pressure type, commonly known as an 'Oyster.' As it was necessary that this type of mine be recovered intact, it was decided to attempt to render the mine safe underwater and on the following day, 8 May 1945, Gosse dived on it again.

Photo Postscript-5

Lt. Comdr. Gosse, RANVR, with 'P' Party 1571 at Bremen, Germany on 8 May 1945.
Collection of Rob Hoole

Photo Postscript-6

Australian Lt. Comdr. George Gosse, RANVR, disarming a GD "Oyster" pressure mine, at Bremen, Germany on 8 May 1945.
Collection of Rob Hoole

Using improvised tools, Gosse eventually succeeded in removing the primer which was followed by a loud metallic crash. The mine was eventually lifted on the quayside when it was found that the detonator had fired immediately after the primer had been removed. During the subsequent ten days, Lieutenant Gosse rendered safe two similar types of mines which were lying in close proximity to shipping and in each instance the detonator fired before the mine reached the surface.

COMMMANDER LIONEL 'BUSTER' CRABB, RNVR

> *While it is the practice for ministers to accept responsibility, I think it is necessary in the special circumstances of this case to make it clear that what was done was done without the authority or knowledge of Her Majesty's ministers. Appropriate disciplinary steps are being taken.*
>
> —Statement by British Prime Minister Sir Robert Anthony Eden on 9 May 1956, during an address to a packed House of Commons in which he refused to reveal the details surrounding the disappearance of a naval diver during a goodwill visit to Portsmouth, England, by the Soviet leadership. This heightened speculation that Comdr. Lionel "Buster" Crabb, RNVR, had been on a secret spying mission for which permission had not been granted.[1]

Lionel Crabb was born in 1909 in South London. (He got the nickname 'Buster' in honour of the American two-time Olympic swimmer and actor, Buster Crabbe.) Crabb tried to join the Navy in 1939. He was refused on medical grounds, was too old to join the Reserves, and so he joined the Merchant Navy instead. When war broke out, he trained as a Merchant Seaman Gunner, and later was able to join the Royal Navy Patrol Service. When he was eventually given a medical examination, Crabb was denied further sea service owing to a weak left eye. Fed up, but undeterred in his desire to serve, he volunteered for Special Duties, and became a mine and bomb disposal expert.[2]

In 1942, Lt. Lionel 'Buster' Crabb, RNVR, was sent to Gibraltar, where he worked in a mine and bomb disposal unit removing Italian limpet mines that enemy divers attached to the hulls of Allied ships. He was one of a group of underwater clearance divers thus employed during a period of Italian frogman and manned torpedo attacks by the Decima Flottiglia MAS, an Italian commando frogman unit of the Regia Marina. (Italian Royal Navy) created during the Fascist regime.[3]

The first operation by the frogmen was in July 1942, when twelve swimming saboteurs damaged three moored merchant ships near Gibraltar. For these and subsequent operations, the unit secretly took up residence in an old derelict Italian tanker, the *Olterra*, moored in a neutral Spanish port just five miles from Gibraltar.[4]

Decima Flottiglia MAS arm badge during the Italian Social Republic (Nazi occupation of Italy)

Motto: Memento Audere Semper (Remember to risk always)

Crabb was awarded the George Medal for his efforts and was promoted to lieutenant commander. He subsequently became Principal Diving Officer for Northern Italy, assigned to clear mines in the ports of Livorno and Venice, which also involved some salvage work. When he finished emptying them of wrecks, Crabb was created an Officer of the Order of the British Empire (OBE). In 1945 he was moved to Palestine to head up an underwater bomb disposal team and he was demobilised in 1947.[5]

A few years later, Crabb was recalled to active duty to assist with salvage of the submarine HMS *Truculent*, which had sunk with all hands in the Thames Estuary on 12 January 1950 after a collision with the Swedish tanker *Divina*. Other operations followed. He was reportedly in the Suez Canal in 1953, and a year later he participated in a search by the British Admiralty for the Tobermory Galleon. An article in *The Times* on 9 August 1953 informed readers of a salvage operation at the galleon wreck site, the expedition being led by Rear Adm. Patrick McLaughlin, RN (retd.) and Lieutenant Commander Crabb.[6]

Legend has it that the wreck of a Spanish galleon, laden with gold, lies somewhere in the mud at the bottom of Tobermory Bay—although its true identity, and cargo, are in dispute. After the defeat of the Spanish armada by the British fleet in 1588, it's believed that a critically damaged Spanish vessel took shelter in the bay of Tobermory on the Isle of Mull, Scotland, and subsequently sank after catching fire. This search, and an earlier one by the Admiralty in 1950, failed to locate any gold.[7]

In 1956, 'Buster' Crabb appeared on the Navy List as Commander Crabb, Special Branch, based at HMS Vernon in Portsmouth—the same year he disappeared under mysterious circumstances, reportedly while in the employment of MI6. On 18 April 1956, Russian leader Nikita

Kruschev and marshall Nikolai Bulganin arrived at Portsmouth aboard the Soviet cruiser *Ordzhonikidze*, for a good will visit.⁸

Photo Postscript-7

Soviet cruiser *Ordzhonikidze* in the Baltic Sea on 1 August 1956.
Naval History and Heritage Command photograph #NH 79977

Crabb and another man, who signed the register as Matthew Smith, had taken rooms the previous night, on the 17th, at the Sally Port Hotel in Portsmouth. The next day the *Ordzhonikidze* and Russian ships in company arrived and tied up at the South Railway Jetty in the dockyard. That night Crabb was seen in Havant having a drink with some old friends before catching the train back to Portsmouth. He reportedly told them that he was "going down to take a dekko [a quick look] at the Russian [ship] bottoms" for which he would earn 60 guineas.⁹

Commander Crabb was last seen leaving the Sally Port Hotel on the morning of 19 April. Both men apparently checked out of the hotel in the afternoon of his disappearance, and several pages of the hotel register were later found to have been removed. Crabb was reported missing, presumed dead, by the Admiralty on 29 April. The official statement declared that he had died following a test dive at Stokes Bay, near Portsmouth, on the Hampshire coast.¹⁰

It appears that Crabb was on a spying mission for MI6, unbeknown to the prime minister, and the statement by the Admiralty an attempt to cover up the mission. When the Soviet Government protested to the

Foreign Office on 4 May that a frogman had been seen in the vicinity of the *Ordzhonikidze*, Sir Robert Anthony Eden was forced to speak out. Sir John Alexander Sinclair, the head of British Intelligence, was subsequently forced to resign.[11]

There was much speculation then, and theories and stories abound since, regarding what befell Commander Crabb. These include (1) his having being killed or captured by Russian sentries posted topside on *Ordzhonikidze* and/or in the water in the event an attempt was made to gather intelligence on her underwater hull and running gear; (2) that he was captured and died during interrogation aboard the ship; or (3) that he was taken back to the Soviet Union and imprisoned. It was reported in 1957 that the headless body of a man in the remains of a diving suit had been found in Chichester Harbour. Crabb's ex-wife, and his current girlfriend were unable to say with any confidence that the body was his owing to the extensive decomposition.[12]

Nevertheless, an inquest recorded an open verdict and the Coroner stated that he was satisfied that the body was that of Commander Lionel Crabb. The remains were buried with Crabb's ornate silver-mounted swordstick after a coroner concluded it was his body.[13]

Based on the following account by Jim Knight, the body believed to be that of Lionel Crabbe was not found in Chichester Harbour as reported, and identification would have been practically impossible:

> The body was not found 'floating around' in Chichester Harbour. It was in fact brought to the surface by 2 net fishermen, from Prinstead or Southbourne, West Sussex. They made some sort of SOS signal that was seen by the lads in the Air traffic control tower at RAF Thorney Island. ATC in turn notified us at the Marine Craft Section in the mid-morning. Myself and two other members of 1107 MCU detachment were duty weekend crew. We had an old World War Two 40 ft assault landing craft which was used for inshore and harbour rescue, and in this craft, we made our way to Pilsey Island which was only about a mile up channel towards the Solent. We found the fishermen with their net tangled up...
>
> Untangling the net, we hitched the body onto the lowered front ramp and got it on board.... The head and hands were missing, and in the cavity where the head had been, were hundreds of small crabs, and other such creatures.... On returning to the MCS (Marine craft Section) we were able to run almost up to West Thorney road, due to it still being high tide. Waiting for us on the bitumen were a mob of RAF Officers, 4 men in long black overcoats, RAF Ambulance, local Police and many unknown onlookers.

The body, still clad in a frogman's suit was removed to Chichester Hospital for examination and identification. Later, we members of the Marine Section discussed the incident. None of us could see how the body was identified. With the advent of DNA, perhaps it could now be proven one way or the other, if the body was that of Cmdr Crabb. Among those on the beach that day was Group Capt. Boxer, Station Commander of RAF Thorney Island. The 4 men in black overcoats must have been MI6 or such.[14]

Photo Postscript-8

Commander Crabb's gravestone at Milton Cemetery, Milton Road, Portsmouth. Courtesy of www.submerged.co.uk/buster-crabb.php

The Cabinet papers concerning the Crabb affair will remain secret until 2057.[15]

CANADIAN AWARDED GEORGE MEDAL FOR MINE DISPOSAL IN THE SUEZ CANAL

Comdr. George Douglas Cook, RCNVR, garnered the George Medal for gallantry and undaunted devotion to duty after assisting Lt. Maurice Griffiths, RNVR with mine disposal in the Suez Canal in May 1941. (Griffiths had earlier been awarded a George Medal for gallantry and undaunted devotion to duty for bomb and mine disposal in London in September 1940). Cook later received a bar to his George Medal for gallantry and undaunted devotion to duty for mine disposal at Haifa, Israel, on 26 July 1941.

Cook's dangerous duty continued in Southeast Asia. While commanding a Landing Craft Obstruction Clearance Unit, he was awarded a Mention in Despatches for courage, tenacity and devotion to duty in operations lasting four months, frequently performed under rapidly changing conditions and with difficult lines of communication, on the Arakan Coast. Arakan, a historic coastal region, faced the Bay of Bengal to the west, the Indian subcontinent to the north, and Burma to the east. This region is now the Rakhine State in Myanmar.

POST-WAR DISBANDMENT OF PORT PARTIES

Work in the European theatre continued until well after VE Day. But eventually, following war's end, most of the Port Parties were disestablished on 30 November 1945. (The exceptions were 'P' Party 2444, which was still operating at Dunkirk, and 'P' Party 2443 formed in June 1945. The latter unit was stood up to deal with residual unexploded ordnance around the UK coast after the war.) Their outstation, HMS Vernon (D), at Brixham in Devon, had closed on 1 October 1945 and 'P' Party 2443 moved its headquarters to HMS Vernon at Portsmouth, but only for a short period.

It was decided to integrate the 'clearance' divers more closely with the mine countermeasures establishment at HMS Lochinvar in Port Edgar on the Firth of Forth and this is where the members of the last 'P' Party on the Continent ('P' Party 2444) were sent in March 1946 after having departed Dunkirk. In the meantime, HMS Vernon became the centre for Deep Diving and the AEDU (Admiralty Experimental Diving Unit) moved there from Siebe Gorman at Tolworth.

This integration had obvious benefits but, as might be expected, given the competitive natures of its members, something of a division sprang up between the deep divers at Vernon (the 'Steamers') and the Clearance Divers at Lochinvar (the 'Corkheads').

CHANGE IN DESIGNATION TO CLEARANCE DIVER

On 3 December 1948, 'P' Party divers were officially renamed Clearance Divers in accordance with Confidential Admiralty Fleet Order 341/1948. On 7 March 1952, the Clearance Diving (CD) specialisation, under the auspices of the TAS Sub-Branch of the Seaman Branch, was officially introduced, although a training nucleus had been set up some two years earlier to take advantage of the few remaining men with wartime experience. These officers and ratings had, in the main, qualified as Shallow Water Divers trained to use the Sladen 'Clammy

Death' diving dress and oxygen breathing apparatus. They were joined by other officers and ratings Qualified in Deep Diving (QDD).

EXPANDED TRAINING RESPONSIBILITIES

In 1950, a Home Station Clearance Diving Team was set up and other Clearance Diving Teams were soon established to support the Mediterranean Fleet and the Far East Fleet. In 1951, clearance diving training moved from Lochinvar to Vernon and a new Clearance Diving School was established combining the training of clearance divers with that of 'deep' divers. This school also took on the new task of training Shallow Water Divers (later called Ships' Divers) so that every ship had a properly equipped air diving capability under its own Diving Officer to conduct ship's bottom searches and undertake simple underwater engineering tasks.

SUB-SPECIALISATION OF THE SEAMAN BRANCH

On 25 February 1966, a new Sub-Specialisation of the Royal Navy's Seaman Branch, Minewarfare and Clearance Diving (MCD), was formed. Seaman Branch officers with CD (clearance diver) and TAS (Torpedo and Anti-Submarine) qualifications automatically became MCD Officers, while those certified only as Clearance Diving Officers (CDOs) undertook a conversion course to be trained in Minewarfare. Minewarfare had been the prerogative of only TAS Sub-Branch officers.

Torpedo and Anti-Submarine ratings (enlisted men) of the Seaman Branch specialised in Weapons (UW) and Control (UC) continued to perform Minewarfare duties until 1975. At that time, two new sub-branches of the Operations Branch were formed: Minewarfare (MW) and Diver (D). In 1981, the MCD Sub-Branch took over responsibility for demolitions training from the Sonar (ex-TAS UC) Sub-Branch.

For purposes of this overview of the evolution and integration of related communities within the Royal Navy, an explanation of progress out to year 1981 is probably sufficient. However, it's important to highlight that post-war progress up to that point, and to the present, was built on the foundation laid by the courageous, innovative, and accomplished Render Mines Safe and Bomb Disposal personnel of the Second World War.

Appendix A: World War II British and German Mines

The most common mines used in World War II were moored mines and ground mines, although drifting mines, oscillating mines, self-burying anti-invasion mines, and limpet mines were also employed.

Moored mines were positively buoyant and were anchored to lie just beneath the surface of the water or at any other chosen depth, e.g., if laid against submerged submarines. Some had delayed riser mechanisms which would only operate after a certain period of time or number of 'ship counts' or minesweeper passes. All mines were meant to self-sterilise if they broke free or after a certain period in accordance with the Hague Convention. The first requirement was usually achieved with a mooring safety lever which broke the firing circuit if tension was released while the second was normally achieved with the use of explosive flooder cutters that pierced the mine shell allowing it to sink to the seabed after a set period.

Most moored mines were detonated by contact. Some were fitted with lead Hertz horns. These were hollow lead protrusions containing a glass ampoule of electrolyte. On being struck by a vessel, the Hertz horn and its glass vial were crushed allowing the electrolyte to come into contact with two terminals in the firing circuit, thus actuating the mine. The Hertz horn was much more reliable than the switch horn which easily corroded, and it provided a long-lasting source of electricity to ignite a detonator, whereas a battery decayed over time. This was also true of chemical horns, each containing a glass ampoule of sulphuric acid held within a mixture of potassium chlorate and fine sugar. When the horn was crushed, its contents produced a chemical reaction producing heat and flame to set off the main charge. Other types of contact firing mechanism included simple switch horns, antennae, Tombac tubes, snaglines, and internal pendulums.

Magnetic Influence moored mines contained coiled rod units, magnetic dip needles, flux meters or other sensors that detected changes in the earth's field caused by the passage of a ferrous hull, while acoustic moored mines contained microphones or 'vibrators' that detected certain noise frequencies produced by a passing ship or submarine.

Most moored mines were laid by surface craft although a few types were designed for lay by aircraft or submarine.

Moored mines tended to contain less explosive than ground mines because they needed a reserve of buoyancy to support their own weight and that of their mooring cable. Mines laid with longer mooring cables tended to have smaller charge weights. Also, ground mines needed to contain greater explosive charges to be effective because they were actuated by influence at some distance from their target.

Photo Appendix A-1

German ground mine, destined for HMS Vernon.
Collection of Rob Hoole

Ground/bottom mines were negatively buoyant and remained on the seabed waiting for targets to approach close enough to be destroyed. They were usually fitted with delayed action or ship count mechanisms, so they would only become 'active' after a certain period of time or number of 'ship counts' or minesweeper passes. Like moored mines, ground mines were meant to self-sterilise after a certain period of time and this was usually achieved through battery decay or the battery was shorted. Some German ground mines were booby-trapped with PSE (Prevent Stripping Equipment) to foil bomb & mine disposal personnel and prevent their secrets becoming known.

Ground mines could contain magnetic and/or acoustic and pressure influence sensors, often used in combination. Depending on type, they could be laid by surface craft, aircraft, or submarine. German parachute ground mines were often used against land targets as blast bombs, exploding in the air, on contact with the ground or after a set delay to kill personnel belonging to the emergency services and bomb disposal units.

Anti-Countermining devices prevented mines from actuating in sympathy with other mines exploding or by the use of explosive mine countermeasures.

Anti-Minesweeping Devices such as moored obstructors, some fitted with chains and cutters, were often sown in moored minefields to complicate the work of minesweepers.

BRITISH MOORED MINES
(The designation of each mine type, and weight of the explosives available with its use are identified in the first column of the table)

Mine Type Explosives	Firing System	Remarks
H2 320 lbs	Hertz horn	World War I mine. Existing stocks used up in early months of World War II.
Mk XIV 320 or 500 lbs	Hertz horn	Emanated from experimental H Type G. First contracts placed in 1920. Superseded by Mk XVII.
Mk XVI 320 lbs	Hertz horn	Laid by *Porpoise*-class submarines.
Mk XVII 320 or 500 lbs	Switch horn or acoustic influence	The general purpose mine of World War II, mass produced from design of Mk XIV. Suitable for laying in depths down to 500 fathoms. Also fitted with a variety of acoustic circuits.
Mk XIX 100 lbs	Switch horn	Small anti-submarine mine not extensively used as such. Also laid offensively by Coastal Forces craft.
Mk XX 320 or 500 lbs	Antenna	Based on the Mk XVII mine case and laid in large numbers as an anti-submarine mine. Instability of the firing circuit was never mastered and most of these mines were laid with sterilised antenna.
Mk XXII 320 or 500 lbs	Switch horn and antenna	Modified Mk XX with floating upper antenna. Used in Northern Barrage during later operations.
Mk XXV 500 lbs	Switch horn with snagline	Modified Mk XXII, used in offensive operations by Coastal Forces craft.
M Mk 1 320 or 500 lbs	Magnetic influence	Moored magnetic mine in service by the beginning of 1941.
Vickers T III 440 lbs	N/K	Used by Allied French and Netherlands submarines. Carried in separate wells, externally, as for the S mine of World War I.

BRITISH GROUND MINES

Mine Type Explosives	Firing System	Remarks
A Mk I-IV 750 lbs	Initially magnetic then acoustic and magnetic acoustic combination	Laid by aircraft or Coastal Forces craft. Mine most used in World War II. Overall weight 2,000 lb. 113½" by 18". These mines were progressively adapted to accommodate acoustic and combination circuits, single steriliser and arming clock or PDM (Period Delay Mechanism). Fitted with parachute if air laid. Superseded in 1944 by the A Mk VI.
A Mk V 600 lbs	Magnetic	Laid by aircraft. Designed in 1940/41 for use in any aircraft capable of carrying a 1,000 lb bomb. Overall weight 1,080 lb. 89" by 15¾". Single and double contact magnetic firing circuits plus arming clock or PDM but not for sterilisation. Continual difficulty with parachute release led to replacement by A Mk VII in 1944.
A Mk VI 950 lbs	Magnetic and acoustic combination	Laid by aircraft or Coastal Forces craft. Approved for service in March 1944 and superseded A Mk I-IV. Overall weight 2,000 lb. 109" by 18¾". Highly versatile. Single steriliser and arming clock or PDM (Period Delay Mechanism). Fitted with parachute if air laid.
A Mk VII 600 lbs	Magnetic	Laid by aircraft. Replaced A Mk V in spring of 1944. 83" by 16½". Overall weight 1,000 lb. Single and double contact magnetic firing circuits plus single steriliser and arming clock or PDM.
M Mk III 1,750 lbs	Magnetic	Laid by Coastal Forces craft from summer of 1942. Poor sensitivity coupled with impossibility of it housing acoustic and combination circuits restricted employment. Redesign cancelled in March 1944.
M Mk II 1,000 lbs	Magnetic	21" diameter mine laid by submarine. Insensitivity, coupled with unsuitability for adaption to take acoustic circuits led to redesign as M Mk V but bulk order of replacement not supplied until after the end of the war.

GERMAN MOORED MINES (MOST REPRESENTATIVE)

Mine Type Explosives	Firing System	Remarks
EMD I/GX 330 lbs	Contact (Hertz horn)	Surface laid. Introduced 1925. EMD II Overall weight 1,860 lbs. Depth of lay 4 to 115 fms (fathoms). Case depth ½ to 10 fms.
EMD II/GX 330 lbs	Contact (Hertz horn)	Surface laid. Introduced 1937/38. Overall weight 2,110 lbs. Depth of lay 4 to 115 fms. Case depth ½ to 10 fms.
EMC I/GY 660 lbs	Contact (Hertz horn)	Surface laid. Introduced 1924. Overall weight 2,375 lbs. Depth of lay 4 to 115 fms. Case depth ½ to 10 fms.
EMC II/GV/GY 550 or 600 lbs	Contact (Hertz horn plus)	Surface laid. Different versions introduced between 1937 and 1943. Overall weight N/K. Depth of lay and case depth varied immensely according to configuration that could include upper and lower antenna, floating antenna, Tombac tube and snagline. Max depth of lay could be 250 fms although mine case had to be 29 fms or less.
FMC/GQ 90 lbs	Contact (Hertz horn)	Surface laid. Introduced 1928. Overall weight 925 lbs. Max depth of lay 72 fms. Mine case ¼ to 3 fms.
UMC/GZ 65 lbs	Contact (Hertz horn)	Surface laid. Different versions introduced between 1928 and 1938. Overall weight 1,800 lbs. Max depth 50 fms. Case depth ¼ to 3 fms.
UMB/GR 90 lbs	Contact (Hertz horn plus Tombac tube and snagline)	Surface laid. Different versions introduced between 1941 and 1944. Overall weight circa 1,400 lbs. Max depth 150 fms depending on version. Case depth ½ to 15 fms. Often fitted with chain and cutter to thwart minesweeping.
BMC/GM 110 lbs	Contact (Leclanché cell horn)	Laid by surface craft or aircraft (without parachute). Developed for the German Navy by the Luftwaffe 1943/44. Its prototypes, the BMA and BMB, were abandoned because of difficulties with the sinker, balance in flight and streamlining. These difficulties were overcome, and the BMC in its final form included a seven-day delay rising clock. It was fitted with four retractable horns which snapped off on impact with a vessel and admitted sea water to a form of dry Leclanché cell thus creating an electrical current sufficient to fire a detonator. Overall weight 1,430 lbs. Max depth 90 fms. Case depth ½ to 7½ fms.
EMF/GO 750 lbs	Magnetic	Surface laid. Introduced 1939-1942. Overall weight circa 2,420 lbs. Max depth 250 fms. Case depth 7½ to 15 fms.

312   Appendix A

Mine Type Explosives	Firing System	Remarks
TMA/GT 460 lbs	Contact plus magnetic	Laid by submarine. Introduced 1944. Overall weight 1,760 lbs. Max depth 135 fms.
LMF/GP 618 lbs	Contact plus magnetic	Laid by aircraft or surface craft. Introduced 1944. Overall weight 2,315 lbs with parachute, 2,182 lbs without parachute. Max depth 164 fms. Case depth 9 to 16 fms.
UMC/GZ 65 lbs	Contact (Hertz horn)	Surface laid. Anti-submarine. Different versions introduced 1928 to 1938. Could be moored at either 25 or 50 fms.

GERMAN GROUND MINES (MOST REPRESENTATIVE)

Mine Type Explosives	Firing System	Remarks
LMB/GC ~1,495 lbs	Magnetic, magnetic acoustic and pressure combination	Laid by aircraft or surface craft. Different versions introduced 1938 (magnetic) to 1941 (acoustic) to 1944/45 (pressure). Overall weight circa 2,155 lbs with parachute, 1,980 lbs without.
LMA/GD ~650 lbs	Magnetic	Laid by aircraft or surface craft. Different versions introduced 1937 (first German magnetic mine rendered safe by Ouvry in 1940) to 1941. Overall weight circa 1,120 lbs with parachute, 1,010 lbs without.
TMB/GS 945 or 1,230 lbs	Magnetic and magnetic acoustic combination	Laid by U-boat (submarine) and E-boat (coastal forces craft). Different versions introduced 1936 to 1944. Overall weight circa 1,500 lbs. Normally laid in waters of 12-15 fms.
TMC/GN 2,045 lbs	Magnetic and magnetic acoustic combination	Laid by U-boat. Enlarged TMB. Different versions introduced 1939 to 1945. Overall weight circa 2,440 lbs. Could be laid in waters up to 20 fms deep.
MTA 1,010 lbs	Magnetic	Delivered by torpedo. Introduced 1942. Overall weight 3,410 lbs.
BM 1000/GG 1,550 lbs	Magnetic and magnetic acoustic combination	Standard Luftwaffe mine laid by aircraft without parachute. Different versions introduced 1941 to 1944. Overall weight 2,200 lbs.

Appendix B: Minewarfare Terms

Acoustic Mine: A mine with an acoustic circuit which responds to the acoustic field of a ship, submarine or sweep.

Actuate: To operate a mine firing system by an influence or a series of influences in such a way that all the requirements of the mechanism for firing, or for registering a target [In NMW; ship] count, are met.

Anchor: A heavy weight to which a mine is moored (see Sinker).

Antenna Mine: A contact mine fitted with antenna which, when touched by a ferrous object, set up galvanic action to fire the mine.

Anti-Countermining Device: A device fitted in a mine to prevent its actuation by shock.

Anti-Submarine Minefield: A field laid specifically against submarines. It may be laid shallow and be unsafe for all craft, including submarines, or laid deep with the aim of being safe for surface ships.

Anti-Sweep Device: Any device incorporated in the mooring of a mine or obstructor, or in the mine circuits to make the sweeping of the mine more difficult.

Armed Mine: A mine from which all safety devices have been withdrawn and, after laying, all automatic safety features and/or arming delay devices have operated. Such a mine is ready to receive a target signal, influence or contact.

Armed Sweep: A sweep fitted with static and/or explosive cutters or other devices to increase its ability to cut mine moorings.

Arming Delay: A device fitted in a mine or any autonomous munition designed to prevent it from being armed for a pre-set time after laying or delivery.

Bottom Mine: A mine that is negatively buoyant; rests on, or can become buried in, the sea bed and is held there by its own mass. (See Ground Mine and Mine).

Bottom Sweep: Two ship wire or chain sweeps used either to sweep mines close to the bottom and to sweep heavy obstructors or to remove such mines and obstructors from a channel by dragging them to a designated area and releasing them.

314   Appendix B

Buoyant Mine: A mine of positive buoyancy held below the surface by a mooring attached to a sinker or anchor on the bottom (see Moored Mine).

Channel: The whole or part of a route specified by a width in which MCM operations will be or have been conducted.

Chemical Horn: A mine horn containing a glass ampoule of sulphuric acid held within a mixture of potassium chlorate and fine sugar. When the horn was crushed, its contents produced a chemical reaction producing heat and flame to set off the main charge.

Clearance Diver: Diver who is qualified to carry out tasks in mine/ordnance search, investigation, disposal, render safe, recovery and removal, underwater and ashore.

Clearance Operations: An MCM operation intended to achieve a high probability of countering any mine in a given area, route or channel.

Photo Appendix B-2

German ground mine disposal continues in 21st Century.
Collection of Rob Hoole

Contact Mine: A mine which is designed to fire by physical contact between the target and the mine case or its appendages.

Controlled Mine: A mine which after laying can be controlled by the user.

Countermine: The process of detonating the main charge in a mine by the shock of a nearby explosion of an independent explosive charge or another mine.

Cutter: In naval mine warfare a device fitted to a sweep wire to cut or part the mooring of mines or obstructors; it may also be fitted in, or to, the mooring of a mine or obstructors to part a sweep.

Deep Water Minefield: An anti-submarine minefield which is safe for surface vessels to cross.

Defensive Mining: A minefield laid in international waters or international straits with the declared intention of controlling shipping in defence of sea lines of communications (primary maritime routes between ports, used for trade, logistics and naval forces).

Degaussing: The process whereby a vessel's magnetic field is reduced by the use of electromagnetic coils, permanent magnets or other means.

Depressor: A device fitted to a sweep which when towed submerges and planes at a predetermined depth without sideways displacement (See 'Kite').

Dip Needle: In naval mine warfare the device within a firing system which responds to a change in the magnitude of the vertical component of the total magnetic field.

Dormant Mine: A mine whose firing system is, by design, prevented temporarily from operating thus preventing actuation.

Drifting Mine: A buoyant or neutrally buoyant mine that is not tethered to the seabed, intentionally laid to be free to move under the influence of wind, waves, current or tide.

Explosive Ordnance Disposal: The detection, identification, onsite evaluation, rendering safe, recovery and final disposal of unexploded explosive ordnance.

Floating Mine: A mine visible on the surface. Whenever possible it should be more exactly defined by the term, Drifting Mine, Free Mine or Watching Mine.

316   Appendix B

Flooder or Flooder Cutter: A device fitted to a buoyant mine which, on operation after a preset time, floods the mine case and causes it to sink to the bottom.

Grapnel: A device fitted to a mine mooring designed to grapple the sweep wire.

Ground Mine: A mine that is negatively buoyant; rests on, or can become buried in, the sea bed and is held there by its own mass (See Bottom Mine and Mine).

Photo Appendix B-2

Royal Naval Volunteer Reserve officer defuzing a parachute land mine.
Collection of Rob Hoole

Horn: A projection from the mine shell of some contact mines which, when broken, or bent by contact, causes the mine to fire.

Influence Mine: A mine actuated by the effect of a target on some physical condition in the vicinity of the mine or on radiations emanating from the mine.

Influence Sweep: A sweep designed to produce influence(s) to actuate mines.

Kite: A device which, when towed, submerges and planes at a predetermined depth without sideways displacement (see Depressor).

Live Mine: A mine with an explosive filling and a means of firing the explosive charge.

Magnetic Mine: A mine with a magnetic influence circuit which responds to the magnetic field of a ship, submarine or sweep.

Married Unit Failure: A moored mine laying on the seabed connected to its sinker from which it has failed to release owing to a defective mechanism.

Mechanical Sweep: Any sweep used with the object of physically contacting the mine or its appendages.

Mine: An explosive device laid in the water with the intention of damaging or sinking vessels or deterring them from entering an area. The term does not include devices attached to the bottoms of vessels or to harbour installations by personnel operating underwater nor does it include devices which explode immediately on expiration of a predetermined time after laying.

Mine Countermeasures (MCM): All methods for preventing or reducing damage or danger from mines.

Mine Disposal: The process of rendering safe, neutralising, recovering, removing or countermining mines.

Minefield: A number of mines laid, or declared to be laid, in a maritime area.

Minesweeping: The technique of countering mines by minesweeping systems using mechanical, explosive or influence gear, which physically removes, destroys or actuates the mine.

Minewarfare: The strategic and tactical use of mines and their countermeasures.

Mining: The strategic and/or tactical use of sea mines.

Mixed Minefield: A minefield containing mines of various types, firing systems, sensitivities, arming delays and ship counter settings.

Moored Mine: A mine of positive buoyancy held below the surface by a mooring attached to a sinker or anchor on the bottom (See Buoyant Mine).

Obstructor: A device laid with the sole object of obstructing or damaging mechanical minesweeping equipment.

Offensive Mining: A minefield laid in enemy territorial waters or waters under enemy control.

Oropesa Sweep: A form of mechanical sweep towed by a single ship.

Otter: A device used in minesweeps which, when towed, displaces itself sideways to a predetermined distance.

Poised Mine: A mine which is ready to detonate at the next actuation.

Pressure Mine: A mine whose circuit responds to the hydrodynamic pressure signature of a target.

Protective Mining: A minefield laid in friendly territorial waters to protect ports, harbours, anchorages, coasts and coastal routes.

Render Safe Procedure (RSP): The action to make a mine inoperative by direct interference with its firing system or explosive train.

Ship Counter: A device in a mine which prevents the mine from detonating until a pre-set number of actuations has taken place.

Ship Influence: The electro-magnetic, acoustic, pressure or other effects of a vessel, or a minesweep simulating a vessel, which is detectable by a mine or other sensing devices.

Sinker: A heavy weight to which a mine is moored (See Anchor).

Snagline Mine: A contact mine with a buoyant line attached to one of the horns or switches which may be caught up and pulled by the hull or propellers of a ship.

Snagline Sweep: Mechanical MCM gear especially fitted to counter Snagline Mines.

Sprocket: An anti-sweep device included in a mine mooring to allow a sweep wire to pass through the mooring without parting the mine from its sinker.

Steriliser: A device included in mines to render the mine permanently inoperative on expiration of a predetermined time after laying.

Swept Channel: The whole or part of a route or a path which has been swept, the width of the channel being specified.

Swept Path: The width of the lane swept down to the sweep depth.

Tactical Mining: Mining conducted in support of a limited military objective generally in a specified area of immediate tactical interest.

Team Sweep: Two or more sweepers linked together by a mechanical sweep.

Watching Mine: A mine secured to its mooring but showing on the surface.

Appendix C: HMS *Adventure* and *Blanche* Casualties

HMS *Adventure* Casualties (Killed)

Alexander Antliffe	Act/Chief Petty Officer
A. R. Butliff	Petty Officer
Charles Gustavus Carlson	Able Seaman
Edward Thomas Clark	Petty Officer Writer
Roy Tyrer Church	Electrical Artificer 1st Class
William Stanley Cock	Able Seaman
Reginald George Ellis	Chief Electrical Artificer
Thomas Augustine Exall	Stoker 1st Class
Ernest G. Fouracre	Joiner
Harold Gough	Able Seaman
Thomas Gray	Chief Cook
Francis Maurice Guilfoyle	Able Seaman
Samuel Hay	Able Seaman
Henry Hugh Horrell	Chief Stoker
Kenneth John Howes	Chief Petty Officer Cook
Samuel McGraham Lowe	Shipwright 3rd Class
George Edgar Merrick	Seaman
William Thomas Passmore	Engine Room Artificer 1st Class
William Charles James Prater	Shipwright 2nd Class
Thomas William Henry Prynne	Seaman
Harry Walter Purssey	Able Seaman
John Risdon	Petty Officer Cook
Henry Victor Salway	Chief Stoker
Joseph Samuel Ronald Wood	Canteen Manager
Thomas Charles Yandell	Shipwright 1st Class

HMS *Blanche* Casualties (Missing, Presumed Killed)

William George Brown	Leading Seaman
	Able Seaman

Bibliography

Brett-James, Antony. *Ball of Fire: The Fifth Indian Division in the Second World War*. Aldershot, UK: Gale & Polden Ltd., 1951.

British Admiralty. *British Minesweeping, Technical Staff Monographs 1939-1945*. London: British Admiralty, 1950.

British Admiralty. *British Mining Operations 1939-1945: Vol 1*. London: Naval Staff History, 1973.

Brown, David. *Battle Summary No. 39, Operation "Neptune" Landings in Normandy June, 1944*. London: HMSO, 1994.

—*The Royal Navy and the Mediterranean: Vol. II: November 1940-December 1941*. London: Routledge, 2002.

Brown, David K. *Nelson to Vanguard: Warship Design and Development 1923-1945*. Annapolis, Md: Naval Institute Press, 2000.

Bruhn, David D. *Ingram's Fourth Fleet: U.S. and Royal Navy Operations Against German Runners, Raiders, and Submarines in the South Atlantic in World War II*. Berwyn Heights, Md: Heritage Books, 2017.

—*We Are Sinking, Send Help! The U.S. Navy's Tugs and Salvage Ships in the African, European, and Mediterranean Theaters in World War II*. Berwyn Heights, Md: Heritage Books, 2015.

—*Wooden Ships and Iron Men: The U.S. Navy's Coastal and Motor Minesweepers, 1941-1953*. Westminster, Md: Heritage Books, 2009.

Bruhn, David D. and Rob Hoole, *Home Waters: Royal Navy, Royal Canadian Navy, and U.S. Navy Mine Forces Battling U-Boats in World War I*. Berwyn Heights, Md: Heritage Books, 2018

—*Nightraiders: U.S. Navy, Royal Navy, Royal Australian Navy, and Royal Netherlands Navy Mine Forces Battling the Japanese in the Pacific in World War II*. Berwyn Heights, Md: Heritage Books, 2018.

Churchill, Winston. *Their Finest Hour. The Second World War, Vol II*. Boston: Houghton Mifflin, 1949.

Cooper, Bryan. *The E-Boat Threat*. Barnsley, UK: Pen & Sword, 2015.

Cowie, J. S. *Mines, Minelayers and Minelaying*. London: Oxford University Press, 1949.

Dönitz, Karl. *The Conduct of the War at Sea*. Washington, DC: Division of Naval Intelligence, 1946.

Gray, Edwyn. *Captains of War: They Fought Beneath the Sea*. London: Leo Cooper Ltd., 1988.

Groner, Eric. *Die deutschen Kriegsschiffe, 1815-1945 Vol. I*. Munchen: J. F.

Lehmanns Verlag, 1966.

Haar, Geirr H. *No Room for Mistakes: British and Allied Submarine Warfare 1939-1940*. Barnsley, UK: Seaforth Publishing, Pen & Sword Books Ltd., 2015.

Hartman, Gregory K. *Mine Warfare History and Technology*. Silver Springs, Md: Naval Surface Weapons Center White Oaks Laboratory, 1975.

Henry, Chris. *Depth Charge: Royal Navy Mines, Depth Charges & Underwater Weapons 1914-1945*. Barnsley, UK: Pen & Sword, 2005.

Hood, Jean. *Submarine: An anthology of firsthand accounts of the war under the sea, 1939-45*. London: Conway Maritime Press, 2009.

Humphreys, Roy. *Hellfire Corner*. Stroud, UK: Sutton Publishing, 1998.

Jane, Fred T., *Jane's Fighting Ships 1939*. London: Macmillan, 1939.

Johnson, Ellis A. and David A. Katcher, *Mines Against Japan*. Washington, DC: Government Printing Office, 1973.

Kennedy, David M. *The Library of Congress World War II Companion*. New York: Simon & Schuster, 2007.

Lott, Arnold S. *Most Dangerous Sea*. Annapolis, Md: U. S. Naval Institute, 1959.

Macpherson, Ken and Ron Barrie. *The Ships of Canada's Naval Forces 1910–2002*. St. Catharines, Ontario: Vanwell, 2002.

McGovern, Terrance and Bolling Smith. *American Coastal Defenses 1885–1950*. Oxford, UK: Osprey Publishing, 2006.

Morison, Samuel Eliot. *The Two-Ocean War*. Boston: Little, Brown, 1963.

Poland, Nicho. *The Torpedomen – HMS Vernon's Story 1872-1986*. Great Britain, Self-Published, 1993.

Roskill, S. W. *History of the Second World War United Kingdom Military Series, War at Sea 1939-1945 Vol. I, The Defensive*. London, HMSO, 1954.

Turner, Mike and Hector Donohue, *Australian Minesweepers at War*. Canberra, Australia: Sea Power Centre – Australia, 2018.

Williams, Jack. *They Led the Way: The Fleet Minesweepers at Normandy – June 1944*. Blackpool, UK: Oropesa, 1994.

Notes

PREFACE NOTES:
[1] Cowie, *Mines, Minelayers and Minelaying*, 166.
[2] Hartman, *Mine Warfare History and Technology*, NSWC/WOL/TR 75-88, Figure 13; Lott, *Most Dangerous Sea*, 264-265; British Admiralty, *British Mining Operations 1939-1945 Vol. 1*, 761.
[3] "Bi-planes smash Italian Fleet at Taranto" (http://ww2today.com/11th-november-1940-italian-fleet-attacked-in-taranto-harbour: accessed 13 April 2018).
[4] "WWII Weapons: The Fairey Swordfish Torpedo Plane a.k.a. Stringbag" (http://warfarehistorynetwork.com/daily/wwii/wwii-weapons-the-fairey-swordfish-torpedo-plane-a-k-a-stringbag/: accessed 26 June 2018).
[5] Ibid.
[6] Ibid.
[7] Ibid.
[8] Cowie, *Mines, Minelayers and Minelaying*, 164-165.
[9] Ibid.
[10] Bruhn and Hoole, *Home Waters: Royal Navy, Royal Canadian Navy, and U.S. Navy Mine Forces Battling U-Boats in World War I*, xxv.
[11] "Abzeichen Für Blockadebrecher / Kriegsmarine Badge for Blockade Runners" (https://www.feldgrau.com/WW2-German-Kriegsmarine-Badge-For-Blockade-Runners: accessed 8 July 2018).
[12] Ibid.
[13] Dwight Messimer correspondence of 4 July 2018.
[14] Groner, *Die deutschen Kriegsschiffe, 1815-1945 Vol. I*, 108-209.
[15] Ibid.
[16] Turner and Donohue, *Australian Minesweepers at War*, xvii.
[17] British Admiralty, *British Minesweeping, Technical Staff Monographs 1939-1945*, 10.
[18] Ibid, 11.
[19] "RNPS Association" (http://rnpsa.co.uk/cms/?History); "Sparrow's Nest, Lowestoft, UK" (https://productforums.google.com/forum/#!msg/gec-history-illustrated-moderated/mtvvUuqzAZs/O0Sjenfm_bEJ: accessed 10 April 2018).
[20] "The Royal Naval Patrol Service at war" (http://www.harry-tates.org.uk/history1.htm: accessed 8 July 2018).
[21] Ibid.
[22] U.S. National Archives Ref PREM 3/314/5
[23] "Invasion of Norway, Battles of Narvik, Blitzkrieg on Western Europe, Dunkirk Evacuation Starts" (http://www.naval-history.net/WW2RN04-194004.htm: accessed 10 April 2018).

[24] "RNPS Association" (http://rnpsa.co.uk/cms/?History: accessed 10 April 2018).
[25] Turner and Donohue, *Australian Minesweepers at War*, xvii, 92.
[26] Cowie, *Mines, Minelayers and Minelaying*, 164-165.
[27] Ibid, 205-206

CHAPTER 1 NOTES:

[1] "Churchill's Sinking of the French Fleet (July 3, 1940)" (https://www.scottmanning.com/content/churchills-sinking-of-the-french-fleet-july-3-1940/: accessed 19 April 2018).
[2] "*Abdiel* class" (https://uboat.net/allies/warships/class/77.html: accessed 3 May 2018).
[3] Rear Adm. Robert Kirk Dickson, DSO, RN, former commanding officer of the minelayer HMS *Manxman*, telling of her exploits in "Ship in Disguise" from the "Now it can be Told" series of radio broadcasts made during his time as Chief of Naval Information 1944-1946 (http://myweb.tiscali.co.uk/hmsmanxman/life/robert_dixon/dixon.htm: accessed 19 April 2018).
[4] Ibid.
[5] Ibid.
[6] Ibid.
[7] Ibid.
[8] Ibid.
[9] Ibid.
[10] Ibid.
[11] Ibid.
[12] Ibid.
[13] Ibid.
[14] Ibid.
[15] Ibid.
[16] Dickson, "Ship in Disguise" from the "Now it can be Told" series of radio broadcasts; "Royal Navy Minelaying Operations" (https://www.naval-history.net/xGM-Ops-Minelaying.htm: accessed 19 May 2018).
[17] Dickson, "Ship in Disguise" from the "Now it can be Told" series of radio broadcasts."
[18] Ibid.
[19] Ibid.
[20] British Admiralty, *British Mining Operations 1939-1945: Vol 1*, 576.
[21] "Robert Kirk Dickson DSO, RN" (https://uboat.net/allies/commanders/733.html: accessed 4 May 2018).

CHAPTER 2 NOTES:

[1] "Royal Navy Minelaying Operations" (https://www.naval-history.net/xGM-Ops-Minelaying.htm: accessed 19 May 2018).

[2] Fraser G. Machaffie, "The Short Life and Sudden Death of HMS *Princess Victoria*," *Sea Breezes*, 18 January 2017 (USAhttp://www.seabreezes.co.im/index.php/features/ships/2370-the-short-life-and-sudden-death-of-hms-princess-victoria: accessed 4 January 2018).
[3] Ibid.
[4] Ibid.
[5] "Royal Navy Minelaying Operations" (https://www.naval-history.net/xGM-Ops-Minelaying.htm: accessed 19 May 2018); "Commonwealth & Dominion Line Port Line" (https://web.archive.org/web/20120708035811/http://www.red-duster.co.uk/PORT7.htm: accessed 4 January 2018).
[6] Cowie, *Mines, Minelayers and Minelaying*, 205-206; "*Adventure* class" (https://uboat.net/allies/warships/class/76.html: accessed 3 January 2018).
[7] Cowie, *Mines, Minelayers and Minelaying*, 205-206; British Admiralty, *British Mining Operations, 1939–1945, Vol. 1*, 665-678; "Auxiliary Minelayers Class" (https://uboat.net/allies/warships/class.html?ID=621&navy=HMS); "M 1 Class" (https://uboat.net/allies/warships/class/624.html: accessed 3 January 2018).
[8] "*Abdiel* class" (https://uboat.net/allies/warships/class/77.html: accessed 3 May 2018.
[9] Cowie, *Mines, Minelayers and Minelaying*, 205-206; "*Abdiel* class."
[10] "*E* class" (https://uboat.net/allies/warships/class.html?ID=13: accessed 3 January 2018).
[11] "*I* class" (https://uboat.net/allies/warships/class.html?ID=17: accessed 3 January 2018).
[12] Cowie, *Mines, Minelayers and Minelaying*, 205-206; "*E* class."
[13] Cowie, *Mines, Minelayers and Minelaying*, 205-206; "*I* class."
[14] "Royal Navy Minelaying Operations" (https://www.naval-history.net/xGM-Ops-Minelaying.htm: accessed 19 May 2018); "*O* class" (https://www.uboat.net/allies/warships/class/26.html: accessed 4 January 2018); British Admiralty, *British Mining Operations, 1939–1945, Vol. 1*, 116-117, 221-222, and 453-454 cites mines laid by O-class minelayers in Operations BS86, 14 October 1942; CH, 8 April 1945; and Trammel, 22 April 1945.
[15] "Admiralty *S* class" (https://www.uboat.net/allies/warships/class/20.html: accessed 3 January 2018).
[16] "Royal Navy Minelaying Operations" (https://www.naval-history.net/xGM-Ops-Minelaying.htm: accessed 19 May 2018).
[17] Cowie, *Mines, Minelayers and Minelaying*, 205-206; "Plover class" https://uboat.net/allies/warships/class.html?ID=622&navy=HMS; "*Plover* Class Coastal Minelayer Leader" (http://gb-navy-ww2.narod.ru/HTM-ML-plover.html: accessed 1 January 2018).
[18] "*Corncrake* Class Coastal Minelayers" (http://gb-navy-ww2.narod.ru/HTM-ML-corncrake.html: accessed 1 January 2018).
[19] Jane, *Jane's Fighting Ships 1939*, 98.

[20] *"Linnet* class" (https://uboat.net/allies/warships/class.html?ID=623&navy=HMS: accessed 4 January 2018).
[21] "Miner Class Coastal Minelayer" (http://gb-navy-ww2.narod.ru/HTML-miner.html: accessed 1 January 2018).
[22] Turner and Donohue, *Australian Minesweepers at War*, 79, 89; "*Atreus* 1911 HMS - Minelayer Base Ship" (http://forums.clydemaritime.co.uk/viewtopic.php?t=25463); "Naval Trawlers" (https://www.battleships-cruisers.co.uk/naval_trawlers.htm: accessed 4 March 2018).
[23] Rob Hoole correspondence, 9 January 2018.
[24] Cowie, *Mines, Minelayers and Minelaying*, 206; British Admiralty, *British Mining Operations, 1939–1945, Vol. 1*, 789; "Royal Navy Coastal Forces" (http://www.unithistories.com/units_british/RN_CoastalForces.html: accessed 6 January 2018).
[25] Cowie, *Mines, Minelayers and Minelaying*, 205-206; "*Porpoise* class" (https://uboat.net/allies/warships/class/49.html); "Chronological Sequence of Submarine Minelaying Operations 1939-45" (http://www.naval-history.net/xGM-Ops-Minelaying.htm: accessed 6 January 2018).
[26] British Admiralty, *British Mining Operations, 1939–1945, Vol. 1*, 788; "T class" (https://uboat.net/allies/warships/class.html?ID=53&navy=HMS: accessed 6 January 2018).
[27] British Admiralty, *British Mining Operations, 1939–1945, Vol. 1*, 788; "S class" (https://uboat.net/allies/warships/class.html?ID=52&navy=HMS: accessed 6 January 2018).
[28] Cowie, *Mines, Minelayers and Minelaying*, 205-206; British Admiralty, *British Mining Operations, 1939–1945, Vol. 1*, 790.
[29] Johnson and Katcher, *Mines Against Japan*, 35.

CHAPTER 3 NOTES:
[1] "Royal Navy Minelaying Operations" (https://www.naval-history.net/xGM-Ops-Minelaying.htm: accessed 19 May 2018).
[2] Cowie, *Mines, Minelayers and Minelaying*, 122.
[3] Ibid, 122-123.
[4] Cowie, *Mines, Minelayers and Minelaying*, 123; Roskill, *History of the Second World War United Kingdom Military Series, War at Sea 1939-1945 Vol. I*, 96.
[5] "HMS *Express* (H 61), later HMCS *Gatineau*" (http://www.naval-history.net/xGM-Chrono-10DD-21E-Express.htm); "HMS *Express*" (https://ww2db.com/ship_spec.php?ship_id=458: accessed 4 May 2018).
[6] Cowie, *Mines, Minelayers and Minelaying*, 139-140.
[7] Ibid.
[8] Dönitz, *The Conduct of the War at Sea*, 3-4
[9] "U-boat types" (https://uboat.net/types/: accessed 19 May 2018).
[10] Dönitz, *The Conduct of the War at Sea*, 2.
[11] Ibid, 4.

[12] "*U-12*" (https://uboat.net/boats/u12.htm); "German U-Boat Casualties in World War Two" (https://www.history.navy.mil/research/library/online-reading-room/title-list-alphabetically/u/united-states-submarine-losses/german-u-boat-casualties-in-world-war-two.html: accessed 5 May 2018).
[13] "*U-40*" (https://uboat.net/boats/u40.htm: accessed 6 May 2018).
[14] "*U-16*" (https://uboat.net/boats/u16.htm: accessed 6 May 2018).
[15] Ibid.
[16] "*U-16*"; "Fate of the crew of HMS *Cayton Wyke*" (http://www.go2war2.nl/artikel/2442/Fate-of-the-crew-of-HMS-Cayton-Wyke.htm: accessed 7 May 2018).
[17] Ut supra.
[18] Cowie, *Mines, Minelayers and Minelaying*, 123; Roskill, *War at Sea 1939-1945 Vol. I*, 96.
[19] "HMS *Express* (H 61), later HMCS *Gatineau*" (http://www.naval-history.net/xGM-Chrono-10DD-21E-Express.htm); Bruhn and Hoole, *Home Waters*, 22, 244.
[20] Bruhn and Hoole, *Home Waters*, 22.
[21] "HMS *Esk* (H 15) - E-class Destroyer" (http://www.naval-history.net/xGM-Chrono-10DD-21E-Esk.htm: accessed 5 May 2018).
[22] "HMS *Esk* (H 15) - E-class Destroyer"; "HMS *Ivanhoe* (D 16) - I-class Destroyer" (http://www.naval-history.net/xGM-Chrono-10DD-29I-Ivanhoe.htm: accessed 5 May 2018)
[23] "HMS *Icarus* (D 03) - I-class Destroyer" (https://www.naval-history.net/xGM-Chrono-10DD-29I-HMS_Icarus.htm); "HMS *Impulsive* (D 11) - I-class Destroyer" (https://www.naval-history.net/xGM-Chrono-10DD-29I-HMS_Impulsive.htm: accessed 5 May 2018).
[24] "HMS *Ivanhoe* (D 16) - I-class Destroyer."
[25] "*U-54*" (https://www.uboat.net/boats/u54.htm: accessed 5 May 2018).
[26] "Unternehmen Wikinger" (http://www.german-navy.de/kriegsmarine/articles/feature4.html: accessed 7 May 2018).
[27] Ibid.
[28] Ibid.
[29] Ibid.
[30] Ibid.
[31] Ibid.
[32] Ibid.
[33] "Unternehmen Wikinger"; "Campaign Summaries of World War II, Actions involving Mine Warfare and Mine Vessels, Part 1 of 2 - 1939-42" (https://www.naval-history.net/WW2CampaignsMineWarfare1.htm: accessed 10 May 2018).

CHAPTER 4 NOTES:
[1] Roskill, *War at Sea 1939-1945 Vol. I*, 98.
[2] Ibid.
[3] Ibid.

[4] Ibid, 98-99.
[5] Ibid, 99.
[6] Ibid, 99-100.
[7] Bruhn and Hoole, *Home Waters*, 307-309; Henry, *Depth Charge: Royal Navy Mines, Depth Charges & Underwater Weapons 1914-1945*, 142.
[8] Henry, *Depth Charge*, 143.
[9] Ibid.
[10] Roskill, *War at Sea 1939-1945 Vol. I*, 100; Dönitz, *The Conduct of the War at Sea*, 2.
[11] Roskill, *War at Sea 1939-1945 Vol. I*, 100; Dönitz, *The Conduct of the War at Sea*, 4-5.
[12] Roskill, *War at Sea 1939-1945 Vol. I*, 100.
[13] Supplementary Report by Capt. Arthur Robert Halfhide, RN, *Adventure*'s commanding officer, on 21 November 1939 (http://www.carlson.eclipse.co.uk/charles-voyage/adventure-report.html: accessed 23 May 2018).
[14] "HMS *Adventure* (M 23)" (https://uboat.net/allies/warships/ship/1267.html: accessed 20 May 2018).
[15] Report by Capt. Arthur Robert Halfhide, RN, *Adventure*'s commanding officer, (179/134) dated 16 November 1939.
[16] Ibid.
[17] Ibid.
[18] Ibid.
[19] Ibid.
[20] Ibid.
[21] Ibid.
[22] Ibid.
[23] Ibid.
[24] Ibid.
[25] "The Men who Died" (http://www.carlson.eclipse.co.uk/charles-voyage/casualty-list.html); "HMS *Blanche* (H 47) – B-class Destroyer" (http://www.naval-history.net/xGM-Chrono-10DD-15B-Blanche); "Casualty Lists of the Royal Navy and Dominion Navies, World War 2" (https://www.naval-history.net/xDKCas1003-Intro.htm#WW2: all accessed 26 May 2018).
[26] "HMS *Adventure* - Cruiser Minelayer" (http://www.naval-history.net/xGM-Chrono-07ML-Adventure.htm: accessed 26 May 2018).
[27] Roskill, *War at Sea 1939-1945 Vol. I*, 100-101.
[28] Ibid, 101-102.
[29] Ibid, 102.

CHAPTER 5 NOTES:
[1] British Admiralty, *British Mining Operations 1939-1945 Vol. 1*, 731.
[2] Ibid, 121, 731-732.
[3] Ibid, 232.
[4] Ibid, 732.

[5] Ibid.
[6] "War Mystery Solved," *Western Morning News*, 22 December 2006.
[7] Ibid.
[8] Ibid.
[9] British Admiralty, *British Mining Operations 1939-1945 Vol. 1*, 732.
[10] British Admiralty, *British Mining Operations 1939-1945 Vol. 1*, 732; Roskill, *War at Sea 1939-1945 Vol. I*, 123.
[11] Cowie, *Mines, Minelayers and Minelaying*, 140.
[12] Ibid.
[13] "HMS *Renown* - *Renown*-class 15in gun Battlecruiser" (http://www.naval-history.net/xGM-Chrono-02BC-Renown.htm: accessed 31 May 2018).
[14] "HMS *Renown* - *Renown*-class 15in gun Battlecruiser"; "HMS *Inglefield* (D 02) - I-class Flotilla Leader" (http://www.naval-history.net/xGM-Chrono-10DD-28I-HMS_Inglefield.htm); "HMS *Hero* (H 99), later HMCS *Chaudiere* - H-class Destroyer" (http://www.naval-history.net/xGM-Chrono-10DD-27H-Hero.htm: accessed 31 May 2018).
[15] "HMS *Renown* - *Renown*-class 15in gun Battlecruiser."
[16] Cowie, *Mines, Minelayers and Minelaying*, 144.
[17] Ibid, 144.
[18] "Now It Can Be Told! - Last Glorious Fight of the *Glowworm*," *The War Illustrated, Volume 9, No. 213*, 236-237, 17 August 1945; "The Comprehensive Guide to the Victoria & George Cross" (http://www.vconline.org.uk/home/4585899435: accessed 1 June 2018).
[19] Ut supra.
[20] Ut supra.
[21] "Now It Can Be Told! - Last Glorious Fight of the *Glowworm*."
[22] Ut supra.
[23] Ut supra.
[24] Ut supra.
[25] Ut supra.
[26] "The Comprehensive Guide to the Victoria & George Cross."
[27] Ibid.
[28] "Narvik Campaign, 9 April-7 June 1940" (http://www.historyofwar.org/articles/campaign_narvik_1940.html: accessed 1 June 2018).
[29] Ibid.
[30] Ibid.
[31] Ibid.
[32] Roskill, *War at Sea 1939-1945 Vol. I*, 123-125.
[33] Cowie, *Mines, Minelayers and Minelaying*, 144-145.
[34] Ibid, 145.
[35] Roskill, *War at Sea 1939-1945 Vol. I*, 125.

CHAPTER 6 NOTES:

[1] "*Willem van der Zaan*" (http://www.netherlandsnavy.nl/Zaan.html); "*Willem van der Zaan* minelayer (1939)"

(http://www.navypedia.org/ships/netherlands/nl_ms_van_der_zaan.htm: accessed 24 July 2018).
[2] *"Willem van der Zaan'."*
[3] *"Willem van der Zaan* History"
(http://www.netherlandsnavy.nl/Zaan_his.htm: accessed 24 July 2018).
[4] "Dutch Navy WW2" (http://www.naval-encyclopedia.com/ww2/netherlands/dutch-navy-ww2/: accessed 25 July 2018).
[5] Ibid
[6] "Dutch Naval forces" (https://uboat.net/allies/ships/dutch.htm); "Ships by name" (http://www.netherlandsnavy.nl/: accessed 25 July 2018).
[7] *"Willem van der Zaan* History."
[8] *"Douwe Aukes*-class"; *"Johan Maurits van Nassau* (I)"; *"G 13"* (http://www.netherlandsnavy.nl: accessed 26 July 2018).
[9] *"Hydra*-class minelayers"; *"Douwe Aukes*-class" (http://www.netherlandsnavy.nl/: accessed 26 July 2018).
[10] Ut supra.
[11] *"Douwe Aukes*-class."
[12] "Minelayer *Nautilus*" (http://www.netherlandsnavy.nl/: accessed 26 July 2018).
[13] Ibid.
[14] *"Franklin K. Lane"* (https://uboat.net/allies/merchants/ships/1766.html: accessed 26 July 2018); *"Jan van Brakel"* (http://www.netherlandsnavy.nl/: accessed 26 July 2018).
[15] *"Willem van der Zaan* History."
[16] Ibid.
[17] Ibid.
[18] *"Willem van der Zaan* History"; "Napoleon's Exile on St Helena" by Lara Jacobs (https://www.historic-uk.com/HistoryUK/HistoryofBritain/Napoleons-Exile-on-St-Helena/: accessed 26 July 2018).
[19] Ibid.
[20] *"Willem van der Zaan* History."

CHAPTER 7 NOTES:
[1] British Admiralty, *British Mining Operations 1939-1945: Vol 1*, 733.
[2] Ibid.
[3] Ibid.
[4] "Sinking of HMS *Esk*," account from the diary of Herbert Vaughan, written whilst in Marlag & Milag Nord POW camp (http://www.scarboroughsmaritimeheritage.org.uk/article.php?article=150: accessed 6 June 2018).
[5] Churchill, *Their Finest Hour. The Second World War, Vol II*, 115.
[6] "Allied Warships in World War Two" (https://uboat.net/allies/warships/: accessed 15 July 2018).
[7] Gray, *Captains of War: They Fought Beneath the Sea.*

[8] "British Submarine HMS *Seal*" by William H. Langenberg (http://www.historynet.com/british-submarine-hms-seal.htm: accessed 15 July 2018).
[9] Ibid.
[10] Ibid.
[11] "U-B (former HMS *Seal*)" (https://uboat.net/boats/foreign_ub.htm: accessed 15 May 2018).
[12] "British Submarine HMS *Seal*"; Gray, *Captains of War: They Fought Beneath the Sea.*
[13] "British Submarine HMS *Seal*"; Haarr; *No Room for Mistakes: British and Allied Submarine Warfare 1939-1940*; "U-B (former HMS *Seal*)."
[14] Ut supra.
[15] "British Submarine HMS *Seal*."
[16] Ibid.
[17] Ibid.
[18] Ibid.
[19] Ibid.
[20] Ibid.
[21] "British Submarine HMS *Seal*"; Haarr; *No Room for Mistakes: British and Allied Submarine Warfare 1939-1940*; Gray, *Captains of War: They Fought Beneath the Sea.*
[22] Ut supra.
[23] Ut supra.
[24] "British Submarine HMS *Seal*."
[25] Ibid.
[26] "British Submarine HMS *Seal*"; "U-B (former HMS *Seal*)"; "Casualty Lists of the Royal Navy and Dominion Navies, World War 2" (https://www.naval-history.net/xDKCas1003-Intro.htm#WW2: accessed 26 May 2018).
[27] "British Submarine HMS *Seal*"; "U-B (former HMS *Seal*)."
[28] Ut supra.
[29] Ut supra.
[30] "British Submarine HMS *Seal*."
[31] Ibid.
[32] Machaffie, "The Short Life and Sudden Death of HMS *Princess Victoria*."
[33] Cowie, *Mines, Minelayers and Minelaying*, 145-146; "HMS *Express* (H 61)" (http://www.naval-history.net/xGM-Chrono-10DD-21E-Express.htm); "HMS *Esk* (H 15)" (https://www.naval-history.net/xGM-Chrono-10DD-21E-HMS_Esk.htm); "HMS *Intrepid* (D 10)" (http://www.naval-history.net/xGM-Chrono-10DD-29I-HMS_Intrepid.htm: accessed 19 July 2018); War Cabinet Weekly Resume (No. 37) of the Naval, Military and Air Situation from 12 noon May 9th to 12 noon May 16th, 1940, 17 May 1940.
[34] "HMS *Express* (H 61)"; "HMS *Esk* (H 15)"; "HMS *Intrepid* (D 10)"; "HMS *Ivanhoe* (D 16)" (http://www.naval-history.net/xGM-Chrono-10DD-29I-Ivanhoe.htm); "HMS *Impulsive* (D 11)" (https://www.naval-history.net/xGM-Chrono-10DD-29I-HMS_Impulsive.htm: accessed 19 July 2018).

[35] Machaffie, "The Short Life and Sudden Death of HMS *Princess Victoria*."
[36] "HMS *Princess Victoria* (M 03)" (https://uboat.net/allies/warships/ship/13416.html); "'Lest We Forget' HMS *Princess Victoria* and war graves in North Norfolk Churchyards" by Richard Jefferson (http://www.bahs.uk/GH-Files/GH8/GH8-Article5.pdf: accessed 20 July 2018).
[37] Ut supra.
[38] "'Lest We Forget' HMS *Princess Victoria* and war graves in North Norfolk Churchyards"; "HMS *Grafton* (H 89) - G-class Destroyer" (http://www.naval-history.net/xGM-Chrono-10DD-25G-HMS_Grafton.htm: accessed 20 July 2018); "HMS *Express* (H 61)"; "HMS *Ivanhoe* (D 16)."
[39] "'Lest We Forge' HMS *Princess Victoria* and war graves in North Norfolk Churchyards."
[40] Ibid.
[41] "Casualty Lists of the Royal Navy and Dominion Navies, World War 2" (https://www.naval-history.net/xDKCas1003-Intro.htm#WW2: accessed 26 May 2018).
[42] "Sunken Royal Navy submarine HMS *Narwhal* found after 77 years" (https://navaltoday.com/2017/10/04/sunken-royal-navy-submarine-hms-narwhal-found-after-77-years/); "HMS *Narwhal* (N 45)" (https://www.uboat.net/allies/warships/ship/3414.html: accessed 20 July 2018).
[43] Ut supra.
[44] Ut supra.
[45] "Sinking of HMS *Esk*."
[46] "HMS *Esk* (H 15)" (https://uboat.net/allies/warships/ship/4381.html); "HMS *Esk* (H 15) - E-class Destroyer"; "HMS *Express* (H 61), later HMCS *Gatineau* - E-class Destroyer" (http://www.naval-history.net/xGM-Chrono-10DD-21E-Express.htm: accessed 4 June 2018); "Sinking of HMS *Esk*."
[47] Ut supra.
[48] Ut supra.
[49] "Sinking of HMS *Esk*."
[50] "HMS *Ivanhoe* (D 16) - I-class Destroyer" (http://www.naval-history.net/xGM-Chrono-10DD-29I-Ivanhoe.htm: accessed 5 June 2018).
[51] "HMS *Ivanhoe* (D 16) - I-class Destroyer"; "HMS *Kelvin* (G.37) - K-class Destroyer" (http://www.naval-history.net/xGM-Chrono-10DD-39K-HMS_Kelvin.htm: accessed 5 June 2018).
[52] "HMS *Express* (H 61), later HMCS *Gatineau* - E-class Destroyer."
[53] "Casualty Lists of the Royal Navy and Dominion Navies, World War 2" (https://www.naval-history.net/xDKCas1003-Intro.htm#WW2: accessed 26 May 2018).
[54] "HMS *Express* (H 61), later HMCS *Gatineau* - E-class Destroyer."
[55] Cowie, *Mines, Minelayers and Minelaying*, 132-133; "Royal Navy Minelaying Operations" (https://www.naval-history.net/xGM-Ops-Minelaying.htm: accessed 19 May 2018); "Organisation of the Royal Navy 1939-1945"

(http://www.naval-history.net/xGW-RNOrganisation1939-45.htm: accessed 21 July 2018).
[56] Cowie, *Mines, Minelayers and Minelaying*, 133.
[57] Cowie, *Mines, Minelayers and Minelaying*, 132-133; "Auxiliary minelayers class."
[58] Cowie, *Mines, Minelayers and Minelaying*, 134.
[59] "Royal Navy Minelaying Operations" (https://www.naval-history.net/xGM-Ops-Minelaying.htm: accessed 19 May 2018).

CHAPTER 8 NOTES:
[1] British Admiralty, *British Mining Operations 1939-1945 Vol 1*, 737.
[2] Ibid.
[3] Bruhn and Hoole, *Nightraiders: U.S. Navy, Royal Navy, Royal Australian Navy, and Royal Netherlands Navy Mine Forces Battling the Japanese in the Pacific in World War II*, 222.
[4] "HMS *Rorqual* (74 M) - *Grampus*-class Minelaying Submarine" (https://www.naval-history.net/xGM-Chrono-12SS-05Grampus-HMS_Rorqual.htm: accessed 11 June 2018).
[5] Ibid.
[6] Kennedy, *The Library of Congress World War II Companion*, 80.
[7] "Italy declares war on France and Great Britain" (https://www.history.com/this-day-in-history/italy-declares-war-on-france-and-great-britain: accessed 11 June 2018).
[8] "HMS *Grampus* (N 56)" (https://uboat.net/allies/warships/ship/3413.html: accessed 11 June 2018).
[9] Ibid.
[10] Ibid.
[11] Ibid.
[12] "*Porpoise* class."
[13] "HMS *Rorqual* (N 74)" (https://uboat.net/allies/warships/ship/3415.html: accessed 12 June 2018).
[14] Ibid.
[15] British Admiralty, *British Mining Operations 1939-1945 Vol 1*, 737.
[16] "HMS *Abdiel* - *Abdiel*-class Fast Cruiser Minelayer" (http://www.naval-history.net/xGM-Chrono-07ML-Abdiel.htm; "HMS *Abdiel* (M 39)" (https://uboat.net/allies/warships/ship/3953.html: accessed 12 June 2018).
[17] Ut supra.
[18] "The Battle for Crete" (https://nzhistory.govt.nz/war/the-battle-for-crete/overview: accessed 13 June 2018).
[19] Ibid.
[20] "HMS *Abdiel* - *Abdiel*-class Fast Cruiser Minelayer"; "HMS *Abdiel* (M 39)."
[21] "HMS *Latona* - *Abdiel*-class Fast Cruiser Minelayer" (http://www.naval-history.net/xGM-Chrono-07ML-HMS_Latona.htm: accessed 13 June 2018).

[22] Brown, *The Royal Navy and the Mediterranean: Vol. II: November 1940-December 1941*, 144; "World War II and Postwar Nationalism" (http://countrystudies.us/cyprus/10.htm: accessed 13 June 2018).
[23] "HMS *Abdiel* - *Abdiel*-class Fast Cruiser Minelayer."
[24] "HMS *Latona* - *Abdiel*-class Fast Cruiser Minelayer."
[25] Ibid.
[26] "HMS *Latona* - *Abdiel*-class Fast Cruiser Minelayer"; "HMS *Hero* (H 99), later HMCS *Chaudiere* - H-class Destroyer"; "Casualty Lists of the Royal Navy and Dominion Navies, World War 2" (https://www.naval-history.net/xDKCas1003-Intro.htm#WW2: accessed 26 May 2018).
[27] Brett-James, *Ball of Fire: The Fifth Indian Division in the Second World War*.
[28] "HMS *Abdiel* - *Abdiel*-class Fast Cruiser Minelayer."
[29] Ibid.
[30] British Admiralty, *British Mining Operations 1939-1945 Vol 1*, 737.
[31] "The Supply of Malta 1940-1942, Part 3 of 3" (http://www.naval-history.net/xAH-MaltaSupply03.htm: accessed 14 June 2018).
[32] "HMS *Cachalot* (N 83)" (https://uboat.net/allies/warships/ship/3416.html: accessed 14 June 2018).
[33] "HMS *Rorqual* (N 74)"; "HMS *Rorqual* (74 M) - *Grampus*-class Minelaying Submarine."
[34] "HMS *Rorqual* (N 74)."
[35] "7 August 1941: Axis Fighter Numbers in Sicily Back to Full Strength" (https://maltagc70.wordpress.com/tag/hms-cachalot/: accessed 14 June 2018).
[36] "HMS *Cachalot* (N 83)."
[37] Ibid.
[38] Ibid.
[39] Ibid.
[40] Ibid.
[41] Ibid.
[42] Ibid.
[43] Ibid.
[44] Ibid.
[45] Ibid.
[46] Ibid.
[47] "General Antonio Cantore destroyers (1921-1922)" (http://www.navypedia.org/ships/italy/it_dd_generali.htm: accessed 15 June 2018).
[48] "HMS *Cachalot* (N 83)"; "HMS *Porpoise* (N 14)" (https://www.uboat.net/allies/warships/ship/3412.html): accessed 14 June 2018).
[49] "HMS *Porpoise* (N 14)"; "HMS *Rorqual* (N 74)."
[50] "HMS *Porpoise* (N 14)".
[51] Ibid.

[52] British Admiralty, *British Mining Operations 1939-1945 Vol 1*, 737.
[53] Ibid.

CHAPTER 9 NOTES:
[1] "Tobruk: The lifting of the siege, 75 years ago" (https://anmm.blog/2016/12/09/tobruk-the-lifting-of-the-siege-75-years-ago/: accessed 6 April 2018).
[2] "World War II: North Africa Campaign" (http://www.historynet.com/world-war-ii-north-africa-campaign.htm: accessed 7 April 2018).
[3] Ibid.
[4] Ibid.
[5] "World War II: North Africa Campaign"; "Battle of Taranto" (https://www.navywings.org.uk/heritage/battle-honours/battle-of-taranto/: accessed 8 April 2018).
[6] Interview with Mitsuo Fuchida, 25 February 1964, Donald M. Goldstein Papers, Archives Service Center, University of Pittsburgh.
[7] "World War II: North Africa Campaign"; "Rats of Tobruk 1941" (https://www.awm.gov.au/visit/exhibitions/tobruk/: accessed 7 April 2018).
[8] "Siege of Tobruk" (http://military.wikia.com/wiki/Siege_of_Tobruk: accessed 9 April 2018).
[9] "Rats of Tobruk 1941"; Leah Riches, "The Patrolling War in Tobruk" (https://www.awm.gov.au/sites/default/files/ssvs-riches-2012-tobruk.pdf: accessed 7 April 2018).
[10] "The Tobruk Run" (https://www.navyhistory.org.au/the-tobruk-run/: accessed 9 April 2018).
[11] Ibid.
[12] Ibid.
[13] Ibid.
[14] Bruhn, *Ingram's Fourth Fleet: U.S. and Royal Navy Operations Against German Runners, Raiders, and Submarines in the South Atlantic in World War II*, 66.
[15] Bruhn, *Ingram's Fourth Fleet*, 66; H. K. Kelly, "HMSAS *Southern Maid*," Military History Journal, Vol 1 No 3 - December 1968.
[16] "South African Naval Forces" (http://www.delvillewood.com/navy.htm); "Royal, Dominion and Indian Navy Ships, January 1941 (Part 2 of 2)" (http://www.naval-history.net/xDKWW2-4101-26RNOverseas-Dominion.htm: accessed 8 April 2018).
[17] "Naval Events, February 1941 (Part 1 of 2) Saturday 1st – Friday 14th" (http://www.naval-history.net/xDKWW2-4102-29FEB01.htm); "Actions involving Mine Warfare and Mine Vessels" (http://www.naval-history.net/WW2CampaignsMineWarfare1.htm: accessed 10 April 2018).
[18] "Trawlers and Drifters" (http://www.wildfire3.com/drifters-and-trawlers.html: accessed 9 April 2018).

[19] "Naval Events, February 1941 (Part 1 of 2) Saturday 1st – Friday 14th."
[20] "'A sole survivor and a ship's crest'; the South African Navy's first loss – HMSAS *Southern Floe*" (https://samilhistory.com/2016/02/16/the-hazards-of-minesweeping-mines-the-sinking-of-the-south-african-navys-southern-floe/: accessed 8 April 2018).
[21] Ibid.
[22] Ibid.
[23] Ibid.
[24] "Naval Events, February 1941 (Part 2 of 2) Saturday 15th – Friday 28th" (http://www.naval-history.net/xDKWW2-4102-29FEB02.htm: accessed 9 April 2018).
[25] "HMSAS *Southern Maid*," Military History Journal, Vol 1 No 3, December 1968; "'A sole survivor and a ship's crest'; the South African Navy's first loss – HMSAS *Southern Floe*."
[26] "Chief Stoker Petty Officer Rene Sethren (1920-1967)" (http://delvillewood.com/sethren2.htm; "Naval Events, June 1941 (Part 2 of 2) Sunday 15th – Monday 30th" (http://naval-history.net/xDKWW2-4106-33JUN02.htm: accessed 8 April 2018).
[27] Brown, *The Royal Navy and the Mediterranean: Vol. II: November 1940-December 1941*, 79; "MS Whalers" (https://uboat.net/allies/warships/class.html?ID=603: accessed 9 April 2018).
[28] "Naval Trawlers" (https://www.battleships-cruisers.co.uk/naval_trawlers.htm: accessed 4 March 2018); "MS Whalers."
[29] "The Royal Canadian Navy and Overseas Operations (1939-1945)" (https://www.canada.ca/en/navy/services/history/naval-service-1910-2010/stepping.html: accessed 6 April 2018); "The Last Canadian VC - Robert Hampton Gray" (http://www.vintagewings.ca/VintageNews/Stories/tabid/116/articleType/ArticleView/articleId/34/The-Last-Canadian-VC--Robert-Hampton-Gray.aspx: accessed 30 May 2018).
[30] "27 January 1942" (http://www.etherit.co.uk/1942/01/27.htm: accessed 6 April 2018); "Ruttan, John McDonald" (http://www.nauticapedia.ca/dbase/Query/Biolist3.php?&name=Ruttan,%20John%20McDonald&id=4181&Page=1&input=Ruttan,%20J: accessed 22 August 2018).
[31] "Casualty Lists of the Royal Navy and Dominion Navies, World War 2" (https://www.naval-history.net/xDKCas1003-Intro.htm#WW2: accessed 26 May 2018); "Telegraph from London dated August 30th, 1941" (http://docs.fdrlibrary.marist.edu/psf/box36/t323a03.html: accessed 6 April 2018).
[32] "World War II Awards for RN Minesweeping" (http://www.mcdoa.org.uk/ww_ii_awards_for_rn_minesweeping_M.htm); "Royal Canadian Navy Citations" (http://www.blatherwick.net/royal%20canadian%20navy%20citations/: accessed 7 April 2018).

[33] "Rats of Tobruk 1941" (https://www.awm.gov.au/visit/exhibitions/tobruk/); "World War II: North Africa Campaign" (http://www.historynet.com/world-war-ii-north-africa-campaign.htm: both accessed 12 April 2018).
[34] "Allies surrender at Tobruk, Libya" (https://www.history.com/this-day-in-history/allies-surrender-at-tobruk-libya: accessed 12 April 2018).
[35] Bjørn L. Basberg, "Keynes, Trouton and the Hector Whaling Company. A personal and professional relationship" (https://brage.bibsys.no/xmlui/bitstream/handle/11250/281948/DP%2008.pdf?sequence=1: accessed 8 April 2018).

CHAPTER 10 NOTES:
[1] British Admiralty, *British Mining Operations 1939-1945: Vol 1*, 389.
[2] Rob Hoole correspondence 2 July 2018.
[3] Cooper, *The E-Boat Threat*, 26; Poland, *The Torpedomen – HMS Vernon's Story 1872-1986*, 112.
[4] Cooper, *The E-Boat Threat*, 26.
[5] "Coastal Forces of World War II" (http://www.coastal-forces.org.uk/history.html: accessed 3 July 2018).
[6] "Royal Navy Coastal Forces" (http://www.unithistories.com/units_british/RN_MTBs.html#MTB Flotillas: accessed 27 June 2018).
[7] "Organisation of the Royal Navy 1939-1945" (http://www.naval-history.net/xGW-RNOrganisation1939-45.htm: accessed 21 July 2018).
[8] "Record of Coastal Force Action during World War Two" (http://collections.rmg.co.uk/archive/objects/1112689.html: accessed 30 June 2018).
[9] British Admiralty, *British Mining Operations 1939-1945: Vol 1*, 292-293.
[10] Ibid, 293.
[11] HMS Beehive" (http://www.maritimeheritageeast.org.uk/exhibitions/hms-beehive/felixstowe-museum-hms-beehive); "MTB HMS Beehive" (http://cfv.org.uk/forum/viewtopic.php?t=466: accessed 29 June 2018).
[12] British Admiralty, *British Mining Operations 1939-1945: Vol 1*, 293, 391.
[13] Ibid, 293-294, 391.
[14] Ibid, 300.
[15] Ibid.
[16] Ibid.
[17] Ibid, 301.
[18] Ibid.
[19] Ibid, 302-303.
[20] Ibid.
[21] Ibid, 303-307.
[22] Cowie, *Mines, Minelayers and Minelaying*, 153; British Admiralty, *British Mining Operations 1939-1945: Vol 1*, 391.

[23] British Admiralty, *British Mining Operations 1939-1945: Vol 1*, 391; "Sperrbrecher" (https://www.german-navy.de/kriegsmarine/ships/minehunter/sperrbrecher/index.html: accessed 1 July 2018).
[24] British Admiralty, *British Mining Operations 1939-1945: Vol 1*, 391.
[25] "Sperrbrecher."
[26] "The Norwegian Navy in the Second World War" (http://www.resdal.org/Archivo/d00000a5.htm: accessed 28 June 2018).
[27] Ibid.
[28] "The Norwegian Navy in the Second World War"; "Camp Norway, Lunenburg, Nova Scotia, World War II" (http://www.skjoldlodge.com/dyk_archives/nova_scotia_WWII.htm): accessed 28 June 2018).
[29] Ut supra.
[30] Ut supra.
[31] British Admiralty, *British Mining Operations 1939-1945: Vol 1*, 326.
[32] Finn Christian Mosgaard Gysler, "The Story of the Norwegian 52 ML Flotilla" (unpublished personal account provided the authors by Colin Stromsoy, son of one of the flotilla members). Pangbourne, identified as a lieutenant in the quoted material, was likely Temporary Acting Gunner (T) Donald James Pangbourne. As a thin stripe Warrant Officer Gunner, he would easily have been mistaken for a Sub. Lt. and would have worn the same single thick stripe as a Sub. Lt. when he became a commissioned gunner.
[33] British Admiralty, *British Mining Operations 1939-1945: Vol 1*, 326.
[34] Ibid.
[35] British Admiralty, *British Mining Operations 1939-1945: Vol 1*, 326; "Casualty Lists of the Royal Navy and Dominion Navies, World War 2" (https://www.naval-history.net/xDKCas1003-Intro.htm#WW2: accessed 26 May 2018).
[36] British Admiralty, *British Mining Operations 1939-1945: Vol 1*, 329-330.
[37] British Admiralty, *British Mining Operations 1939-1945: Vol 1*, 330; Gysler "The Story of the Norwegian 52 ML Flotilla."
[38] Ut supra.
[39] British Admiralty, *British Mining Operations 1939-1945: Vol 1*, 330; "Casualty Lists of the Royal Navy and Dominion Navies, World War 2"; Gysler "The Story of the Norwegian 52 ML Flotilla."
[40] British Admiralty, *British Mining Operations 1939-1945: Vol 1*, 330; Gysler "The Story of the Norwegian 52 ML Flotilla."
[41] British Admiralty, *British Mining Operations 1939-1945: Vol 1*, 396.

CHAPTER 11 NOTES:

[1] British Admiralty, *British Mining Operations 1939-1945 Vol. 1*, 734-738.
[2] Ibid, 738.
[3] Ibid.
[4] Ibid, 739.
[5] Ibid.

[6] Ibid.
[7] Ibid.
[8] Ibid, 739-740.
[9] "No. 825 Squadron Fleet Air Arm & The Channel Dash 12 February 1942" (https://www.airshowspresent.com/no-825-squadron-fleet-air-arm-8203-the-channel-dash-820312-february-1942.html: accessed 14 July 2016).
[10] Ibid.
[11] "KMS *Scharnhorst*" (https://www.militaryfactory.com/ships/detail.asp?ship_id=KMS-Scharnhorst: accessed 12 July 2018).
[12] "No. 825 Squadron Fleet Air Arm & The Channel Dash 12 February 1942."
[13] "KMS *Scharnhorst*."
[14] Ibid.
[15] "Lieutenant Commander Eugene Kingsmill Esmonde VC DSO" https://www.navywings.org.uk/heritage/heroes/lieutenant-commander-eugene-kingsmill-esmonde-vc-dso/: accessed 12 July 2018.
[16] "No. 825 Squadron Fleet Air Arm & The Channel Dash 12 February 1942."
[17] "No. 825 Squadron Fleet Air Arm & The Channel Dash 12 February 1942"; "Lieutenant Commander Eugene Kingsmill Esmonde VC DSO."
[18] "KMS *Scharnhorst*"; "No. 825 Squadron Fleet Air Arm & The Channel Dash 12 February 1942."
[19] "No. 825 Squadron Fleet Air Arm & The Channel Dash 12 February 1942"; "With Gallantry and Determination" The Story of the Torpedoing of the Bismarck by Mark E. Horan (http://www.kbismarck.com/article2.html: accessed 17 July 2018).
[20] "Fairey Swordfish" (http://aviation-history.com/fairey/swordfish.html: accessed 12 July 2018)
[21] British Admiralty, *British Mining Operations 1939-1945 Vol. 1*, 740.
[22] Ibid.
[23] Ibid.

CHAPTER 12 NOTES:

[1] "The Cross of the Liberation" (www.ordredelaliberation.fr/en/ordre-de-la-liberation/un-ordre-national/la-croix-de-la-liberation: accessed 23 July 2018).
[2] British Admiralty, *British Mining Operations 1939-1945 Vol 1*, 328; "FR *Rubis*" (https://uboat.net/allies/warships/ship/6112.html); "In Honour and by Victory, sixteenth episode: FFS *Rubis*" (http://actualites.musee-armee.fr/expositions-en/dans-lhonneur-et-par-la-victoire-episode-16-le-sous-marin-rubis/?lang=en: accessed 22 July 2018).
[3] "*Saphir* class" (https://uboat.net/allies/warships/class/297.html); "FS *Rubis* (1933)" (https://www.militaryfactory.com/ships/detail.asp?ship_id=FS-Rubis-1933: accessed 22 July 2018); "The *Rubis* – Free French Submarine" by Stefen

Styrsky (http://armchairgeneral.com/the-rubis-free-french-submarine.htm: accessed 22 July 2018).
[4] Ut supra.
[5] "FR *Rubis*."
[6] "FR *Rubis*"; "FS *Rubis* (1933)"; "The *Rubis* – Free French Submarine" by Stefen Styrsky (http://armchairgeneral.com/the-rubis-free-french-submarine.htm: accessed 22 July 2018).
[7] "The *Rubis* – Free French Submarine"; "FR *Rubis*."
[8] "The *Rubis* – Free French Submarine."
[9] "The *Rubis* – Free French Submarine"; "FR *Rubis*."
[10] Ibid.
[11] Ibid.
[12] Ibid.
[13] Ibid.
[14] "FR *Rubis*."
[15] British Admiralty, *British Mining Operations 1939-1945 Vol 1*, 328.
[16] Ibid, 328-329.
[17] Ibid, 329.
[18] "FR *Rubis*."
[19] Ibid.
[20] Ibid.

CHAPTER 13 NOTES:
[1] "Badges and Battle Honours" (http://readyayeready.com/tradition/badges-and-battle-honours.php: accessed 9 July 2018).
[2] Warrant Officer Program, United States Army Combined Arms Center (http://usacac.army.mil/organizations/cace/wocc/woprogram: accessed 15 April 2017).
[3] Ibid.
[4] Ibid.
[5] Ibid.
[6] Tim Colton, "Neafie & Levy, Philadelphia PA" (http://www.shipbuildinghistory.com/shipyards/19thcentury/neafie.htm: accessed 16 April 2017); Army Appropriation Bill, 1916.
[7] Army Appropriation Bill, 1916; Tim Colton, "Pusey & Jones, Wilmington DE" (http://www.shipbuildinghistory.com/shipyards/large/pusey.htm: accessed 15 April 2017).
[8] McGovern and Smith, *American Coastal Defenses 1885–1950*.
[9] Tim Colton, "Marietta Manufacturing, Point Pleasant WV" (http://www.shipbuildinghistory.com/shipyards/large/marietta.htm: accessed 17 April 2017).
[10] "USS *Oglala* (CM-4, later ARG-1), 1917-1965" (http://www.shipscribe.com/usnaux/ww1/ships/cm4.htm; "Class: *Terror* (CM-5)" (http://www.shipscribe.com/usnaux/CM/CM05.html: accessed 23 December 2017).

[11] "Class: *Terror* (CM-5)".
[12] "Class: *Terror* (CM-5)"; Robert S. Egan, "USS *Terror* and Her Family," *Warship International*, Vol 48, Issue 4, December 2011.
[13] "Class: *Terror* (CM-5)"; *Terror, DANFS*.
[14] *Keokuk, DANFS*.
[15] *Monadnock* and *Miantonomah, DANFS*.
[16] *Salem, DANFS*.
[17] *Weehawken, DANFS*.
[18] *Salem* War Diary, April-May, 1944; *Monadnock* War Diary, December 1941-July 1943.

CHAPTER 14 NOTES:
[1] Bruhn, *We Are Sinking, Send Help! The U.S. Navy's Tugs and Salvage Ships in the African, European, and Mediterranean Theaters in World War II*, 1.
[2] Ibid, 1-2.
[3] Ibid, 2.
[4] Ibid, 3-4.
[5] Commander Task Force Thirty-four, Torch Operation, preliminary report of, 28 November 1942.
[6] Ibid.
[7] Ibid.
[8] *Hamilton* War Diary, November 1942; Lott, *Most Dangerous Sea*, 107; Bruhn, *We Are Sinking, Send Help!*, 8; ONI Combat Narratives, The Landings in North Africa 1942, 1944.
[9] Bruhn, *We Are Sinking, Send Help!*, 7.
[10] Bruhn, *We Are Sinking, Send Help!*, 8-9; ONI Combat Narratives, The Landings in North Africa 1942, 1944; Algeria-French Morocco, US Army Campaigns in World War II, CMH Pub 72-11 (https://www.ibiblio.org/hyperwar/USA/USA-C-Algeria/index.html: accessed 27 May 2017).
[11] Lott, *Most Dangerous Sea*, 108.
[12] Bruhn, *We Are Sinking, Send Help!*, 9; Lott, *Most Dangerous Sea*, 108.
[13] Lott, *Most Dangerous Sea*, 108; ONI Combat Narratives, The Landings in North Africa 1942, 1944.
[14] ONI Combat Narratives, The Landings in North Africa 1942, 1944.
[15] Ibid.
[16] Ibid.
[17] ONI Combat Narratives, The Landings in North Africa 1942, 1944; Algeria-French Morocco, US Army Campaigns in World War II, CMH Pub 72-11.
[18] Commanding Officer, USS *Hogan*, Action Report, Encounter with French Corvette on 8 November, 1942, 9 December 1942.
[19] Commanding Officer, USS *Hogan*, Action Report, Encounter with French Corvette on 8 November, 1942, 9 December 1942; Lott, *Most Dangerous Sea*, 109.

[20] Commanding Officer, USS *Hogan*, Action Report, Encounter with French Corvette on 8 November, 1942, 9 December 1942; Commander Task Group 34.9, Operation Torch – Report on, 30 November 1942.
[21] Commander Task Force Thirty-four, Torch Operation, preliminary report of, 28 November 1942.
[22] Commander Task Force Thirty-four, Torch Operation, preliminary report of, 28 November 1942; ONI Combat Narratives, The Landings in North Africa 1942, 1944.
[23] Commander Task Group 34.9, Operation Torch – Report on, 30 November 1942; *Miantonomah*, *Tillman* War Diary, November 1942.
[24] *Miantonomah* War Diary, November 1942; Lott, *Most Dangerous Sea*, 109-110.
[25] Commander Task Group 34.9, Operation Torch – Report on, 30 November 1942; *Tillman* War Diary, November 1942.
[26] *Tillman* War Diary, November 1942.
[27] Commander Task Group 34.9, Operation Torch – Report on, 30 November 1942; *Tillman* War Diary, November 1942.
[28] Algeria-French Morocco, US Army Campaigns in World War II, CMH Pub 72-11.
[29] Ibid.
[30] Lott, *Most Dangerous Sea*, 110; *Hamilton* War Diary, November 1942.
[31] Morison, *The Two-Ocean War*, 232.
[32] ONI Combat Narratives, The Landings in North Africa 1942, 1944.
[33] Bruhn, *We Are Sinking, Send Help!*, 17-18.
[34] Commanding Officer, USS *Terror*, USS *Terror* (CM5); report of, in connection mining operations off Casablanca, Morocco, 28 November 1942.
[35] Commanding Officer, USS *Terror* (CM-5), Operation Order No. 1-42, 15 November 1942.
[36] *Monticello*, *Terror* War Diary, November 1942; Code Names Operations of World War 2 (http://codenames.info/operation/ug/: accessed 1 May 2017).
[37] *Terror* War Diary, November 1942.
[38] Commanding Officer, USS *Terror*, USS *Terror* (CM5); report of, in connection mining operations off Casablanca, Morocco, 28 November 1942.
[39] Ibid.
[40] Ibid.
[41] Ibid.
[42] Ibid.
[43] Ibid.
[44] Ibid.

CHAPTER 15 NOTES:
[1] Bruhn, *We Are Sinking, Send Help!*, 53.
[2] Dwight Messimer correspondence of 31 July 2018.
[3] Ibid
[4] Ibid.
[5] Bruhn, *We Are Sinking, Send Help!*, 53.

[6] Ibid, 53-54.
[7] Ibid, 54.
[8] Ibid, 59.
[9] Ibid.
[10] Ibid.
[11] Ibid, 63.
[12] Ibid, 63-64.
[13] Lott, *Most Dangerous Sea*, 116-117; Morison, *The Two-Ocean War*, 254.
[14] Lott, *Most Dangerous Sea*, 117; Morison, *The Two Ocean War*, 254-255; "Bitter Fight at Gela" (https://warfarehistorynetwork.com/daily/wwii/bitter-fight-at-gela/: accessed 31 July 2018).
[15] Lott, *Most Dangerous Sea*, 117; Morison, *The Two Ocean War*, 254-255.
[16] Morison, *The Two Ocean War*, 260.
[17] Bruhn, *We Are Sinking, Send Help!*, 81.
[18] Ibid, 81-82
[19] Ibid.
[20] Ibid, 82.

CHAPTER 16 NOTES:
[1] "Motor Gun Boats Spitfire of the Seas" by Mike Kemble and Ray Holden (https://www.39-45war.com/mgb.html: accessed 5 August 2018).
[2] British Admiralty, *British Mining Operations 1939-1945 Vol 1*, 398.
[3] Ibid, 406.
[3] Ibid.
[4] Ibid.
[5] Ibid, 398-399.
[6] British Admiralty, *British Mining Operations 1939-1945 Vol 1*, 332; "Royal Navy Minelaying Operations" (https://www.naval-history.net/xGM-Ops-Minelaying.htm: accessed 19 May 2018).
[7] British Admiralty, *British Mining Operations 1939-1945 Vol 1*, 332.
[8] "Norwegian Navy and Dover" (https://doverhistorian.com/2013/05/12/norwegian-navy-and-dover/); "Lord Warden Hotel/House" (https://doverhistorian.com/2013/10/02/lord-warden-hotel-house/: accessed 30 July 2018).
[9] Ut supra.
[10] Hood, *Submarine: An anthology of firsthand accounts of the war under the sea, 1939-45*.
[11] Ibid.
[12] Ibid.
[13] Finn Christian Mosgaard Gysler, "The Story of the Norwegian 52 ML Flotilla" (unpublished personal account provided the authors by Colin Stromsoy, son of one of the flotilla members).
[14] "Lord Warden Hotel/House"; "Oil Mill Caves, Dover" (http://www.subterraneanhistory.co.uk/2007/08/oil-mill-tunnels-

dover.html: accessed 5 August 2018); Gysler, "The Story of the Norwegian 52 ML Flotilla."
[15] "Norwegian Navy and Dover."
[16] British Admiralty, *British Mining Operations 1939-1945 Vol 1*, 334-335); Gysler, "The Story of the Norwegian 52 ML Flotilla."
[17] Gysler, "The Story of the Norwegian 52 ML Flotilla."
[18] British Admiralty, *British Mining Operations 1939-1945 Vol 1*, 334.
[19] Ibid.
[20] Ibid, 406, 411.
[21] Ibid, 334.
[22] "Casualty Lists of the Royal Navy and Dominion Navies, World War 2" (https://www.naval-history.net/xDKCas1003-Intro.htm#WW2: accessed 26 May 2018).
[23] British Admiralty, *British Mining Operations 1939-1945 Vol 1*, 335.
[24] "Royal Navy Minelaying Operations" (https://www.naval-history.net/xGM-Ops-Minelaying.htm: accessed 19 May 2018); British Admiralty, *British Mining Operations 1939-1945 Vol 1*, 335.
[25] British Admiralty, *British Mining Operations 1939-1945 Vol 1*, 337-338.
[26] Ibid, 337.
[27] Ibid.
[28] Ibid, 340.
[29] Ibid, 399.
[30] Ibid, 399-400.
[31] Ibid, 400.
[32] Ibid.
[33] British Admiralty, *British Mining Operations 1939-1945 Vol 1*, 343; Gysler, "The Story of the Norwegian 52 ML Flotilla."
[34] "Royal Norwegian Navy Ship Histories, Convoy Escort Movements, Casualty Lists 1940-1947" (http://www.naval-history.net/xDKCas3000-Norwegian.htm: accessed 7 July 2018).
[35] Humphreys, *Hellfire Corner*, 136; Hood, *Submarine: An anthology of firsthand accounts of the war under the sea, 1939-45*.
[36] Gysler, "The Story of the Norwegian 52 ML Flotilla."
[37] Ibid.

CHAPTER 17 NOTES:

[1] "HMS *Manxman* - *Abdiel*-class Fast Cruiser Minelayer" (http://www.naval-history.net/xGM-Chrono-07ML-Manxman.htm: accessed 12 August 2018).
[2] Ibid.
[3] "HMS *Manxman* (M 70)" (https://uboat.net/allies/merchants/ship/2481.html: accessed 12 August 2018).
[4] Ibid.
[5] "HMS *Welshman* (M 84)" (https://uboat.net/allies/warships/ship/3956.html); "HMS *Welshman* (M 84)" (https://uboat.net/allies/merchants/ship/2614.html): "HMS *Welshman* -

Fast *Abdiel*-class Cruiser Minelayer" (http://www.naval-history.net/xGM-Chrono-07ML-Welshman.htm: accessed 12 August 2018).
[6] "HMS *Welshman* (M 84)"
(https://uboat.net/allies/merchants/ship/2614.html).
[7] "HMS *Welshman* (M 84)"
(https://uboat.net/allies/warships/ship/3956.html).
[8] "Operation Slapstick, the Taranto Landings, 9 September 1943" (http://www.historyofwar.org/articles/operation_slapstick_taranto.html: accessed 12 August 2018).
[9] Ibid.
[10] "Operation Slapstick, the Taranto Landings, 9 September 1943"; "HMS *Abdiel* (M 39)."

CHAPTER 18 NOTES:
[1] Brown, *Battle Summary No. 39, Operation "Neptune" Landings in Normandy June, 1944*, 47.
[2] Ibid, 47-48.
[3] "HMS *Apollo* - Abdiel-class Fast Cruiser Minelayer" (http://www.naval-history.net/xGM-Chrono-07ML-Apollo.htm: accessed 10 August 2018).
[4] British Admiralty, *British Mining Operations 1939-1945 Vol. 1*, 401; "HMS *Plover* (M 26) - Coastal Minelayer" (https://www.naval-history.net/xGM-Chrono-08ML-HMS_Plover.htm: accessed 10 August 2018).
[5] British Admiralty, *British Mining Operations 1939-1945 Vol. 1*, 750.
[6] Ibid, 356.
[7] Ibid.
[8] Ibid, 347.
[9] Ibid, 348.
[10] Ibid, 366.
[11] Ibid.
[12] "Channel Clashes" (http://www.netherlandsnavy.nl/Special_mtb.htm: accessed 10 August 2018).
[13] British Admiralty, *British Mining Operations 1939-1945 Vol. 1*, 750.
[14] Ibid.
[15] Ibid, 402.
[16] Ibid, 402-403.
[17] Ibid, 403.
[18] "WWII Awards for RN Minelaying" (http://www.mcdoa.org.uk/MCD_History_Frames.htm: accessed 10 August 2018); "Royal Navy Minelaying Operations" (https://www.naval-history.net/xGM-Ops-Minelaying.htm: accessed 19 May 2018).

CHAPTER 19 NOTES:
[1] *Barricade, Chimo, Planter* War Diaries, April-May 1944.
[2] Ut supra.
[3] *Planter* War Diary, May 1944.
[4] Ibid.

[5] *Barricade* War Diary, May-June 1944.
[6] *Planter, Chimo* War Diary, May 1944.
[7] *Planter* War Diary, May 1944.
[8] *Chimo* War Diary, May-June 1944.
[9] Ibid, June 1944.

CHAPTER 20 NOTES
[1] "Royal Canadian Navy" (https://www.thecanadianencyclopedia.ca/en/article/royal-canadian-navy/: accessed 19 June 2018).
[2] "Canadian Naval Reserve History" (http://military.wikia.com/wiki/Special:Upload?wpDestFile=HMCS_Pictou_K146_MC-2774.jpg: accessed 21 June 2018).
[3] "Canada in the Second World War, Minesweepers" (https://www.junobeach.org/canada-in-wwii/articles/the-ships/minesweepers/); "Canadian Navy Ships (1910-1939)" (http://www.navy.gc.ca/project_pride/history/history_e.asp?section=4&category=11&title=1: accessed 20 June 2018); Macpherson and Barrie, *The Ships of Canada's Naval Forces 1910–2002*, 32-33.
[4] "HMCS *Fundy* (1st) (J88)" (http://www.readyayeready.com/ships/shipview.php?id=1144&ship=FUNDY%20(1st): accessed 21 June 2018)
[5] "*Fundy* class" (https://uboat.net/allies/warships/class/239.html: accessed 20 June 2018).
[6] "Canada in the Second World War, Minesweepers."
[7] Richard Oliver Mayne, "The little boat has just put her lights out:" The Life, Fate and Legacy of HMCS *Bras d'Or* (https://www.cnrs-scrn.org/northern_mariner/vol14/tnm_14_4_25-40.pdf: accessed 20 June 2018).
[8] "HMCS *Bras D'Or* J06 Former Lightship *No. 25*" (http://www.forposterityssake.ca/Navy/HMCS_BRAS_D_OR.htm: accessed 22 June 2018).
[9] Mayne, "The little boat has just put her lights out."
[10] Ibid.
[11] Mayne, "The little boat has just put her lights out:" "HMCS *Bras D'Or* (1st)" http://www.readyayeready.com/ships/shipview.php?id=1047&ship=BRAS%20D'OR%20(1st): accessed 21 June 2018).
[12] Mayne, "The little boat has just put her lights out."
[13] "101-GO minesweepers (1944)" (http://www.navypedia.org/ships/japan/jap_ms_ex_brit.htm: accessed 20 June 2018).
[14] "*Bangor* class" (https://uboat.net/allies/warships/class.html?ID=132: accessed 20 June 2018).
[15] Ibid.
[16] Brown, *Nelson to Vanguard: Warship Design and Development 1923-1945*.

[17] "Canada in the Second World War, Minesweepers."
[18] "*Isles* class" (https://uboat.net/allies/warships/class/339.html); "Western Isles Class A/S Trawlers" (http://www.forposterityssake.ca/RCN-SHIP-INDEX.htm#TRAWLERS: accessed 22 June 2018).
[19] Ut supra.
[20] Ut supra.
[21] "HMCS *Sankaty* Then Prince Edward Island Ferry *Charles A. Dunning*" (http://www.forposterityssake.ca/Navy/HMCS_SANKATY.htm: accessed 22 June 2018).
[22] "*Llewellyn* class" (https://uboat.net/allies/warships/class.html?ID=138&navy=HMCS: accessed 21 June 2018).
[23] "HMCS *Llewellyn* (J278)" (http://www.readyayeready.com/ships/shipview.php?id=1224&ship=LLEWELLYN: accessed 21 June 2018).
[24] "*Llewellyn* class."
[25] "Lake Class Minesweepers" (http://www.forposterityssake.ca/Navy/LAKE_CLASS_MINESWEEPERS.htm: accessed 15 August 2018).
[26] "Suderøy Factory & Suderøy Whale Catchers" (http://www.warsailors.com/singleships/suderoy.html: accessed 22 June 2018).
[27] Ibid.
[28] Ibid.
[29] "Allied War Losses" (https://uboat.net/allies/warships/war_losses.html?navy=HMCS: accessed 17 June 2018).
[30] "Royal Canadian Warships that Participated in the Battle of the Gulf of St. Lawrence" (http://www.veterans.gc.ca/eng/remembrance/history/second-world-war/battle-gulf-st-lawrence/canwarship); "Commemoration" (http://www.veterans.gc.ca/eng/remembrance/history/second-world-war/battle-gulf-st-lawrence/commemoration: accessed 22 June 2018).
[31] Ibid.

CHAPTER 21 NOTES:
[1] Brown, *Battle Summary No. 39, Operation "Neptune" Landings in Normandy June, 1944*, 49.
[2] "Naval Guns at Normandy" (https://www.history.navy.mil/content/history/nhhc/research/library/online-reading-room/title-list-alphabetically/n/naval-guns-normandy.html: accessed 15 August 2018).
[3] Morison, *The Two-Ocean War*, 386.
[4] Ibid.
[5] "Naval Guns at Normandy."
[6] Lott, *Most Dangerous Sea*, 184.

[7] Lott, *Most Dangerous Sea*, 184, 186; Brown, *Battle Summary No. 39, Operation "Neptune" Landings in Normandy June, 1944*, 51-53.
[8] Lott, *Most Dangerous Sea*, 186-187.
[9] Ibid, 187.
[10] Michael Whitby, "There Must Be No Holes in Our Sweeping": The 31st Canadian Minesweeping Flotilla on D-Day (http://scholars.wlu.ca/cgi/viewcontent.cgi?article=1132&context=cmh: accessed 15 August 2018).
[11] Whitby, "There Must Be No Holes in Our Sweeping."
[12] Whitby, "There Must Be No Holes in Our Sweeping"; Williams, *They Led the Way: The Fleet Minesweepers at Normandy - June 1944*, 14-15.
[13] Williams, *They Led the Way: The Fleet Minesweepers at Normandy - June 1944*, 14-15; Macpherson and Barrie, *The Ships of Canada's Naval Forces 1910–2002*, 192.
[14] Whitby, "There Must Be No Holes in Our Sweeping."
[15] Whitby, "There Must Be No Holes in Our Sweeping"; Williams, *They Led the Way: The Fleet Minesweepers at Normandy - June 1944*, 14-15.
[16] Whitby, "There Must Be No Holes in Our Sweeping."
[17] Brown, *Battle Summary No. 39 Operation "Neptune" The Landing in Normandy 6th June, 1944*, App.A(1).
[18] "HMCS *Mulgrave* (J 313)" (https://uboat.net/allies/warships/ship/2694.html: accessed 17 June 2018).
[19] Whitby, "There Must Be No Holes in Our Sweeping."
[20] Ibid.
[21] Ibid.
[22] Ibid.
[23] Ibid.
[24] Ibid.
[25] Ibid.
[26] Ibid.
[27] Whitby, "There Must Be No Holes in Our Sweeping"; "Rear Admiral RCN (ret'd) Antony H. G. Storrs, DSC & Bar" (https://www.blatherwick.net/documents/Royal%20Canadian%20Navy%20Citations/S%20-%20RCN%20-%20WW2%20.pdf: accessed 15 August 2018).
[28] Ut supra.
[29] Whitby, "There Must Be No Holes in Our Sweeping."
[30] "Rear Admiral RCN (ret'd) Antony H. G. Storrs, DSC & Bar."
[31] "Canadian Orders Decorations and Medals," https://www.blatherwick.net/royal%20canadian%20navy%20citations/); "World War II Awards for RN Minesweeping" (http://www.mcdoa.org.uk/ww_ii_awards_for_rn_minesweeping_M.htm).
[32] Ut supra.
[33] Brown, *Battle Summary No. 39, Operation "Neptune" Landings in Normandy June, 1944*, 51.
[34] Ibid, Appendices A(1) and B(1).

[35] Lott, *Most Dangerous Sea*, 184-185.
[36] Brown, *Battle Summary No. 39, Operation "Neptune" Landings in Normandy June, 1944*, App.B(1).
[37] Bruhn, *Wooden Ships and Iron Men: The U.S. Navy's Coastal and Motor Minesweepers, 1941-1953*, 93.
[38] Ibid, 93-94.
[39] USS *Chimo*, Action Report – 6 June to 20 June 1944 – Allied expenditionary Force, Western Naval Task Force, Bai de la Seine, France, 4 September 1944.
[40] Lott, *Most Dangerous Sea*, 187.
[41] Lott, *Most Dangerous Sea*, 187; Bruhn, *Wooden Ships and Iron Men: The U.S. Navy's Coastal and Motor Minesweepers, 1941-1953*, 94-95.
[42] Bruhn, *Wooden Ships and Iron Men: The U.S. Navy's Coastal and Motor Minesweepers, 1941-1953*, 95.
[43] Fact Sheet: Normandy Landings (https://obamawhitehouse.archives.gov/the-press-office/2014/06/06/fact-sheet-normandy-landings: accessed 18 August 2018).
[44] Lott, *Most Dangerous Sea*, 189.
[45] USS *Chimo*, Action Report – 6 June to 20 June 1944 – Allied expenditionary Force, Western Naval Task Force, Bai de la Seine, France, 4 September 1944.
[46] Lott, *Most Dangerous Sea*, 189.
[47] Ibid.

CHAPTER 22 NOTES:
[1] Lott, *Most Dangerous Sea*, 189
[2] Brown, *Battle Summary No. 39, Operation "Neptune" Landings in Normandy June, 1944*, 114.
[3] Ibid.
[4] Ibid, Appendix J.
[5] "Mines used in D-Day" (https://www.historyonthenet.com/mines-used-in-d-day/: accessed 19 August 2018).
[6] Brown, *Battle Summary No. 39, Operation "Neptune" Landings in Normandy June, 1944*, 107.
[7] Bruhn, *Wooden Ships and Iron Men: The U.S. Navy's Coastal and Motor Minesweepers, 1941-1953*, 99.
[8] Ibid, 99-100.
[9] Ibid, 100.
[10] Ibid.
[11] Ibid, 100-101.
[12] Ibid, 101.
[13] Ibid.
[14] *Chimo* War Diary, June 1944.
[15] Ibid.

[16] *Chimo* War Diary, June 1944; USS *Chimo*, Action Report – 6 June to 20 June 1944 – Allied expeditionary Force, Western Naval Task Force, Bai de la Seine, France, 4 September 1944.
[17] Bruhn, *Wooden Ships and Iron Men: The U.S. Navy's Coastal and Motor Minesweepers, 1941-1953*, 49, 54.
[18] *Chimo* War Diary, June and July 1944; Commander "Y" Squadrons, Special Action Report – Submission of, 3 August 1944.
[19] Commander "Y" Squadrons, Special Action Report – Submission of, 3 August 1944.
[20] Ibid.
[21] Ibid.
[22] Ibid.
[23] Ibid.
[24] Commander "Y" Squadrons, Special Action Report – Submission of, 3 August 1944; Lott, *Most Dangerous Sea*, 193.

CHAPTER 23 NOTES:
[1] Monograph, Official Study of Port of Cherbourg, 1944-1945.
[2] Bruhn, *We Are Sinking, Send Help!*, 192; Lott, *Most Dangerous Sea*, 194.
[3] Bruhn, *We Are Sinking, Send Help!*, 192.
[4] Ibid, 197.
[5] "The Loss of *YMS-350*: As Seen From HMML 137" by Brendan A. Maher, California State Library Foundation Bulletin, No. 77, Spring 2004; Lott, *Most Dangerous Sea*, 193.
[6] Maher, "The Loss of *YMS-350*: As Seen From HMML 137."
[7] Maher, "The Loss of *YMS-350*: As Seen From HMML 137"; Lott, *Most Dangerous Sea*, 195.
[8] Maher, "The Loss of *YMS-350*: As Seen From HMML 137."
[9] "Memories of a Survivor from the Sinking of the USS *YMS-350*, July 2, 1944" by Mead B. Kibbey, California State Library Foundation Bulletin, No. 77, Spring 2004.
[10] Ibid.
[11] Ibid.
[12] Bruhn, *We Are Sinking, Send Help!*, 198.
[13] Ibid.
[14] Ibid, 199.
[15] Ibid, 199-201.
[16] Ibid, 203.
[17] Ibid, 207.
[18] Ibid, 208.
[19] Ibid, 209.
[20] *Miantonomah, DANFS*.
[21] Commander, Eleventh Amphibious Force War Diary, June 1944; Brown, *Battle Summary No. 39, Operation "Neptune" Landings in Normandy June, 1944*, 31.
[22] *Miantonomah, DANFS*.

[23] Narration by Commander Austin E. Rowe, USNR., Office of Naval Records and Library
[24] Ibid.
[25] *Miantonomah*, *DANFS*; Narration by Commander Austin E. Rowe, USNR.
[26] Ut supra.

CHAPTER 24 NOTES:
[1] Bruhn, *We Are Sinking, Send Help!*, 211.
[2] Ibid, 211-212.
[3] Ibid, 212.
[4] Ibid.
[5] Ibid.
[6] Ibid, 213.
[7] Ibid.
[8] Ibid, 213-214.
[9] Ibid, 214.
[10] British Admiralty, *British Mining Operations 1939-1945: Vol 1*, 754.
[11] Commander Task Group 84.5, Report of Action Operation [word redacted, replaced by handwritten "S. France"] for Period 5 August 1944 until 0503B 17 August 1944, 15 September 1944.
[12] "HMS *Aries* (J 284) ex-AM-327" (http://www.navsource.org/archives/11/02327.htm: accessed 26 August 2018).
[13] "HMS *Aries* (J 284)" (https://uboat.net/allies/warships/ship/3807.html); "Carnall, Herbert Bruce" (http://www.nauticapedia.ca/dbase/Query/Biolist3.php?&name=Carnall%2C%20Herbert%20Bruce&id=16719&Page=4&input=1: accessed 26 August 2018).
[14] "Carnall, Herbert Bruce"; "Casualty Lists of the Royal Navy and Dominion Navies, World War 2" (https://www.naval-history.net/xDKCas1003-Intro.htm#WW2: accessed 26 May 2018).
[15] *The London Gazette*, Tuesday, 22 January, 1946.
[16] "HMS *Aries* (J 284) ex-AM-327."
[17] Commander Task Group 84.5, Report of Action Operation.
[18] Commander Task Force Eighty Four, Report of action, Operation Dragoon, by CTG 84.8 (Commander Minesweeping Group), 3 October 1944.
[19] Commander Task Force Eighty Four, Report of action, Operation Dragoon, by CTG 84.8, 3 October 1944; Commander U.S. Eighth Fleet, Report of action, Operation Dragoon, 28 November 1944.
[20] Commander U.S. Eighth Fleet, Report of action, Operation Dragoon, 28 November 1944; Commander Task Group 84.5, Report of Action Operation.
[21] Brown, *Invasion of the South of France, Operation "Dragoon", 15th August, 1944* (London: HMSO, 1994), 20-21.
[22] Ibid, 21.

[23] Ibid.
[24] Commander Task Force Eighty Four, Report of action, Operation Dragoon, by CTG 84.8.
[25] *Barricade* War Diary, August 1944.
[26] Ibid.
[27] Bruhn, *We Are Sinking, Send Help!*, 219-220.
[28] Ibid, 225-226.

CHAPTER 25 NOTES
[1] "Illustrating victory on VE Day" (https://blog.nationalarchives.gov.uk/blog/ve-day-posters/: accessed 12 August 2018).
[2] Bruhn, *We Are Sinking, Send Help!*, 237.
[3] Ibid, 237-238.
[4] Ibid, 248-249.
[5] Ibid, 249.
[6] Ibid.
[7] Ibid, 252.
[8] Cowie, *Mines, Minelayers and Minelaying*, 194.
[9] British Admiralty, *British Mining Operations 1939-1945: Vol 1*, 768.
[10] Ibid, 762-763.

POSTSCRIPT NOTES:
[1] "1956: Mystery of missing frogman deepens" (http://news.bbc.co.uk/onthisday/hi/dates/stories/may/9/newsid_4741000/4741060.stm: accessed 6 July 2018).
[2] "Commander Lionel 'Buster' Crabb" (https://www.submerged.co.uk/buster-crabb.php: accessed 6 July2018).
[3] Ibid.
[4] "Odd Fighting Units: Italian Frogmen and the Human Torpedoes of the Decima Flottiglia MAS, 1940-1943" (http://warfarehistorian.blogspot.com/2014/03/odd-fighting-units-italian-frogmen-and.html: accessed 6 July 2018).
[5] "Commander Lionel 'Buster' Crabb."
[6] "The Tobermory Treasure Galleon" (https://www.detecting.org.uk/tobermory/index.html: accessed 6 July 2018).
[7] Ibid.
[8] "Commander Lionel 'Buster' Crabb."
[9] "1956: Mystery of missing frogman deepens"; "Commander Lionel 'Buster' Crabb."
[10] Ut supra.
[11] "1956: Mystery of missing frogman deepens."
[12] Ibid.
[13] Ibid.
[14] "Commander Lionel 'Buster' Crabb."
[15] "1956: Mystery of missing frogman deepens."

Index

Adams, William Leslie Graham (Rear Adm., OBE, RN), 202
Africa
  Algeria, Oran, Mers-el-Kebir, 1, 6, 8, 162, 164-165, 177, 206, 286
  Libya
    Bardia, 104, 107, 119
    Beda Fomm, 107
    Benghazi, 20, 92, 100, 111-112, 118
    Misrata, 92
    Ras Azzaz, 13, 97
    Tobruk, 13, 96-98, 105-120, 163, 205-207
    Tripoli, 104, 107
    Tripolitania, 87, 164
    Tulmaythah, 92
  Morocco
    Casablanca, 159-182
    Fedala, 165-182
    Mehdia, 164-171
    Port Lyautey, 164-170
    Rabat-Sale, 169
    Safi, 164-177
  Senegal, Dakar, 167
  Tunisia, Bizerte, 147, 207, 218, 220
Alexander, Harold Rupert Leofric George (Field Marshall, 1st Earl Alexander of Tunis, KG, GCB, OM, GCMG, CSI, DSO, MC, CD, PC, BA), 185
Allan, Peter (T/A/Lt. Comdr., RNR), 85
Allen, Henry T., 169
Allen, William Donald, 250
Allison, John Hamilton (Capt., DSO, RN), 83
Arnold, C. H., 114
Ashdown, Thomas Aubrey (T/A/Lt. Comdr., DSO, Hkn, RNR), 194
Australia/Australian
  2/10th Infantry Battalion, 108
  9th Infantry Division ("Rats of Tobruk"), 96, 119
Badoglio, Marshal Pietro, 184
Bailey, Hugh, 198
Baker, J. P. (Sub. Lt.), 114
Baldwin, Charles E. (CPO, DSM, RN), 40
Barnard, Raymond (AB), 215
Barnes, Maurice (Sub. Lt., RNVR), 283
Barraclough, Edward Murray Conrad (Capt., CBE, OBE, RN), 86
Barrett, James Herbert (AB Gunner, DSM), 194
Barten, Wolfgang, 27

Bateman, John Robert, 214
Bateson, Stuart Latham (Rear Adm., CB, CBE, RN), 95
Beale, Colin R., 97
Beckham Jr., William E., 254
Beet, Trevor Agar (Capt., OBE, RN), 73, 77-78
Belgium
  Antwerp, 298
  Ostend, 66
  Zeebrugge, 196
Belknap, Robert, 281, 291
Bergum, Kermit Luther, 276
Beynon (Sub. Lt., RNVR), 145
Bickford, Jack Grant (Capt., DSO, DSC, RN), 72, 83, 85
Blackwell, Cecil Llewellyn, 182
Bligh (Sub. Lt., RNVR), 145
Blomberg, Alfred Julius, 201
Bloxham, John T. (T/Midshipman, RNR), 117
Bottcher, Heintz, 76
Boulnois, Philip William Hubert (Lt. Comdr., RN), 73
Bouvet, Georges-Régis, 280
Brantley, Daisy Louis, 250
Brandi, Albrecht, 205, 207
Brennan, Leo, 190
Bretherton, John, 97
Bridge, John (Lt. (Sp), GC, GM & Bar, RNVR), 295
Brock, Neville Brevoort Carcy (Capt., RN), 86
Broili, Nikolaus, 77
Brown, Henry J., 97
Brown, R. C., 218
Brown, William George (Leading Seaman), 318
Bruce, George F. W., 97
Brunskill, William (Able Bodied Seaman), 298
Bryan, Lewis Calvin, 253-254
Buckley, Franklin Duerr, 254
Bulganin, Nikolai, 302
Bulkeley, John, 261, 280
Bunce (Petty Officer Airman), 145
Burch, Ronald James (Lt. Comdr., DSO, RN), 71, 82, 91
Burfeind, Otto, xxx
Burgess, Stanley R., 198
Burnett, Robert Lindsay (Adm. Sir, GBE, DSO, CStJ, RN), 3-5, 85
Burney, Charles Dennison (Comdr. Sir, 2nd Baronet, CMG, RN), 29
Burns, Wallace Bruce (Lt., RN), 241
Burrows, James W., 81
Bush, Elizabeth Kauffman, 296
Bush, George H. W., 296

Butler, Terence Brinsley John Danvers (Lt., RN), 73-77
Cabanier, Georges E. J., 149
Calamei, Marco (Italian admiral), 98
Cameron, Roger Stannard (Comdr., DSC & Bar, RNZNVR), 283
Campbell, Hugh (A/Lt., DSC, RCNR), 242, 248
Canada/Canadian
  Halifax, 132-133, 222, 225, 230-234, 240
  Royal Canadian Army
    1st Infantry Division, 184
    3rd Canadian Division, 241
  Royal Canadian Navy
    Mine Force, 221-236
      31st Minesweeping Flotilla, 240-249, 267, 271
      Volunteer Reserve (RCNVR), 116-117, 222
        Port Arthur Half Division, Winnipeg Division, 116
  Quebec
    Clarke City, Rimouski, 225
    Sorel, 223, 225
  St. Lawrence River, 223-225, 234-236
  St. John's, 231, 233
  Sydney, 225, 230-233
  Victoria, 222
Carnall, Herbert Bruce (Capt., DSC, CD, RCN), 283
Cartwright, Thomas Nelson (T/A Lt. Comdr., DSC & 2 Bars, RNVR), 216
Cazalet, Peter Grenville Lyon (Vice Adm., Sir, KBE, CB, DSO, DSC, RN), 98
Chandler, Theodore E., 280
Churcher, Francis John, 194
Churchill, Winston, xxxv, 1, 23, 53, 70, 183, 188, 191, 273, 277, 287, 289
Ciliax, Otto, 143
Clark, Ronald Harrison Senior (Lt., RN), 73
Clinton, William Johnson (Petty Officer Airman), 144
Close, Anthony (T/Lt. (E), MBE, RNR), 283
Cochrane, James, 111
Cockey, Richard Kendall, 250
Cockshutt, John F., 81
Conolly, Richard L., 186
Cook, George Douglas (Lt., GM & Bar, RCNVR), 3-4-305
Cook, John, 28
Cook Jr., Albert G., 182
Cookson, Ernest G. W., 81
Cooper, Joshua Winfred, 182
Couch, Richard John Hollis (Lt. Comdr., DSC, RN), 72, 83
Coupethwaite, James, 81
Cox, Robert Arthur Neville (Lt., DSC, RNR), 115, 118
Crabb, Lionel Kenneth Phillip (Comdr., OBE, GM, RNVR), 300-304
Crane, George, 261

Cranefield, Donald Lewis (T/Sub. Lt.), 214
Crisp, Arthur F. L., 97
Crosscurth, David J., 198
Cunliffe, Robert Lionel Brooke (Cdre., CBE, RN),
Cunningham, Andrew Browne (Adm. Sir, 1st Viscount Cunningham of Hyndhope, KT, GCB, OM, DSO & 2 Bars, RN), 3, 105, 185
Dahlquist, John E., 280
Darby, Sydney N., 81
Dargie, John Clunie (Telegraphist, DSM), 215
Davidson, Lyal A., 190, 279
Davidson, Thomas, 81
Davies, John Mortimer (Lt. Comdr., DSC, RCNVR), 115
Davis, Donald (Lt., RCNVR), 242, 248
Dean, W. H., 114
de Back, Johannes Adolph, 65
De Gaulle, Charles, 147, 152
De Lattre la Tassigny, Jean, 279
Denmark, 23, 53, 61, 69, 73-74, 150, 211
   Aalborg, 74
   Frederikshavn, 77
Dewhurst, Ronald Hugh (Comdr., DSO, RN), 88, 92, 99
Dickson, Robert Kirk (Rear Adm., CB, DSO, RN), 3-8, 205
Dietl, Eduard (Gen., German Army), 58
Dönitz, Karl, 291
Doorman, Karl, 64
Drake, Walter Charles (T/Lt., DSC, RNZNVR), 215
Dudley, Sydney W., 81
Duffy (Lt.), 291
Dymond, R. P. D., 113
Eagles, William W., 280
Eatherall, James (Leading Wireman), 283
Eburah, Robert George (T/Lt., DSC & Bar, RNVR), 215
Ede, E. P. (A/Sub Lt., RNR), 113
Eden, Robert Anthony (1st Earl of Avon, KG, MC, PC), 300, 303
Edmundson, D. V., 281, 291
Edwards, J. (Gunner), 113
Edwards, Raymond Dorsey, 154, 182
Edwards, Thomas Bottrell (Lt., RCNR), 243
Egypt
   Alexandria, 88, 90, 93-116, 164, 205-207
   El Alamein, Battle of, 3, 97
   Mersa Matruh, 109, 114
   Port Said, 100, 118, 195,
   Sidi Barrani, 107
Eisenhower, Dwight D., 165, 185, 188, 238, 277
Ellis, Robert A. L., 250

Emmanuel III, King Victor, 184
Esmonde, Eugene Kingsmill (Lt. Comdr., VC, DSO, RN), 143-144
Evans, C. L. (Lt.), 114
Evans, Henry F., 81
Fagg, Maurice O. (Telegraphist), 117
Fairbanks Jr., Douglas E., 270-280
Ferrill, Homer Ewald, 250
Fiebach Jr., Ralph P., 254
Fife Jr., James, 98
Fildew, Frederick Charles (A/Leading Seaman, DSM), 214
Finsen (surgeon commander), 201
Fitch, Howard W., 178-180
Fitzmaurice, W. V. (Sub. Lt., RNR), 113
Flaherty, Patrick, 81
France/French
  Arcachon, 151
  Army
    1st and 7th Regiments Moroccan Tirailleurs, French Foreign Legion, 169
  Bayonne, 147, 150, 200
  Boulogne, 24, 128-129, 196, 210, 298
  Brest, xxv, 122, 139-147, 192, 200, 210-213
  Brittany, 123, 191-192, 197, 199, 210, 213
  Calais, 23, 27, 29, 128, 136, 144, 210, 298
  Cannes, Hyères, 277
  Cap Gris Nez, 198
  Cavalaire, 278-286
  Channel Islands, Guernsey and Herm, 197, 199
  Cherbourg, 147, 191-192, 197-199, 209-211, 238, 264-278, 298
  Corsica, 6-7, 278
  Dieppe, 123, 283, 298
  Dunkirk, 24, 27-28, 66, 70-71, 106, 125-128, 198-199, 238, 305
  Dyck Shoal, 24, 128
  Etaples, 128-129, 192, 199, 211, 213
  Fecamp, 192
  Grandcamp, 274
  Gravelines, 193, 198
  Ile de bas, 197, 200
  La Rochelle, 99
  Le Havre, 160-162, 209-211, 238, 243, 273-275, 292, 298
  Les Heaux de Brehat, 140
  Les Sept Iles, 200
  Marseilles, 122, 273, 278, 287, 290
  Navy/Marines
    1er Régiment de Fusiliers Marins, 152
    5th and 7th Submarine Squadrons, 147
    Vickers T III mines, 150-151

Ouistreham, 238
Reims, 291
St. Marcouf Islets, 252
St. Nazaire, xxxiii, 123
St. Vaast, 252, 263-264
Toulon, 6-8, 147, 278, 287, 290
Ushant Island, 213
Vieux-Boucau, 151
Fraser, Charles Robertson (T/A/Comdr., DSC & Bar, RNR), 291
Friedberger, Wiliam Howard Dennis (Capt., DSO, RN), 86, 206
Fuchida, Mitsuo, 107
Fuller-Wright (A/Sub. Lt., RNVR), 145
Fulton, Albert E., 198
Gauw, Jan August, 65
Gay, James (Seaman), 117-118
Germany/German
  Army, Afrika Korps, 87, 107, 120
  B-Dienst (Observation Service), 82
  Brunsbuttel, 144
  DFS 230 gliders, 184
  Elbe River, 51, 144
  Heligoland Bight, xxxiii, 25-26, 32-33, 51, 54, 62, 69-70, 79-82, 213
  Kiel, 27, 78, 213
  Kriegsmarine (Navy)
    1st Minehunter Flotilla, 36
    8th Flotilla, 193
    12th UJ Flotilla, 77
    Mines
      GC pressure type ('Oyster'), 40-41, 275-276, 298-199, 311
      TMA submarine-launched moored mine, 26-28
      TMB submarine-launched ground mine, 27, 311
    Security West (Befehlshaber der Sicherung West), 193
  Luftwaffe (Air Force)
    706 Coastal Defence Wing, 76
    Aircraft
      Arado Ar-196 seaplane, 76-78
      Dornier Do 17 light bomber, 7, 82, 91, 195
      Heinkel
        He111 medium bomber, 35
        He 115 seaplane/torpedo bomber, 74, 77
      Junkers
        Ju 87 ("Stuka") dive bomber/ground attack aircraft, 13, 97, 114, 116
        Ju 88 multi-role combat aircraft, 14
      Messerschmitt Bf 109 fighter aircraft, xxvi, 144
    Küstenfliegergruppe 76, 74
    X Fliegerkorps (10th Air Corps), 35

Varel, 61
Wilhelmshaven, 27
Glenny, John Edward Maxwell (Capt., DSO, DSC, RN), 40
Goldsmith, Frederick O., 154, 182
Goodall, Robert McLean, 215
Gordon, Roderick Cosmo (Comdr., DSO, RN), 83
Gosse, George (Lt. Comdr., GC, RANVR), 298-200
Gough, Thomas I., 198
Graham, William, 81
Grainger, William, 198
Grant, Alexander (T/Lt., RCNVR), 243
Granville-Smith (Leading Airman), 145
Graziani, Rodolfo, 106
Great Britain
  British Army
    1st Airborne Division, 207-208
    5th Indian Infantry Division, 97
    6th Royal Welsh Battalion, 208
    8th Army, 185, 190, 207
    11th Indian Brigade, 120
    50th (Northumbrian) Infantry Division, 96
    South African Division, 120
    Western Desert Force, 107
  British Protectorate, St. Helena Island, 68
  England
    Bletchley Park (Milton Keynes, Buckinghamshire), 50
    Blyth, 13-14, 81, 91
    Bristol, 50-51, 238, 274
    Chatham, 9, 47, 72, 85
    Cornwall, 50-51, 150-151
    Cromer, 80
    Dartmouth, 17, 151
    Dover, 10, 23-29, 49, 67, 124-128, 134-137, 142-144, 192-212, 238, 272
    Falmouth, 66, 151
    Farne Islands, 32
    Felixstowe, 122-125
    Folkestone, 23-24
    Gosport, 123-124, 216
    Great Yarmouth, 124, 272
    Grimsby, 31, 44, 65, 272
    Hartlepool, 66
    Harwich, 31-32, 80, 122, 272
    Immingham, 25, 31, 67, 79-82
    Isle of Wight, 239
    Kent, 28, 144, 298
    Kingston-upon-Hull, 85

Margate, 43, 46
Newcastle-upon-Tyne, 81
Norfolk, 48, 80
Plymouth, 47, 124, 191-192, 197-200, 209-220, 241, 255, 263, 274-275
Portsmouth, 17-18, 25, 32, 40, 44, 65-67, 70, 78, 123-124, 133, 151, 192, 195, 210-211, 241, 243, 271-272, 294, 298-305
Saltfleet, 66
Seaton, xxx
Sheerness, 46-47, 66, 68, 271
Shoeburyness, 39-41, 46
Southampton, 66, 68
St. Margaret's Bay, 28
Thames Estuary, 21, 31, 39, 42-48, 65-68, 85, 238, 301
Weymouth, 124, 195
Ireland
  Belfast, 219
  Londonderry, 68
  Rathlin Island, 219
Isle of Man, 2
Royal Air Force
  Aircraft
    Beaufort, xxv, 59-61, 143
    Catalina PBY, 22
    Halifax III, Lancaster III, Liberator I/II, Manchester I, Marauder, Mosquito XIB, Stirling I, Ventura, 22
    Hampden xxv, I, 59-61
    Wellington IC/X, 22, 104
  Bomber Command, 59, 141, 209, 211-213
  Coastal Command, 59, 61, 141-142, 147
  RAF Manston (in Kent on the Isle of Thanet), 144
  RAF Thorney Island, 303-304
Royal Navy
  15th Anti-Submarine Striking Force, xxxvi
  Destroyer Flotilla
    1st, 34
    5th, 83
    10th, 109
  Eastern Fleet, 98, 140
  Fleet Air Arm, xxv, xxiv, 22, 118, 143
    Aircraft
      Albacore, Avenger II, Barracuda II, 22
      Fairey Swordfish, xxiv, xxv, xxvi, xlvi, 22, 58, 62, 87, 104-106, 141-146
    No. 815 Naval Air Squadron, 59
    No. 825 Naval Air Squadron, xxv, 143
    No. 828 Naval Air Squadron, 104
    No. 830 Naval Air Squadron, 87, 104

Frater Armament Depot, 198
Gibraltar (British Colony), xxxvii, 2, 13, 26, 99, 104, 167, 206, 300-301
HMS Beehive, HMS Black Bat, HMS Grasshopper, HMS Midge, 123-128
HMS Hornet (HQ), 123-124, 216
HMS King Alfred, 116
HMS Lochinvar, 305
HMS Lynx, 196
HMS Vernon, 16-19, 38-40, 121, 123, 135, 241, 294-295, 301, 305-306
HMS Wasp, 124, 127, 194-196, 202
Malta (British Colony), xxxvii, 2, 13, 17, 87-104, 205-207
Mines (extensive details about mines may be found in Appendix A)
    A Mk I (ground-magnetic), 8, 128-129, 199, 309
    H Mk II (moored), 24
    A Mk I-IV (ground), 128-129, 199, 309
    Mk XVI (self-mooring type), 72, 309
    Mk XIX (moored-contact), 127, 129, 201, 309
    Mk XXVII (snagline), 192-93, 201, 267-270, 309-310, 317
Minesweep gear
    'LL Sweep,' 41
Minelaying Squadron
    1st, 47, 79, 85-86
(Minelaying) Destroyer Flotilla
    20th, 31-32, 51, 67, 70-71, 79, 82-86
Minesweeping Flotilla
    4th, 242, 249
    9th, 267, 271, 283
    14th, 242, 245, 249, 262
    15th, 102nd, 131st, 271
    16th, 243, 249
    104th, 248-249
    109th, 115
    132nd, 249
    139th, 157th, 163rd, 165th, 168th, 169th, 170th, 206th, 272
    159th, 248, 263, 267
    167th, 249, 272
Motor Gunboat Flotilla
    9th, 198
Motor Launch Flotilla
    4th, 195
    8th, 281
    10th, 191, 197, 199, 214
    50th, 127-136, 194-202
    51st, 128, 215
Motor Torpedo Boat Flotilla
    1st, 91, 121-122
    3rd, 122

362   Index

    4th, 125-128, 215
    21st, 215
    22nd, 52nd, 261
    66th, 275
  Naval Auxiliary personnel (Merchant Marine), 80
  Port Clearance Parties ('P' Parties),
    1571, 294, 298-299
    1572, 1573, 1574, 1575, 294
    2443, 305
    2444, 294, 305
  Royal Naval Air Station Lee-on-the-Solvent, 144
  Royal Naval Patrol Service, xxxiii-xxxvi, 117, 293
    HMS Europa (Sparrow's Nest) at Lowestoft, xxxiv,
    Minesweeping course at Lochinvar, Scotland, 116
  Submarine Flotilla,
    1st, 4th, 88
Scotland
  Aberdeen, 20, 71, 82, 91
  Ardrossan, 241, 272
  Cromarty Firth, 21
  Dalmuir, 13, 21
  Dumbarton, 79
  Dundee, 132, 148, 150-151
  Faslane, 21
  Firth of Forth, 2, 24, 68, 305
  Greenock, 18, 241
  Isle of Mull, 301
  Kyle of Lochalsh, 3, 8, 12, 47, 72, 85-86
  Loch Striven, 21
  Methil, 68
  Scapa Flow, 32, 47, 53, 85
  Skye, 3
  Troon, 13-14, 21
Wales
  Cardiff, 50
  Holyhead, 66
  Milford Haven, 5, 14, 21, 210
Greece/Greek
  Cephalonia, 104
  Crete, 3, 94-95
  Navarino (Pylos), 104
Green, G. L., 282
Greenland, William (A/Leading Stoker, DSM), 215
Greene, James Henry (Lt., RCNR), 242
Gregg, Henry (Engine Room Artificer, CGM, RN), 57
Gresham, Ronald (Lt. Comdr., DSC, RNVR), 241

Griffiths, Maurice Walter (Lt. Comdr., GM, RNVR), 304
Grøneng, Asbjørn Harr, 201
Gross, John Winston, 154, 217, 220
Guest, Eric (Able Bodied Seaman), 216
Gysler, Finn Christian Mosgaard, 134-136, 194-201
Haber, Charles Percy, 154, 200, 288
Haderlie, Gene, 296-297
Hadow, Philip Henry (Comdr., RN), 72, 83
Halfhide, Arthur Robert (Capt., CBE, RN), 42-46
Hall, David Alfred, 110, 114
Hall, John Buller Edward (Capt., RN), 71-81
Hall Jr., John L., 186, 238
Hall, Kenneth William Newman (Lt. Comdr., Croix de Guerre, RCNR), 242
Hallifax, Guy Waterhouse (Rear Adm. CMG, RN),
Hamilton, John, 110
Hardwich, N. M., 114
Harmon, Ernest N., 176
Harrelson, George D., 254
Harries, James Leslie (Comdr., OBE, GM & Bar, RCNVR), 298
Hart, R. D. C. (Lt., RNVR), 101
Henderson, Alexander Robert Lushington (Sub. Lt., RN), 73
Hennecke, Walter, 265
Hewitt, Henry K., 165, 173, 177-178, 183, 185-189, 279-280, 287
Heye, Helmuth Guido Alexander (Vice Adm., Kriegsmarine), 57
Heys, Norman (Able Bodied Seaman), 216
Heyward, Allard Barnswell, 250
Heyward, Allen B., 261
Hitler, Adolf, 90, 142, 184, 255, 262, 265, 291
Hoare, Dudley (T/A Comdr, DSC, RNR), 241
Hogendoorn, Jacobus Johannes, 65
Homewood, Robert Henry (Able Bodied Seaman), 194
Hopgood, Joseph G. W., 97
Howitt, Charles William Eaton (T/Surg. Lt., RCNVR), 283
Howlett, Ronald Stanley (Lt. Comdr., DSC, RN), 83
Hughes, Sidney, 97
Humphreys, Henry A., 81
Hunt, Douglas Eric James (T/Lt., MBE, DSC & Bar, RNVR), 216
Hupton, T. S., 114
Ingersoll, Royal Eason, 166
Italy/Italian
  10th Army, 107
  Bari, Grottaglie, 207
  Brindisi, 87, 92, 103-104, 207
  Capraia Island, 281
  Gorgona Island, 7-8
  Leghorn (Livorno), 2-8, 104, 301

364 Index

Regia Marina (Navy), 90, 103, 300
  1st Torpedo boat Flotilla, 13th Torpedo boat Squadron, 91
  Decima Flottiglia MAS, 300-301
Sardinia, 3, 8, 99, 104, 106, 187
Sicily
  Augusta, 90-91
  Catania, Licata, Scoglitti, 185-186
  Gela, 161-162, 183-188
  Messina, 185, 190
  Palermo, 103, 185, 189, 190
  Syracuse, 20, 71, 90-91, 184-188
Social Republic (a German puppet state), 184, 301
Somaliland, 105
Sparviero S.79 three-engined bomber, 97
Taranto, xxi, xxiv, xlvi, 13, 106, 205-208
Venice, 301
Jenkins (Lt.), 291
Jennings, Richard Borthwick (A/Capt., DSO, DSC & Bar, RN), 249
Jessop, Thomas Harry (Seaman, DSM, RNPS), 118
Jewell, H. M., 114
Jodl, Alfred, 291
Johnson, Ambrose Lawrence (Airman Petty Officer, RN), 145
Johnson, E. S., 176
Johnson, Henry C., 279
Johnson, Howard, 281
Johnston, Thomas (Able Seaman), 283
Jones, C. J. (Leading Stoker), 112
Jones, Ronald, 97
Karapiner, Sheres (Turkish admiral), 98
Kauffman, Draper L., 295-296
Kauffman, James L., 295
Keeping, Edward C., 198
Keitel, Wilhelm, 291
Kekewich, Piers Keane (Rear Adm., CB, RN), 123
Kelly, Arthur Vincent, 214
Kelly, Monroe, 168
Kemsley, Harold Thomas (T/A/Lt. Comdr., DSC & Bar, RNVR), 214
Kennedy, Ronald, 97
Kibbey, Mead B., 265-269
Keppler, R. H. T., 114
Kincaid, John (T/Lt., RCNR), 243
King George VI, 39-40
Kingsmill (A/Sub. Lt., RNVR), 145
Kirk, Alan G., 186, 237-238, 249, 287
Kirkwood, Frederick T., 97
Klopper, Henrik Balzazar (Gen., DSO, SAA), 120

Knapp, Richard Albert, 154, 288
Kockett, P. D., 114
Könenkamp, Jürgen, 205-206
Kruschev, Nikita, 302
Ladds, Charles Robert (Wireman, DSM), 118
Lamb, James Barrett (T/Lt., RCNVR), 242
Lambert, Louis A., 81
Lang, Otto, 77
Lawler, Paul A. (Leading Wireman), 117
Lee (A/Sub. Lt.), 145
Leigh-Mallory, Trafford (Air Chief Marshal Sir, KCB, DSO & Bar, RAF), 238
Lewis, John E. J., 110
Lewis, P. L., 114
Lewis, Roger Curzon (Capt., DSO, OBE, RN), 40
Lewis, Spencer S., 280
Lindner, Hendrik Dirk, 63, 64
Lipscombe, Albert E. J., 97
Little, Alexander, 81
Little, James, 81
Little-Kalsøy, Alf August, 201
Logan, Alfred C., 81
Logger, Johannes Marius, 65
Lonsdale, Rupert Philip (Comdr., RN), 71-78
Lowe, Douglas Wilson (Lt., RCNVR),
Lowry, Frank J., 242
Lowther, Louis C., 81
Lundgren, Oscar Berndherd, 250
MacDonald, Albert H., 81
MacKay, Herman Dwight (Comdr., Croix de Guerre avec Etoile en Vermeil, RCNR), 242, 248
Mackenzie, John, 81
MacMillan, Robert Cunningham (Lt., DSC & Bar, RCNVR), 115, 118
Madden, John Francis, 250
Maguire, Edward Henry (Lt., RCNVR), 242
Maher, Brendan A. (Lt., RNVR), 268
Maher, Joseph Benedict, 182
Mahn, Bruno, 78
Main, Douglas Wharrie (Lt. Comdr., RCNR), 243
Man, Frank Otto Stoe (Comdr., OBE, DSC, MRN[V]R), 214
Mantle, Harold J., 87
Marston, Joseph Charles (Lt. Comdr., DSC, RCNR), 242-248
Martin, Alister Angus, (Capt., DSO, DSC & 2 Bars, RD, RNR), 281, 290
Martinsen, Ivar, 201
Matson, A. C., 114
Maud, Colin Douglas (Capt., DSO, DSC, RN), 83
Mayes (Leading Seaman), 77

McCalmont, James, 81
McCarroll, Edward (Ord. Seaman), 118
McInally, John (AB Seaman, DSM), 215
McIsaac, Alexander, 81
McKenzie, Thomas (Commodore, CB, CBE, RNR), 266
McLaughlin, Patrick Vivian (Rear Adm., CB, DSO, RN), 301
McLean, Ian G., 81
McLean, William, 214
McNab, Alexander, 81
Mehrens, Gunther, 76
Meredith, Ralph Morton (Lt. Comdr., DSC, RCNR), 242, 248
Merritt, Reginald (AB, CGM, RN), 57
Miles, R. C. E., 114
Mills, Robert Edwin, 154, 190
Mitchell, Jack, 81
Montgomery, Bernard Law (Field Marshal, 1st Viscount Montgomery of Alamein, KG, GCB, DSO, PC, DL, BA), 185, 189-190, 238
Moon, Don P., 238
Mooney, A., 114
Morley, Frank, 282, 291
Morley, Norman Eyre (T/Comdr., DSC & 3 Bars, RNVR), 290
Mornu, Louis (French admiral), 98
Morris, William Harold (Leading Stoker, DSM), 194
Morrison, Thomas (Ord. Seaman), 118
Morshead, Leslie (Lt. Gen. Sir, KBE, CMG, DSO, ED, RAA), 108
Mountbatten, Louis Francis Albert Victor Nicholas (Adm., 1st Earl Mountbatten of Burma, KG, GCB, OM, GCSI, GCIE, GCVO, DSO, PC, FRS, RN), 98
Müller, Bernhard, 82
Murphy, Wilfred H., 97, 174
Murray, Felix T. Murray, 81
Murray, Leonard Warren (Rear Adm., CB, CBE, RCN), 222
Murray, Michael (Petty Officer, DSM), 216
Muscat, Giuseppe, 103
Mussolini, Benito, 90, 106, 184
Naito, Takeshi, 106
Napier, Joseph Henry, 250
Napier, Lennox William (Capt., DSO, DSC, RN), 99
Nelmes, Ronald Charles (Telegraphist, DSM), 194
Netherlands/Dutch
  Ameland Island, 28
  Callantsoog, Den Helder, 64-65
  East Indies, xxvii, 64, 68, 98
  Hook of Holland, 127, 129
  Koninklijke (Royal Netherlands Navy),
    9th Motor Torpedo Boat Flotilla, 212, 215

203rd Minesweeping Flotilla, 67, 211
Nieuwpoort (Nieuport), 127
Sint Philipsland, 65
Terschelling Island, 33, 141
Texel Island, 14, 70, 72, 82, 191, 213
Vlissingen (Flushing), 66, 298
West Indies, 67
Ymuiden, 79
Zuydcote Pass, 125, 128
Netterburg, John Charles, 110, 114
Newton, Hugo Rowland Barnwell (Capt., DSC, RN), 99-103
Nicholas, John Harold (Lt. Comdr. (L), RNVR), 241
Nolte, Heinz, 77
Norway/Norwegian
Bud, 53-54
"Camp Norway" in Nova Scotia, 133
Egerøy, 149-150
Egersund, 148, 152
Feiestein-Rinne, 152
Finnmark, 232
Haugesund, Herdla, 148
Karmøy Island, 141
Kristiansund, 54, 81, 91, 132
Namsos, xxxvi
Narvik, Ofotfjord, 52-59
Rombak's Fjord, 58
Royal Norwegian Navy, 131-133, 194-195
52nd Motor Launch Flotilla, xxxviii, 133-136, 194-202, 216
Port Clearance Party ('P' Party) 3006, 294
Stadtlandet, 53-54
Stavanger, 52, 152
Trondheim, 57
Vestfjord, 53-54
O'Connor, Richard Nugent (Gen. Sir, KT, GCB, DSO & Bar, MC, BA), 107
O'Daniel, John W., 219, 280
Offer, H., 14
O'Kelly, Maurice H., 81
Oldford, Augustus Albert (Leading Seaman, RNPS), 117-118
Operation
ANTAGONIZE (post-war mine clearance in the Adriatic Sea), 283
CERBERUS (German plan to break out of RN-blockaded Brest), 142
CULTIVATE (replacing 9th Australian Division with British soldiers), 97
DYNAMO (evacuation of Allied troops from Dunkirk), 70
FULLER (Plan for RN and RAF to prevent breakout of Brest), 142, 144
HUSKY (invasion of Sicily), 183-190
JUBILEE (Dieppe raid), 283

368   Index

MAPLE (mining for the Normandy landings), 192, 209-216, 281
MERKUR (German parachutists/glider-borne troops assault of Greece), 95
MINCEMEAT (disguising HMS Manxman as a French cruiser), 1-8
MINCEMEAT (corpse dressed as a British officer to fool Germans), 186
NEPTUNE (English Channel crossing to Normandy beaches), 237-264
OVERLORD (Allied invasion at Normandy), 237-264
ROMEO, SITKA (Support of the invasion of Southern France), 279
SLAPSTICK (British 1st Airborne Division at Taranto), 207
TORCH (Allied invasion of North Africa), xxxvii, 159, 163-182
WIKINGER (Ill-fated German destroyer operations), 33-36
WILFRED (British mining of Norwegian waters), 53-59
Oram, Alfred, 117
Orr-Ewing, David (Capt., DSO, RN), 207
Ouvry, John Garnault Delahaize (Comdr., DSO, RN), 39-40, 311
Overton, T. E. E., 114
Pangborne, Donald James (T/A Gunner T), 134, 196
Parker, Frederick Arthur (Ordinary Signalman), 216
Parker, Thomas R., 81
Parkinson (A/Sub. Lt., RN), 144
Patch, Alexander M., 280
Patton, George S., 165, 177, 185, 188-190, 277
Pead, Joseph Henry (Ordinary Signalman, DSM), 194
Pearson, James William (Leading Telegraphist, DSC), 216
Peedle, Charles A., 97
Pierce, Stanley (T/Lt., RCNR), 242
Pillinger, Eric W., 97
Pizey, Edward Fowle (Comdr., DSC, RN), 104
Plander, Henry, 249-250, 266
Pleydell-Bouverie, Edward (Capt., the Hon, MVO, RN), 93
Plunkett, Reginald Aylmer Ranfurly (Adm., Sir, DSO, JP, DL, RN), 84
Poland, Albert Lawrence (Vice Adm., DSO, DSC, RN), 112
Portugal, Azores, 166-167, 218-219, 241
Pound, Alfred Dudley Pickman Rogers (Adm. Sir, GCB, OM, GCVO, RN), 129, 143,
Pratt, Harold Irving, 250
Preston, Henry Francis Morrison (T/Lt., DSC, RNR), 241
Prince Philip, 195
Proudfoot, Harry, 81
Ramsay, Bertram Home (Adm. Sir., KCB, KBE, MVO, RN), 185, 188, 213, 238, 240
Ramsey, Robert (Lt., DSO, RN), 57
Reid, Howard Emmerson (Vice Adm. CB, RCN), 226
Ribbink, Louis Botha, 110, 114
Rickett, Thomas, 81
Rhoades, Rodney (Commodore, DSC, RAN), 109
Rich, C. I., 252

Richards, George D. K. (Lt.), 198
Riley, J., 114
Robinson, William J., 81
Rodgers, Bertram J., 280
Rommel, Erwin, 87, 107, 120, 163
Roope Gerard Broadmead (Lt. Comdr., VC, RN), 55-57
Roosevelt, Franklin Delano, 90, 163, 277
Ropp, Dietrich von der, 27
Rose (A/Sub. Lt., RNVR), 145
Rosica, Gino, 103
Rossi, Aldo, 91
Rossiter, Albert Cecil (AB Seaman, DSM), 215
Rousselot, Henri L. G., 149-150
Rowe, Austin Edward, 154, 275-276
Rowe, Charles Alexander (Lt. Comdr., RN), 90
Rucker, Colby G., 182
Ruge, Friedrich, 121
Ruitenschild, Joost, 65
Russell, Benjamin Thomas Robert (Lt. Comdr., RCNVR), 242
Rutherford, John, 97
Ruttan, John McDonald (Lt., DSC, RCNVR), 116, 118
Ryan, William David, 182
Sackritz (unteroffizier), 76
Samples (Sub. Lt.), 145
Sampson, Robert Ray, 182
Sayer, Guy Bourchier (Vice Adm. Sir, KBE CB DSC, RN), 122
Schmidt, Karl, 76
Scott, Walter (PO, CGM, RN), 56-57
Scudder Jr., Theodore Thomas, 154, 217-220, 288
Scutt, Walter C., 81
Self, Denis A., 81
Sethren, Rene, 114
Sharp Jr., Ulysses S. G., 172-173, 182

Ships and Craft
  Australian
    Australia, 88
    Bungaree, xxxix
    Nizam, 95, 119
    Orora, xii
    Parramatta, Stuart, Voyager, Waterhen, Yarra, 119
    Vampire, 15, 112, 119
    Vendetta, 109, 119
  British
    Abdiel, xxxvi, xl, 6, 13, 93-98, 104, 140, 205-207, 292
    Adventure, xxxix, 9-11, 24, 42-47, 86, 319
    Agamemnon, xxxix, 10, 86

Ajax, 93
Alsey, Atreus, Skylark (later *Vernon*, and *Vesuvius*), 18
Aphis, 104
Apollo, xl, 13, 209-210, 213
Arab, xxxvi
Ardrossan, Blackpool, Bootle, Boston, Bridlington, Bridport, Bryher, Calvay, Dalmatia, Dorothy Lambert, Dunbar, Fort York, Fraserburgh, Eastbourne, Green Howard, Gunner, Ijuin, James Lay, Llandudno, Lyme Niblick, Regis, Sigma, Tenby, Worthing, ML 345, ML 454, ML 465, ML 473, MMS 15, MMS 19, MMS 40, MMS 44, MMS 45, MMS 56, MMS 59, MMS 71, MMS 78, MMS 113, MMS 115, MMS 181, 271
Ariadne, xl, 1, 13
Aries, 281-283, 291
Ark Royal, 6, 8, 104
Arthur Cavanagh, Gloxinia, Milford Countess, Salvia, 111, 119
Auckland, 94
Bangor, 227
Basilisk, 44-46
Belfast, 42
Belvoir, Tetcott, 207
Birmingham, Greyhound, Hyperion, Ilex, Imogen, Inglefield, Isis, 54
Black Prince, 252
Blanche, 42-47, 319
Boreas, Brazen, Codrington, 32
Brixham, Bude, Polruan, Stornoway, 281
Bruiser, 277
Bustler, Eskimo, Restive, Pathfinder, 206
BYMS 2032, BYMS 2035, BYMS 2038, BYMS 2039, BYMS 2041, BYMS 2047, BYMS 2050, BYMS 2051, BYMS 2052, BYMS 2055, BYMS 2058, BYMS 2061, BYMS 2069, BYMS 2071, BYMS 2076, BYMS 2078, BYMS 2079, BYMS 2141, BYMS 2154, BYMS 2155, BYMS 2156, BYMS 2157, BYMS 2167, BYMS 2182, BYMS 2188, BYMS 2202, BYMS 2205, BYMS 2206, BYMS 2210, BYMS 2211, BYMS 2213, BYMS 2214, BYMS 2221, BYMS 2230, BYMS 2233, BYMS 2234, BYMS 2252, BYMS 2255, BYMS 2256, Conway Castle, Courtier, Georgette, MMS 1002, MMS 1004, MMS 1020, MMS 1047, MMS 1048, MMS 1077, Northcoates, Perdrant, Probe, Proctor, Prowess, Sir Agravaine, Sir Gareth, Sir Geraint, Sir Kay, Sir Lamorak, Sir Tristram, 272
BYMS 2034, 270, 272
BYMS 2070, BYMS 2173, 263, 272
Cachalot, 20, 73, 93, 99-103, 292
Cayton Wyke, Puffin, 28
Chakla, Fareham, Gloucester, 112
City of Paris, 39
Corncrake, Redshank, 16
Cricket, 114

Crista, Decoy, Hydrangea, 111
Devis, Ulster Prince, 110
Dittisham, Flintham, 20
Empire Broadsword, Minister, Swift, Wrestler, MGB 17, MGB 326, MMS 229,
    258
Empire Garden, Marsdale, War Hindoo, 219
Empire Henchman, 66
Empress of Asia, 195
Encounter, 6, 97
Esk, xl, 13-14, 25, 29, 31-32, 51, 54, 67-72, 79-85, 292
Express, xl, 13-14, 25-29, 31-32, 51, 68-72, 79-85
Fabia, Matra, Ponzano, 47
Flamingo, 94, 114
Foresight, 6, 80
Forester, Fury, Hermione, Nelson, Nestor, 6
Franklin, 275
Garth, St. Cyrus, MTB 15, 85
Glowworm, 53-57
Grampus, 69-73, 81-91, 292
Hampton, xxxix, 9, 12, 24,
Hardy, Hostile, 58
Hartland, Walney, ML 564, 286
Havock, Hunter, 54, 58
Hero, 54, 95, 97
Hotspur, 54, 58, 95, 97
Hyacinth, 111-112, 119
Icarus, xl, 13-14, 31-32, 51, 54, 68, 71, 82-83
Illustrious, xxiv, xxv, xlvi, 106
Impulsive, xl, 13-14, 31-32, 51, 54, 67-71, 79-80
Intrepid, xl, 13-14, 31-32, 67-71, 79-83, 92, 292
Jackal, Phoebe, 95
Jupiter, 83, 85
Ivanhoe, xl, 13-14, 31-32, 54, 68-72, 79-85, 292
Kelvin, 14, 83, 85
Kimberley, 95, 287
Kingston Amber, 220
Kung Wo, 10
Lady Brassey, Northumberland, 68
Latona, xxxvi, xl, 13, 93-99, 205, 292
LBO 68, MMS 279, 270
Linnet, Redstart, Ringdove, 9, 16
Lord Kelvin, 234
M1 (Miner I), M2 (Miner II), M3 (Miner III), M4 (Miner IV), M5 (Miner V),
    M6 (Miner VI), Miner VII, Miner VIII, 17
Manxman, xl, 1-13, 25, 93, 104, 139-140, 205-206
Mao Yeung, 10, 12

Medway, Proteus, Salmon, 88
Mendip, Norbo, 256
Menestheus, Port Quebec, xxxix, 10, 12, 86
MGB 110, 198
ML 100, 194, 215
ML 101, 129, 136, 194, 211
ML 102, 193-194
ML 103, 135-136
ML 104, 193-194, 199, 202
ML 105, ML 106, MTB 224, MTB 232, MTB 233, MTB 238, 215
ML 107, 193, 201, 211
ML 108, 136, 194, 199
ML 110, MTB 30, MTB 44, MTB 45, MTB 47, 128
ML 117, Poole, Romney, Seaham, Whitehaven, 262
ML 125, 133-137, 194, 201, 211
ML 137, 267-269
ML 139, ML 142, ML 275, 266
ML 143, 268
ML 157, ML 159, ML 180, ML 181, ML 184, ML 186, 199, 214
ML 213, 136-137, 193
ML 220, 127
ML 257, 266, 268
ML 259, ML 488, MTB 38, MTB 219, MTB 221, 199
MMS 8, 258, 271
MMS 110, 128, 271
MMS 181, 199, 214, 271
MMS 1019, 268-272
MMS 1084, xxxv
MTB 14, MTB 16, MTB 17, MTB 18, 122
MTB 23, 211
MTB 31, 125
MTB 32, 125, 128
MTB 34, 124
MTB 88, MTB 93, MTB 245, MTB 263, MTB 497, MTB 673, 216
MTB 235, 212, 215
MTB 240, 199, 212
Murrayfield, 65-66
Narwhal, 20, 71, 81-82, 91-92
Neptune, xxx
Nightingale, 10, 18-19
Obdurate, Obedient, Opportune, Orwell, 14
Osiris, Otus, Perseus, Regent, Rover, Taku, Tetrarch, Torbay, Triumph, Truant, 99
Ouse, 113
Pandora, Parthian, 88, 99
Peony, 111-112
Persian, 282

Phoenix, 88, 95
Plover, xxxix, 9-10, 15, 24, 67-68, 146, 209-213
Porpoise, 20, 73, 88, 103-104
Port Napier, xxxix, 10, 12, 69, 72, 85-86, 292
Princess Victoria, xl, 10, 12, 32, 69-71, 78-81, 292
Renown, 54-55
Rorqual, xv, 9, 20, 73, 87-104
Rothesay, 218
*Scout, Stronghold, Tenedo*s, *Thanet, Thracian*, 10, 14-15
Sea Rover, Stoic, Surf, Tactician, Tally-Ho, Tantalus, Tantivy, Taurus, Templar, Thorough, Thule, Tradewind, Trenchant, Trespasser, Tudor, 21
Seal, 69-78, 91, 292
Shepperton, xxxix, 9, 12, 24
Sidmouth, 271, 283
Skudd III, Skudd IV, Skudd V, Sotra, 115-119
Sobkra, xxxv
Southern Prince, xxxix, 10, 12, 85-86
Stuart, 112, 119
Swona, 115
Teviot Bank, xxxix, 9, 21, 54, 281
Tongue Light Vessel, 44
Truculent, 21, 301
Valiant, 93, 147
Vortigern, 83
Warspite, 59, 93
Welshman, xl, 13, 103, 139-140, 205-207, 292
Canadian
  Anticosti, 228-229
  Baffin, Miscou, Whitethroat, 229, 235
  Bayfield, Blairmore, Caraquet, 227, 235, 242-248, 271
  Bellechasse, Chignecto, Courtenay, Kelowna, Miramichi, Outarde, Quatsino, Transcona, 227
  Bras d'Or, 221-234
  Brockville, Burlington, Lockeport, Mahone, Medicine Hat, Melville, 227, 235
  Cailiff, Ironbound, Liscomb, Magdalen, Stonechat, 229
  Canso, 227, 235, 243
  Chedabucto, 227, 233-235
  Clayoquot, 227, 234, 236
  Comox, 222-223, 225
  Coquitlam, Cranbrook, Daerwood, Kalamalka, Lavallee, Revelstoke, Rossland, St. Joseph, 231
  Cowichan, 227, 235, 242-243, 271
  Digby, Drummondville, Gananoque, Goderich, Granby, Grandmere, Ingonish, Kentville, Lachine, Nipigon, Noranda, Port Hope, Quinte, Red Deer, Sarnia, Stratford, Swift Current, Trois Rivières, Truro, Ungava, 227, 236
  Elk, Husky, Reindeer, Vison, 226

Erica, 92
Esquimalt, 227, 234, 236
Fort William, 227, 236, 242-248, 271
Fundy, 221-223,
Gaspé, 222-223, 225
Gatineau, 14
Georgian, 227, 236, 243, 245
Guysborough, 227, 234, 236, 242, 245, 262
Kenora, 226-227, 236, 242, 245
Llewellyn, 221, 230-231, 235
Lloyd George, 231, 235
Malpeque, Mulgrave, 227, 235, 241-248, 271
Milltown, Minas, 227, 235, 242, 244, 271
Nootka, 222-223
Rayon d'Or, Reo II, Ross Norman, Standard Coaster, Venosta, 233
Sankaty, 229-230
Thunder, 227, 236, 242, 248, 266
Vegreville, 227, 236, 243, 245
Venture, 116
Wasaga, 227, 236, 242-243
Westmount, 226, 236

Dutch
  O-19, 68
  Adida, 111
  Dorus Rijkers, Douwe Aukes, G-13, G-15, Hydra, Jan van Brakel, Johan Maurits van Nassau, Medusa, 65-67
  Jan van Gelder, 150
  MTB 202, MTB 203, MTB 204, MTB 229, MTB 231, MTB 235, 211-212
  Nautilus, Van Meerlant, 63-68
  Willem van der Zaan, xxxviii, xxxix, 63-68

Finnish, *Hogland*, 149-150

French (French, Free French and Vichy French)
  Aconit, Chasseur 8, 151-152
  Dahomey, Foudrayant, Lorrain, Loup de Mer, Simon Duhamel, Strasbourgois, 176
  Estefette, 175
  George Leygues, Montcalm, 247
  Léopard, 4, 92
  Mistral, 258
  Quand Meme, 151
  Rubis, xv, xxxviii, 141, 147-152, 200, 212
  Victoria, 171-173

German
  Admiral Hipper, 55-57
  Adolph Woermann, xxx
  Antoine Henriette, 155
  Bernd Von Arnim, Paul Jakobi, 55

Bismarck, xxv
Cläre Hugo Stinnes, Grönland, Lesum, R-402, Seehund, 152
Erich Koeller, Friedrich Eckoldt, Richard Beitzen, Theodor Riedel, 34
Gauleiter A. Meyer, Hans Loh, Imbrim, Marie Frans, 151
Gneisenau, xxv, 58, 76, 139, 142-44
Hans Ludemann, Hermann Kunne, Karl Galster, 43
Helene, 93
Hugo Olendorff, xxxi
Kybfels, Marburg, 94
Leberecht Maass, Max Schultz, 33, 36
Prinz Eugen, xxv, 142-144
Rauenfels, 58
S-54, S-61, 13, 205, 207
Scharnhorst, xxv, 58, 139, 142-144
Seeteufel, 77-78
Spezia, 92
U-12, 27-28
U-13, 39
U-16, 27-28
U-40, 27
U-48, U-52, 33
U-54, 33, 36
U-130, U-173,177
U-190, 234
U-325, U-400, U-1021, 50
U-360, 14
U-375, 205-206
U-431, 115, 119
U-502, 67
U-559, 119
U-617, 13, 205, 207
U-652, 12
U-702, 150
U-806, U-868, 234
U-B, 78
UJ-128, 77
UJ-1106, UJ-1113, UJ-1116, UJ-1702, 152
V-1101, 33
Weichselland, 152
Wilhelm Heidkamp, 43
Wolf, 125, 129
Z-25, Z-29, Z-39, 142-143
Greek
  Antiklia, Miranda, 114
  Armatolos, 283
Icelandic, *Hekla*, 66

Italian
  Andrea Doria, Caio Duilio, 207
  Aldebaran, Altair, Salpi, 99
  Cadamosto, Calipso, Celio, Fratelli Cairoli, Ischia, Intrepido, Leopardi, Pascoli, Peppino C., Rina Croce, 92
  Calliope, Clio, Polluce, 91
  Capo Orso, 103
  Carlo Mirabello, 94
  Circe, 20, 71, 91
  Cleo, 20, 71
  Generale Achille Papa, 20, 102-103
  Generale Antonio Cantore, Generale Antonio Cascino, Generale Carlo Montanari, 102-103
  Generale Antonio Chinotto, 92, 103
  Generale Marcello Prestinari, 103, 207
  Olterra, 301
  Pellegrino Matteucci, 94
  Procellaria, 207
  Rodi, 111
  S-54, S-61, 13, 205, 207
  Sebastiano Venier, 104
  Sirio, 281
  Ticino, Verde, 93
Japanese
  Asagiri, Fubuki, Shirayuki, Yugiri, 15
  W-101, W-102, 227
Norwegian
  Almora, Argo, Blaamannen, Jadarland, Kem, 148
  Castor, Knute Nelson, 152
  Eidsvoll, 58
  Hector VI (whale-catcher), 120
  ML 123, 133, 211
  ML 125, 133, 136-137, 194, 201, 211
  ML 128, 133
  ML 210, 133-137, 193, 195, 201
  ML 573, 194, 201
  Norge, 58
  Star XVI, Suderøy I, Suderøy II, Suderøy IV, Suderøy V, Suderøy VI (whale-catchers), 232-233
  Suderøy (factory ship), 232
  Sverre Sigurdssøn, Vansø, 148
Polish, *Orzel*, 82
Romanian
  Carmen Sylva, 93
  Inginer N. Vlassopol, 224-225
South African

Bever, 119-120
Boksburg, Gribb, Imhoff, Langlaate, Seksern, Treern, 119
General Botha, 110
Johannesburg, 233
Parktown, 119, 233
Protea, 113, 119
Southern Floe, Southern Isles, Southern Maid, Southern Sea, 110-119
Soviet, *Ordzhonikidze*, 302-303
Swedish, *Divina*, 21
United States
  Army
    Gen. E. O. C. Ord, 156
    ST-253, ST-344, 270
  Merchant Marine
    Franklin K. Lane, 67
    John Fairfield, 218
  Navy
    amphibious/transports
      Ancon, 173
      Bayfield, 256-257, 261, 280
      Edward Rutledge, Electra, Hugh L. Scott, Joseph Hewes, Lakehurst, Leonard Wood, Tasker H. Bliss, 175-177
      Henry T. Allen, 169
      LCG-1062, 270
      LCI-489, LCI-496, 256
      LCI(L)-219, 262
      LST-391, 270
      Monrovia, 188
      Panamint, 160
      Susan B. Anthony, 169, 256, 262
    Auxiliaries/tugs
      Brant, Cherokee, 170
      Cowanseque, 218
      Dekanawida, 156
      Pinto, 256
      Winooski, 177
    combatant ships
      aircraft carrier, *Ranger*, 168
      battleships
        Arkansas, 246-247
        Massachusetts, 170
        New York, 177
      cruisers
        Augusta, 288
        Boise, 207-208
        Brooklyn, 174

378   Index

    Philadelphia, 177, 190
    Savannah, 190
    Tuscaloosa, Wichita, 170
  destroyers
    Butler, Cowie, Herndon, Shubrick, 190
    Corry, Meredith, Rich, 258
    Dallas, Jenkins, Mayrant, Rhind, Wainwright, 170
    DD-939, 142
    Glennon, 190, 258
    Hambleton, Ludlow, Murphy, Swanson, Tillman, Wilkes, 173-177
    MacLeish, Thornhill, 218
    McCormick, 220
  patrol boat, *PT-509*, 260
  submarine chasers
    SC-498, SC-535, SC-655, SC-770, SC-978, SC-979, 281-282
minelayers
  Barbican, Bastion, Buttress, Obstructor, Picket, Trapper, 157-158
  Barricade, 154, 157-158, 217-218, 274, 281-282, 286-288
  Chimo, 154, 157-158, 217-220, 249, 255-256, 262-264, 274, 276
  Keokuk, 160-162, 187, 190
  Miantonomah, 153-154, 158-167, 174-182, 256, 274-276, 292
  Monadnock, 153-161, 166-167, 176-182
  Oglala (ex-*Shawmut*), 158-161
  Planter, 154, 157-158, 217-220, 274, 282, 286, 288
  Salem, Weehawken, 154, 161-162, 187-190
  Terror, 158-160, 178-182
minesweepers
  Auk, 166-176, 182, 249-250, 266
  Broadbill, 250, 266
  Chickadee, 250-251, 266
  Dextrous, Pioneer, Prevail, Seer, 281
  Hamilton, 166-167, 176-177, 182
  Hogan, 166-167, 171-173, 182
  Howard, 166-167, 176-178, 182
  Nuthatch, 266
  Osprey, 166-169, 182, 250-251, 258, 262
  Palmer, 166-167, 171, 182
  Pheasant, 256, 260262, 266
  Raven, 166-169, 182, 149-250
  Staff, 250
  Stansbury, 166-167, 171, 182
  Steady, Sustain, 187
  Swift, Threat, 250, 256, 260-261, 266
  Tide, 250-251, 256-262
  YMS-13, YMS-20, YMS-27, YMS-199, YMS-251, 282
  YMS-18, YMS-21, YMS-34, YMS-82, YMS-355, 281

YMS-62, *YMS-69*, *YMS-207*, *YMS-208*, *YMS-226*, *YMS-227*, 187
YMS-231, 252, 263, 271
YMS-247, *YMS-346*, *YMS-348*, *YMS-351*, *YMS-352*, *YMS-358*, 252, 271
YMS-304, 252, 258, 263-264, 271
YMS-305, 252, 255, 263
YMS-347, 252, 263, 270-271
YMS-349, 252, 263, 275
YMS-350, 252, 258, 263-271
YMS-356, 252, 262, 271
YMS-375, 252, 263, 271
YMS-377, 252-253, 255, 271
YMS-378, 252, 255, 258, 262-264, 271
YMS-379, 252, 255, 271
YMS-380, 252, 255, 263-264, 271
YMS-381, *YMS-382*, 252, 263-264, 271
YMS-406, 252-253, 255, 271

Shiret, James Frederick (AB), 214
Silvester, Christopher A., 97
Sinclair, John Alexander (Sir, KCMG, CB, OBE), 303
Sketchley, Frank Godfrey, 214
Skorzeny, Otto, 184
Somerville, James Fownes (Adm. Sir, GCB, GBE, DSO, DL, RN), 2-5
South Africa/South African
   22nd Anti-Submarine Group, 110
   Royal Naval Volunteer Reserve (SA), 110
   Seaward Defence Force (SDF), 110, 113
   Simon's Town, 110
Smith, Charles C., 77
Smith, Einar, 201
Smith, Richard Leigh (Lt. Comdr., RCNVR), 115
Smuts, Jan Christiaan (Gen. PC, OM, CH, DTD, ED, KC, FRS, SAA), 110
Snowling, Frederick H., 81
Spain
   Balearic Islands, 6
   Canary Islands, 167
   Cape Finisterre, 33, 234
   Corunna, 122
   Galicia, 33
   Punta Umbría, 187
Spinage, Frederick T. (Able Seaman), 283
Sprigg, John Leslie (Telegraphist), 194
Sri Lanka (formerly Ceylon), 15, 64, 98
   Trincomalee, 98
Stagg, Arthur R. (Sub. Lt., RNVR), 198
Stanley, Edwin George (Stoker First Class), 241

Stannard, Richard Been (Capt., VC, DSO, RD, RNR), xxxvi
Stern, B. B., 286
Storrs, Antony Hubert, Gleadow (Rear Adm., DSC & Bar, Legion of Merit, Croix de Guerre avec Palme en Bronze, Croix de la Legion d'Honneur, RCNR), 241-248
Stromsoy, Olav Martin, 137-138
Sublette, William H., 172
Sullivan, William A., 266
Sweden, Vinga Island, 74
Swift, Edward Rhodes (T/Lt., RNVR), 117-118
Swimm, Charles H., 250-251
Sworder, Edward Robert Denys (T/Lt. Comdr. MBE, DSC, RNVR), 286-288
Syme, Hugh Randall (Lt., GC, GM & Bar, RANVR), 295
Tait, John Norman (Capt., CBE, DSC, RN), 72, 86
Tapping (Leading Airman), 144
Tarling, B. D., 114
Tatham, Charles (Leading Telegraphist, DSM), 215
Temple, John Bruce Goodenough (Comdr., DSC & 2 Bars, RN), 241, 249, 255-256
Theobald, Frederick A., 81
Tholfson, Kjell (Lt. Comdr., DSC, RNR), 115
Thomas, Allan, 114
Thomas, Alwyn,
Thomas, Kenneth Lewis (AB, DSM), 194
Thomas, Reginald G., 198
Thomson, Roger William David (Comdr., RN), 266
Thompson, John Chute (Sub. Lt., RN), 144
Thomsen, William H., 117
Trew, A. F., 110
Truscott Jr., Lucian K., 168, 280, 288
Underwood, William, 81
United States
  Army
    1st Army, 265
    3rd Army, 277
    7th Army, 185, 189-190, 279-280, 290
    1st Infantry Division, 183, 186
    3rd Infantry Division, 171, 186, 190, 219, 279-280
    9th Infantry Division, 168, 176
    36th Infantry Division, 279-280
    45th Infantry Division, 186, 279-280
    47th Infantry Division, 176
    60th Infantry Division, 168
    79th Infantry Division, 266
    Mine Planter Service, 16, 153-157
  Coast Guard, Lighthouse Service, 156

Navy
  Bomb Disposal School, 296
  Construction Battalion ("Seabee"), 28th, 275
  Destroyer Squadron
    8, 189
    13, 166
    15, 166
  Fleet
    8th, 218-219,
    12th, 218, 220, 274
  Mine Division
    16, 281, 285
    19, 166, 182
    21, 182
    50, 161-162, 182
  Mine Squadron
    6, 218
    7, 182
  Minesweeping
    "A" Minesweeping Flotilla, 249-250, 263
    "Y1" Squadron, 251-252, 254, 263, 271
    "Y2" Squadron, 249, 251-252, 271
  Motor Torpedo Squadron 15, 189
  U.S. Advanced Naval Base, Cherbourg, 272
Valentine, John Thomas (Leading Seaman, DSM), 215
van Asbeck, Thomas Karel, 65
Vaughan, Herbert, 70, 82, 84
Vearncombe, Archibald L. (AB, DSM, RN), 40
Vian, Philip Louis (Adm. Sir, GCB, KBE, DSO & 2 Bars, RN), 238
von Rabenau, Wolf-Rüdiger, 150
Wake-Walker, William Frederic (Adm. Sir, CBE, RN), 40, 85
Walker, Edwin A., 280
Wallis, A. V., 281
Warburton-Lee, Bernard Armitage (Capt., VC, RN), 58
Ward, Percy, 97
Waterman, William (T/Sub. Lt., RNVR), 198
Watson, C. B. B., 114
Watson, Edward Clifford (Capt., DSO, RN), 86
Wavell, Archibald Percival (Field Marshall Sir, 1st Earl Wavell, GCB, GCSI, GCIE, CMG, MC, KStJ, PC, BA), 106
Wear, Charles W., 81
Welsh, Daniel (PO Motor Mechanic, DSM), 216
Wheeler, Henry Thomas Albert (Leading Airman), 145
Whincup, John H., 97
White, Henry J., 252, 254
Wilkes, John E., 271-275

Wilkinson, Stephen Austin (Sub. Lt., RNVR), 296, 298
Willats, Philip T., 81
Williams (Lt., RN), 144
Williams, Henry Goodman, 154
Winch, Theodore G. B. (Comdr., RN), 97
Winter, Robert William (T/Lt., RNVR), 214
Wolfe, John Alfred (T/Lt., DSC & Bar, RNVR), 215
Wolfe, William Jackson (Lt. Comdr., RNVR), 115
Wood (Sub. Lt., RN), 145
Wood, Andrew, 194
Wood, Geoffrey Ernest (Leading Telegraphist, DSM), 215
Wyburd, Derek Bathurst (Comdr., DSO, DSC, RN), 83
Zepos, George (Greek admiral), 98
Zhukov, Georgi, 291
Zondorak, Charles Joseph, 182

About the Author

Rob Hoole joined the Royal Navy as a 'seaman officer' (later termed 'warfare officer') in 1971. He qualified as a Ships' Diving Officer in 1975 and as a Minewarfare & Clearance Diving Officer in 1976. In 1991, he co-founded the Royal Naval Minewarfare & Clearance Diving Officers' Association (MCDOA) of which he is Vice Chairman. Rob acquired an MBA from Oxford Brooks University in 2000 and left the Royal Navy in late 2002 after serving 32 years in all manner of operational and training roles at sea and ashore including Command of diving teams and mine countermeasures vessels. Rob lives in Waterlooville near Portsmouth with his wife Linda. They have three grown children, two young grandsons and a granddaughter.

Aside from his family, his passions include country pubs & real ale, sailing, reading and maritime history, particularly the development of naval & military diving, minewarfare and explosive ordnance disposal (EOD) otherwise known as bomb & mine disposal. He has contributed to many books, journals and other publications covering different aspects of his favourite subjects and was the editor of the book *Last of the Wooden Walls – An Illustrated History of the TON Class Minesweepers & Minehunters*. Last but not least, he is a fierce supporter of Project Vernon, the campaign to erect a monument at Gunwharf Quays in Portsmouth, Hampshire, to commemorate the minewarfare and diving heritage of HMS Vernon which previously occupied the site.

About the Author

Commander David D. Bruhn, U.S. Navy (Retired) served twenty-two years on active duty and two in the Naval Reserve, as both an enlisted man and as an officer, between 1977 and 2001.

Following completion of basic training, he served as a sonar technician aboard USS *Miller* (FF 1091) and USS *Leftwich* (DD 984). He was commissioned in 1983 following graduation from California State University at Chico. His initial assignment was to USS *Excel* (MSO 439), serving as supply officer, damage control assistant, and chief engineer. He then served in USS *Thach* (FFG 43) as chief engineer and Destroyer Squadron Thirteen as material officer.

After graduation from the Naval Postgraduate School, Commander Bruhn was assigned to Secretary of the Navy and Chief of Naval Operation staffs as a budget analyst and resources planner before attending the Naval War College in 1996, following which he commanded the mine countermeasures ships USS *Gladiator* (MCM 11) and USS *Dextrous* (MCM 13) in the Persian Gulf.

Commander Bruhn's final assignment was executive assistant to a senior (SES 4) government service executive at the Ballistic Missile Defense Organization in Washington, D.C.

Following military service, he was a high school teacher and track coach for ten years, and is now a USA Track & Field official. He lives in northern California with his wife Nancy and has two sons, David and Michael.

www.ingramcontent.com/pod-product-compliance
Lightning Source LLC
Chambersburg PA
CBHW071435300426
44114CB00013B/1444